FUZZY LOGIC CONTROL
ADVANCES IN APPLICATIONS

WORLD SCIENTIFIC SERIES IN ROBOTICS AND INTELLIGENT SYSTEMS

Published:

Vol. 1: Genetic Algorithms and Robotics — A Heuristic Strategy for Optimization (*Y Davidor*)

Vol. 2: Parallel Computation Systems for Robotics: Algorithms and Architectures (*Eds. A Bejczy and A Fijany*)

Vol. 3: Intelligent Robotic Planning Systems (*P C-Y Sheu and Q Xue*)

Vol. 4: Computer Vision, Models and Inspection (*A D Marshall and R R Martin*)

Vol. 5: Advanced Tactile Sensing for Robotics (*Ed. H R Nicholls*)

Vol. 6: Intelligent Control: Aspects of Fuzzy Logic and Neural Nets (*C J Harris, C G Moore and M Brown*)

Vol. 7: Visual Servoing: Real-Time Control of Robot Manipulators Based on Visual Sensory Feedback (*Ed. K Hashimoto*)

Vol. 8: Modelling and Simulation of Robot Manipulators: A Parallel Processing Approach (*A Y Zomaya*)

Vol. 9: Advanced Guided Vehicles — Aspects of the Oxford AGV Project (*Eds. S Cameron and P Probert*)

Vol. 10: Cellular Robotics and Micro Robotic Systems (*T Fukuda and T Ueyama*)

Vol. 11: Recent Trends in Mobile Robots (*Ed. Y F Zheng*)

Vol. 12: Intelligent Assembly Systems (*Eds. M Lee and J J Rowland*)

Vol. 14: Intelligent Supervisory Control: A Qualitative Bond Graph Reasoning Approach (*H Wang and D A Linkens*)

Vol. 15: Neural Adaptive Control Technology (*Eds. R Zbikowski and K J Hunt*)

Vol. 17: Applications of Neural Adaptive Control Technology (*Eds. J Kalkkuhl, K J Hunt, R Zbikowski and A Dzielinski*)

Vol. 20: Obstacle Avoidance in Multi-Robot Systems: Experiments in Parallel Genetic Algorithms (*M A C Gill and A Y Zomaya*)

Vol. 21: High-Level Feedback Control with Neural Networks (*Eds. F L Lewis and Y H Kim*)

Vol. 23: Fuzzy Logic Control: Advances in Applications (*Eds. H B Verbruggen and R Babuška*)

World Scientific Series in Robotics and Intelligent Systems – Vol. 23

FUZZY LOGIC CONTROL
ADVANCES IN APPLICATIONS

EDITED BY

HENK B. VERBRUGGEN
ROBERT BABUŠKA

Faculty of Information Technology and Systems
Delft University of Technology
Delft
The Netherlands

World Scientific
Singapore • New Jersey • London • Hong Kong

Published by

World Scientific Publishing Co. Pte. Ltd.

P O Box 128, Farrer Road, Singapore 912805

USA office: Suite 1B, 1060 Main Street, River Edge, NJ 07661

UK office: 57 Shelton Street, Covent Garden, London WC2H 9HE

British Library Cataloguing-in-Publication Data
A catalogue record for this book is available from the British Library.

First published 1999
Reprinted 2001

World Scientific Series in Robotics and Intelligent Systems — Vol. 23
FUZZY LOGIC CONTROL
Advances in Applications

ISBN 981-02-3825-8

This book is printed on acid-free paper.

Printed in Singapore by Uto-Print

PREFACE

This book presents a number of applications of fuzzy control. The application areas are the process industry, biotechnology, manufacturing, electromechanical systems, traffic control, avionics and biomedical systems. To position these applications in a wider perspective, we have also included two tutorial chapters presenting the main concepts and directions of fuzzy modeling and control. For those who are new to the field of fuzzy control, a chapter is included on the design procedure for a simple fuzzy controller. It is also a good introduction to a course on the Internet, which enables the user to design a fuzzy controller for the same problem as treated in the chapter. Moreover, we have included in this book chapters on advanced methodologies which seem to be most interesting for future applications.

This book is the result of the International School 'Fuzzy Logic Control – Advances in Applications,' which was held at the Delft University of Technology, Delft, the Netherlands in April 1999. This school was funded by the TMR program of the European Commission and consisted of two events, coordinated by Prof. Claudio Bonivento, University of Bologna, one held in Ferrara, Italy and one held in Delft, the Netherlands. Proceedings of the first event, 'Fuzzy Logic Control – Advances in Methodology,' were published by World Scientific in 1998 (ISBN 981-02-3506-2).

This book contains 16 chapters clustered into three parts: tutorials, surveys on advanced methodologies, and applications. The two tutorials in Part I address the position of fuzzy systems in modeling, identification and control.

The first tutorial, by Babuška, gives an overview of the role of fuzzy modeling, identification and of control based on fuzzy models. This chapter also contains an introductory section on fuzzy sets and fuzzy logic.

The second tutorial, by Årzén and Johansson, addresses fuzzy control from the viewpoint that fuzzy systems are a special form of nonlinear mappings and that fuzzy control can be seen as a special class of nonlinear control. Both the heuristic design and the model-based design of fuzzy controllers are addressed.

Part II of the book contains four surveys of advanced methodologies related to neuro-fuzzy modeling, complexity reduction in fuzzy systems, fault detection and decision making.

The survey by Nauck is devoted to the acquisition of neuro-fuzzy systems from training data. Examples are given in the area of function approximation, control and classification.

In the next chapter, Setnes deals with the reduction of complexity in fuzzy and neuro-fuzzy systems, focusing on aspects such as pruning of rules, merging of linguistic terms and rules, fusion of inputs and modular structuring of the rule base.

In the chapter by Kaymak, the relationship between fuzzy control and fuzzy decision making is addressed. It is shown that these two approaches are closely related when supervisory control and task-oriented control are considered. Applications include the selection of discrete control actions in model-based predictive control and the design of objective functions from expert knowledge.

The chapter by Frank and Marcu addresses the use of fuzzy techniques in fault detection, isolation and diagnosis. Fuzzy models for symptom generation and fuzzy decision methods for symptom evaluation are described. A number of successful applications are outlined.

The third part of the book contains 10 chapters describing interesting applications and case studies. In the previous two parts, illustrative examples have were already given. This part of the book, however, concentrates on real applications and possible future application areas. A wide range of application domains is covered: process industry, biotechnology, packaging in the food industry, manufacturing, traffic management and control, aircraft control and applications in the medical field.

Østergaard, one of the pioneers who introduced fuzzy control in Europe, addresses the application of fuzzy control techniques in the cement industry. He describes a design methodology for the development and implementation of high-level control strategies using fuzzy systems. Practical problems of installing these systems are described by Østergaard as well, and he presents some typical results.

Hellendoorn describes the problems arising in the control of complex machines in the paper and pulp industry. The enormous amount of data which the many sensors produce in this process contains useful information, which can be used by the company. Hellendoorn delineates a methodology for the extraction of information from large industrial data sets. He gives examples of industrial applications where fuzzy-logic-based data analysis plays an important role.

The chapter by Setnes presents an advisory fuzzy control system in the manufacturing of washing powders. The system consists of several fuzzy rule bases that model the control actions of experienced process operators. A fuzzy supervisor activates the various rule bases when corrective actions are needed and advises the operator to take the most rewarding actions, those which result in better quality control. The development, implementation and test results of this PC-hosted fuzzy advisory system are reported.

Te Braake, Roubos and Babuška describe a semi-mechanistic modeling approach for biochemical processes. This approach combines known white-box models (first-principle approach) with black-box models (fuzzy or neural models). The procedure is described for an enzymatic conversion of Penicillin-G. It is shown that by means of the semi-mechanistic model one can obtain useful data to construct a complete neural model for the process under consideration. This model is useful for real-time control.

The presented method is also well suited for the use in other application areas.

Filling systems of packaging machinery, as described in the chapter by Fantuzzi, are quite common in the food industry. It is necessary to accurately control the quantity of the product dispensed despite the high speed of the filling procedure, rapid changes in product properties and many unforeseen disturbances. A fuzzy controlled filling system is introduced and the test results on a real plant show a good performance.

Isermann and Fuessel present a fault detection and diagnosis system for electro-mechanical drives and actuators. Fault detection can benefit from transparent nonlinear fuzzy models, which can be learnt from experiments or from simulated data. It is also easy to incorporate rules which are known a priori. The approach is illustrated on the diagnosis of a valve and a DC motor.

Hellendoorn addresses applications of fuzzy logic in traffic management and control. Traffic systems exhibit complex behavior and a great extent of human influence. A fuzzy system for traffic-flow control and incident detection is described. This system has already been operational for some time. Another example Hellendoorn describes is the use of fuzzy logic in forecasting the occupancy of parking garages in an urban environment.

For flight safety, a flight control system is needed that has good performance under nominal as well as under exceptional, perhaps even hazardous conditions. The design of a flight control system is very time consuming and hence expensive. Applications of fuzzy systems with the aim of reducing the design effort and enhancing the flight performance are reviewed in the chapter by Schram, Oosterom and Verbruggen.

The chapter by Linkens and Abbod is concerned with medical applications of fuzzy logic. Fuzzy systems seem to be a natural paradigm for dealing with the complexity and measurement problems present in the monitoring and control problems in the human body. The application of fuzzy logic in anaesthesia is described, especially the fuzzy control for unconsciousness management of patients undergoing surgery.

Control engineers are mainly familiar with designing and analyzing linear controllers. It is also possible to design fuzzy controllers where the starting point is a linear controller, which is usually easy to design. This fuzzy controller can be adjusted for the given application, and its nonlinearity can be exploited. This design approach is the subject of the last chapter, by Jantzen. This chapter is also a good introduction to an Internet course on fuzzy logic control developed by the same author.

Finally, the editors wish to thank the authors for their contributions, which were produced within a tight time schedule of a half year.

Henk B. Verbruggen
Robert Babuška
Delft, the Netherlands

CONTENTS

PART I: TUTORIALS

AN OVERVIEW OF FUZZY MODELING AND MODEL-BASED FUZZY CONTROL

ROBERT BABUŠKA

Control Engineering Laboratory, Faculty of Information Technology and Systems Delft University of Technology, Mekelweg 4, 2628 CD Delft, The Netherlands e-mail: r.babuska@et.tudelft.nl

This tutorial addresses fuzzy modeling and control based on an explicit use of fuzzy models within the control scheme. First, a brief overview of fuzzy sets and fuzzy set operations is given. Then, the concept of a fuzzy system is explained and three types of rule-based fuzzy systems are described: the linguistic, relational, and Takagi–Sugeno model. For each model, the structure of the rules, the inference and defuzzification methods are presented. Fuzzy modeling of dynamic systems is addressed, as well as various methods to construct fuzzy models from knowledge and data. Finally, control schemes including inverse control, internal model control and predictive control based on fuzzy models are reviewed.

1 Introduction

Since its introduction in 1965, fuzzy sets and fuzzy logic have found applications in a wide variety of disciplines. Automatic control is a field in which these techniques have received considerable attention, not only from the scientific community but also from industry [1,2,3]. While most of the early design methods for fuzzy control were based on heuristic considerations [4], recent research has focused on the development of model-based fuzzy control techniques [5,6,7]. In the model-based approach, a fuzzy model is first developed to approximate the behavior of a complex process to be controlled. Based on this model, a controller can be designed.

Fuzzy set techniques have been recognized as a powerful tool for the development of models for systems that are not amenable to conventional modeling approaches due to the lack of precise, formal knowledge about the system, due to strongly nonlinear behavior or time varying characteristics. The rule-based nature of fuzzy models allows the use of information expressed in the form of natural language statements. This makes the models transparent to qualitative interpretation and analysis. At the computational level, fuzzy models can be regarded as flexible mathematical structures, similar to neural networks, that can approximate a large class of complex nonlinear systems to a desired degree of accuracy [8,9,10].

The first section of this tutorial briefly reviews several basic concepts of fuzzy set theory. In the subsequent sections, different types of fuzzy rule based systems are presented and several procedures for the building of fuzzy models are outlined. The last section of the tutorial is devoted to nonlinear control schemes that explicitly include fuzzy models, such as predictive control or internal model control. A companion tutorial in this book [11] focuses on direct fuzzy controllers obtained either though heuristic design or through model-based design by means of Lyapunov methods.

2 Fuzzy Sets and Fuzzy Systems

A static or dynamic system which makes use of fuzzy sets or fuzzy logic and of the corresponding mathematical framework is called a *fuzzy system*. Most common are fuzzy systems defined by means of *if-then* rules with fuzzy predicates. Fuzzy systems can serve different purposes, such as modeling, data analysis, prediction or control. Historically, automatic control was among the first technical application domains of fuzzy systems [4,1]. The basic idea of a fuzzy logic controller (FLC) is to formulate the control strategy of a human operator, which can be represented as a collection of if–then rules, in a way tractable for computers and for mathematical analysis. As an example, consider a control rule from a handbook for cement kiln operators [2,12]:

If temperature in the burning zone is *OK*, **and**
oxygen percentage in the exhaust gases is *Low*, **and**
temperature at the back-end is *High*,
then reduce fuel *Slightly* **and** reduce fan speed *Moderately*.

The first part of the rule, called the *antecedent*, specifies the conditions under which the rule holds, while the second part, called the *consequent*, prescribes the corresponding control action. Both the antecedent and the consequent contain linguistic terms (*OK*, *High*, *Low*, *Slightly* etc.). These terms represent in an approximate, quantized, way the magnitude of different variables involved.

When the if-then rules are to be represented in a form tractable for computers, one needs to define the linguistic terms, the logical connectives that operate on them (**and** in the above example), and the representation of the **if–then** relations. These concepts are explained below.

2.1 Fuzzy Sets

Consider the variable "temperature in the burning zone" in the previous example. The operator's knowledge about the magnitude of this variable may be represented, for instance, by three linguistic terms (*Low*, *OK*, *High*). In the given context, a set of high temperatures may be defined such that a temperature classified as *High* if it is, for instance, greater or equal to 600°C. In the conventional set theory, this set can be represented by a characteristic function $\chi_{High}(T)$ defined as

$$\chi_{High}(T) = \begin{cases} 1 & T \geq 600 \\ 0 & \text{otherwise.} \end{cases}$$

This characteristic function is depicted in Fig. 1.

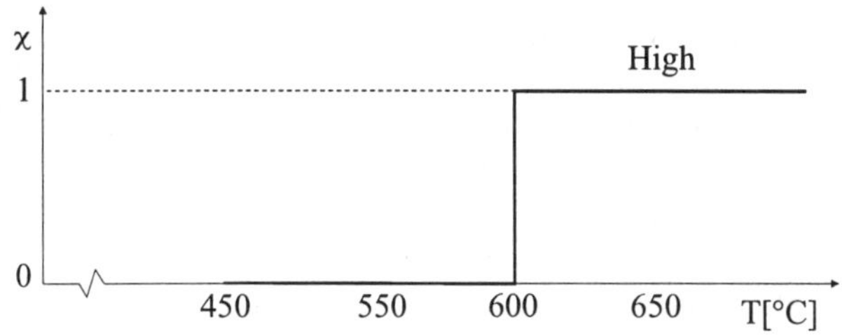

Figure 1. Characteristic function of the set of high temperatures in the burning zone.

According to this definition, a particular temperature is either an element of the set *High* or not. For example, 600°C is classified as a high temperature, while 599.5°C is not. It is, however, much more reasonable to assume that the human operator has a general idea of what a high temperature is, without defining an abrupt transition. One can imagine that there is a gradual transition from non-high temperatures, to high temperatures. This concept can be represented by a fuzzy set [13], which is defined by generalizing the characteristic function of conventional sets. The characteristic function of a fuzzy set is called the *membership function*.

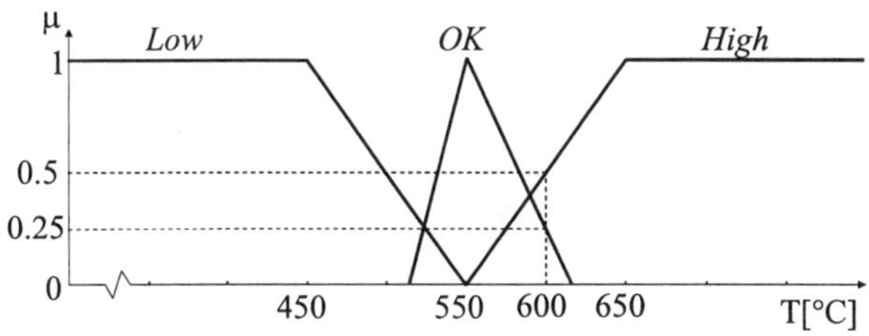

Figure 2. Membership functions of the fuzzy sets defined for burning zone temperatures.

The fuzzy set of high temperatures can now be defined as depicted in Fig. 2. For high temperatures, the membership value $\mu_{High}(T)$ equals 1, while it equals 0 for temperatures that are definitely not *High* and is between 0 and 1 for other temperatures.

$$\mu_{High}(T) = 1 \quad \text{temperature is definitely } High$$
$$\mu_{High}(T) \in (0,1) \quad \text{temperature is partially } High$$
$$\mu_{High}(T) = 0 \quad \text{temperature is not } High$$

In this example, temperatures above 650°C definitely belong to the set of high temperatures. There is a gradual transition from non-membership to full-membership from 550°C to 650°C. Because of this gradual transition between the sets, the temperatures can be partitioned into overlapping intervals as shown in Fig. 2. It is thus possible that a particular point in the domain belongs to more than one set with various degrees of membership, such as 600°C which belongs to the set of high temperatures with membership 0.5 and to the set of *OK* temperatures with membership 0.25. Because fuzzy sets are used to denote linguistic terms, the variable 'burning zone temperature' is also called a linguistic variable. It takes on different linguistic values such as *Low*, *OK* and *High* with different membership degrees.

In computer implementations, fuzzy sets are usually represented in two ways – on discrete domains as a list of ordered pairs $\{x_i, \mu(x_i)\}, x_i \in X$, and on continuous domains as an analytical formula of the membership function, e.g., $\mu(x) = 1/(1+x^2), x \in \mathbb{R}$. The position and the shape of the membership functions depend on particular applications and on the context. Fig. 3 shows some commonly used types of membership functions. The fuzzy singleton depicted in this figure is a degenerated fuzzy set: the fuzzy representation of a crisp (non-fuzzy) number.

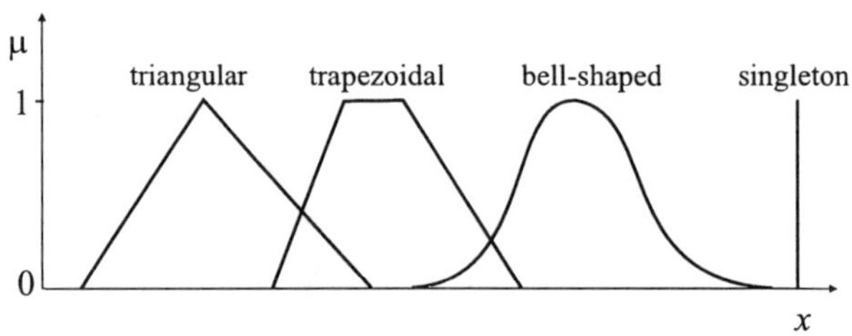

Figure 3. Various types of membership functions.

2.2 Fuzzy Set Operations

In order to work with fuzzy rules, one needs to combine fuzzy sets using logical connectives such as AND (conjunction), OR (disjunction) or NOT (complement). For this purpose, the logical connectives from conventional Boolean logic have been extended to their fuzzy equivalents. The generalization of conjunction (intersection) to fuzzy sets is done by functions called *t*–norms. The most commonly used conjunction operators are the minimum and the product operators. In terms of membership degrees, for $C = A \cap B$ we have

$$\mu_C(x) = \min\big(\mu_A(x), \mu_B(x)\big) . \tag{1}$$

The minimum operator is also denoted by '$\wedge$', i.e., $\mu_C(x) = \mu_A(x) \wedge \mu_B(x)$. The disjunction (union) is generalized by *t*–conorms. Usually the maximum operator is used for the disjunction. In terms of membership degrees, for $C = A \cup B$ we have

$$\mu_C(x) = \max\big(\mu_A(x), \mu_B(x)\big) . \tag{2}$$

The maximum operator is also denoted by '$\vee$', i.e., $\mu_C(x) = \mu_A(x) \vee \mu_B(x)$.

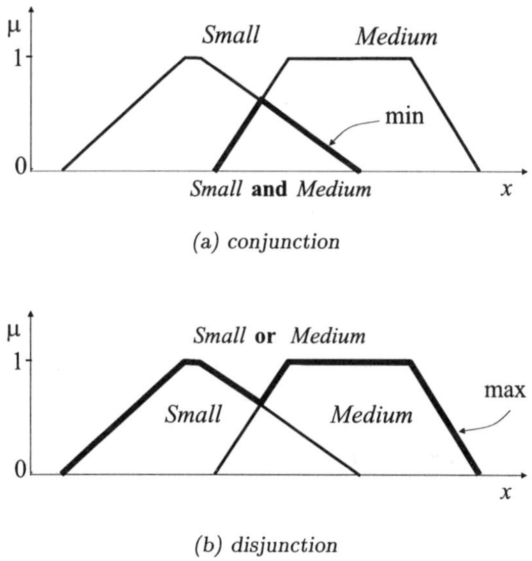

Figure 4. Conjunction and disjunction of two fuzzy sets.

Figure 4 shows the conjunction and disjunction of two fuzzy sets using the minimum and the maximum operators. For the complement, $(\mu_{\bar{A}}(x) = 1 - \mu_A(x))$ is usually used.

If the fuzzy sets are defined in different domains, such as X and Y, the operator has to be applied in the Cartesian product space of X and Y, i.e., for all possible pairs of x and y.

2.3 Fuzzy Relations

A fuzzy relation is a fuzzy set in a Cartesian product space of several domains, $X_1 \times X_2 \times \cdots \times X_n$. The membership grades represent the degree of association (correlation) among the elements of the different domains X_i. An n-ary fuzzy relation is a mapping

$$R: X_1 \times X_2 \times \cdots \times X_n \to [0, 1], \tag{3}$$

which assigns membership grades to all n-tuples $(x_1, x_2, \ldots, x_n)$ from the Cartesian product $X_1 \times X_2 \times \cdots \times X_n$. We will use fuzzy relations later to represent fuzzy if-then rules.

2.4 Types of Rule-Based Fuzzy Models

We have seen that in rule-based fuzzy systems, the relationships between variables are represented by means of fuzzy if–then rules of the following general form:

If *antecedent proposition* **then** *consequent proposition.*

The antecedent proposition is always a fuzzy proposition of the type "$\tilde{x}$ is A" where $\tilde{x}$ is a linguistic variable and A is a linguistic term. Depending on the form of the consequent, three main types of rule-based fuzzy models are distinguished:

- *Linguistic fuzzy model*: both the antecedent and the consequent are fuzzy propositions.
- *Fuzzy relational model*: generalization of the linguistic model, the relation between the antecedent and consequent terms is fuzzy.
- *Takagi–Sugeno (TS) fuzzy model*: the antecedent is a fuzzy proposition, the consequent is a crisp function.

These three types of fuzzy models are detailed in the subsequent sections.

3 Linguistic Fuzzy Model

The linguistic fuzzy model [14,15] has been introduced as a way to capture available (semi-)qualitative knowledge in the form of if–then rules:

$$\mathcal{R}_i\text{: } \textbf{If } \tilde{x} \text{ is } A_i \textbf{ then } \tilde{y} \text{ is } B_i, \qquad i = 1, 2, \ldots, K\,. \tag{4}$$

Here $\tilde{x}$ is the input (antecedent) linguistic variable, and A_i are the antecedent linguistic terms (constants). Similarly, $\tilde{y}$ is the output (consequent) linguistic variable and B_i are the consequent linguistic terms. The values of $\tilde{x}$, $\tilde{y}$ and the linguistic terms A_i, B_i are fuzzy sets defined in the domains of their respective base variables:[a] $\mathbf{x} \in X \subset \mathbb{R}^p$ and $\mathbf{y} \in Y \subset \mathbb{R}^q$. The membership functions of the antecedent and consequent fuzzy sets are the mappings: $\mu(\mathbf{x})\colon X \to [0,1]$ and $\mu(\mathbf{y})\colon Y \to [0,1]$, respectively. Fuzzy sets A_i define fuzzy regions in the antecedent space, for which the respective consequent propositions hold. The linguistic terms A_i and B_i are usually selected from sets of predefined terms, such as *Small*, *Medium*, etc. By denoting these sets by $\mathcal{A}$ and $\mathcal{B}$ respectively, we have $A_i \in \mathcal{A}$ and $B_i \in \mathcal{B}$. The rule base $\mathcal{R} = \{\mathcal{R}_i | i = 1, 2, \ldots, K\}$ and the sets $\mathcal{A}$ and $\mathcal{B}$ constitute the knowledge base of the linguistic model.

In order to be able to use the linguistic model, we need an algorithm which allows us to compute the output value, given some input value. This algorithm is called the *fuzzy inference* algorithm. For the linguistic model, the inference algorithm can be derived by using fuzzy relational calculus, as shown in the following section.

3.1 Relational Representation of Linguistic Models

Each rule in (4) can be regarded as a fuzzy relation (fuzzy restriction on the simultaneous occurrences of values $\mathbf{x}$ and $\mathbf{y}$): $R_i\colon (X \times Y) \to [0,1]$. This relation can be computed in two basic ways: by using fuzzy conjunctions (Mamdani method) and by using fuzzy implications (fuzzy logic method), see for instance [16]. Fuzzy implications are used when the if-then rule (4) is strictly regarded as an implication $A_i \to B_i$, i.e., "A_i implies B_i". In classical logic this means that if A holds, B must hold as well for the implication to be true. Nothing can, however, be said about B when A does not hold, and the relationship also cannot be inverted.

When using a conjunction, $A_i \wedge B_i$, the interpretation of the if-then rules is "it is true that A_i and B_i simultaneously hold". This relationship is symmetrical and can be inverted. For simplicity, in this text we restrict ourselves

[a] Base variable is the domain variable in which fuzzy sets are defined.

to the Mamdani (conjunction) method. The relation R_i is computed by the minimum ($\wedge$) operator:

$$R_i = A_i \times B_i, \quad \text{that is,} \quad \mu_{R_i}(\mathbf{x},\mathbf{y}) = \mu_{A_i}(\mathbf{x}) \wedge \mu_{B_i}(\mathbf{y})\,. \tag{5}$$

Note that the minimum is computed on the Cartesian product space of X and Y, i.e., for all possible pairs of $\mathbf{x}$ and $\mathbf{y}$. The fuzzy relation R representing the entire model (4) is given by the disjunction (union) of the K individual relations R_i:

$$R = \bigcup_{i=1}^{K} R_i, \quad \text{that is,} \quad \mu_R(\mathbf{x},\mathbf{y}) = \max_{1 \le i \le K} \left(\mu_{A_i}(\mathbf{x}) \wedge \mu_{B_i}(\mathbf{y})\right). \tag{6}$$

Now the entire rule base is encoded in the fuzzy relation R and the output of the linguistic model can be computed by the relational max-min composition ($\circ$) [17]:

$$\tilde{y} = \tilde{x} \circ R\,. \tag{7}$$

In terms of membership degrees, the composition operator is detailed in the next paragraph. Since the relation R can be visualized as a "fuzzy" graph in $X \times Y$, the linguistic fuzzy model is sometimes called a *fuzzy graph*. The relational composition (7) can be regarded as a function evaluation on the fuzzy graph.

3.2 Max-Min (Mamdani) Inference

We have seen that a rule base can be represented as a fuzzy relation. The output of a rule-based fuzzy model is then computed by the max-min relational composition. In this section, it will be shown that the relational calculus can be by-passed. This is advantageous, as the discretization of domains and storing of the relation R can be avoided. To show this, suppose an input fuzzy value $\tilde{x} = A'$, for which the output value B' is given by the relational composition:

$$\mu_{B'}(\mathbf{y}) = \max_X \left(\mu_{A'}(\mathbf{x}) \wedge \mu_R(\mathbf{x},\mathbf{y})\right), \quad y \in Y\,. \tag{8}$$

After substituting for $\mu_R(\mathbf{x},\mathbf{y})$ from (6), the following expression is obtained:

$$\mu_{B'}(\mathbf{y}) = \max_X \left[\mu_{A'}(\mathbf{x}) \wedge \max_{1 \le i \le K} \left(\mu_{A_i}(\mathbf{x}) \wedge \mu_{B_i}(\mathbf{y})\right)\right]. \tag{9}$$

Since the max and min operation are taken over different domains, their order can be changed as follows:

$$\mu_{B'}(\mathbf{y}) = \max_{1 \le i \le K} \left[\max_X \left(\mu_{A'}(\mathbf{x}) \wedge \mu_{A_i}(\mathbf{x})\right) \wedge \mu_{B_i}(\mathbf{y})\right]. \tag{10}$$

Note that the expression $\max_X (\mu_{A'}(\mathbf{x}) \wedge \mu_{A_i}(\mathbf{x}))$ represents the degree of match (similarity) between the input fuzzy set A' and the antecedent fuzzy set A_i in the ith rule. We will call this quantity the *degree of fulfillment* of the ith rule and denote it by

$$\beta_i = \max_X (\mu_{A'}(\mathbf{x}) \wedge \mu_{A_i}(\mathbf{x})) .$$

The output fuzzy set of the linguistic model is thus given by:

$$\mu_{B'}(\mathbf{y}) = \max_{1 \le i \le K} (\beta_i \wedge \mu_{B_i}(\mathbf{y})), \quad \mathbf{y} \in Y . \tag{11}$$

This algorithm, called the *max-min* or the *Mamdani inference*, is summarized in Algorithm 3.1 and visualized in Fig. 5.

Algorithm 3.1 *Mamdani (max-min) inference.*

1. Compute the degree of fulfillment by:

$$\beta_i = \max_X [\mu_{A'}(\mathbf{x}) \wedge \mu_{A_i}(\mathbf{x})] , \quad i = 1, 2, \dots, K .$$

2. Derive the output fuzzy sets B_i':

$$\mu_{B_i'}(\mathbf{y}) = \beta_i \wedge \mu_{B_i}(\mathbf{y}), \quad \mathbf{y} \in Y, \quad i = 1, 2, \dots, K .$$

3. Aggregate the output fuzzy sets B_i' into a single fuzzy set B':

$$\mu_{B'}(\mathbf{y}) = \max_{1 \le i \le K} \mu_{B_i'}(\mathbf{y}), \quad \mathbf{y} \in Y .$$

Note that for a singleton fuzzy set[b] A', Step 1 of the above algorithm simplifies to $\beta_i = \mu_{A_i}(\mathbf{x}')$.

The Mamdani inference method does not require the discretization of the domains and can work with analytically defined membership functions. It can also make use of learning algorithms, as discussed in Section 7.4.

3.3 Multivariable Systems

So far, the linguistic model was presented in a general manner covering both the SISO and MIMO cases. In the MIMO case, all fuzzy sets in the model are defined on vector domains by multivariate membership functions. It is,

[b] A singleton fuzzy set is a fuzzy representation of a crisp number. Suppose a crisp number $\mathbf{x}'$, the corresponding fuzzy singleton A' is defined by: $\mu_{A'}(\mathbf{x}') = 1$ and $\mu_{A'}(\mathbf{x}) = 0$, $\forall \mathbf{x} \in X$, $\mathbf{x} \neq \mathbf{x}'$.

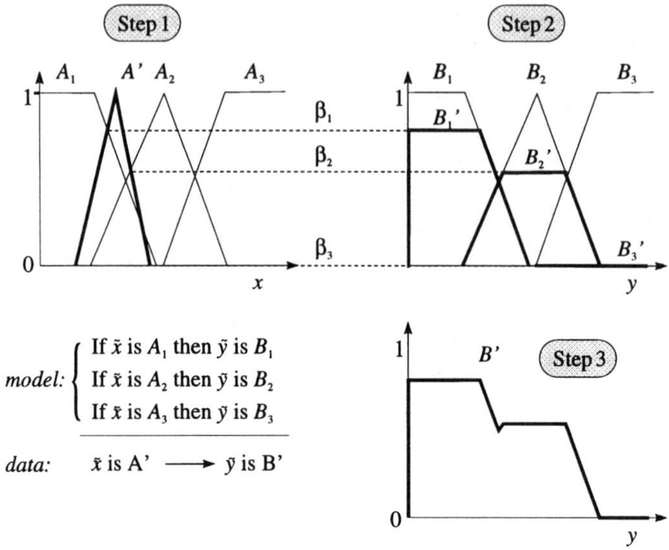

Figure 5. A schematic representation of the Mamdani inference algorithm.

however, often more convenient to write the antecedent and consequent propositions as logical combinations of fuzzy propositions with univariate membership functions. Fuzzy logic operators, such as the conjunction, disjunction and negation (complement) can be used to combine the propositions. Furthermore, a MIMO model can be written as a set of MISO models. Therefore, for the ease of notation, we will write the rules for MISO systems. Most common is the *conjunctive form* of the antecedent, which is given by:

$$\mathcal{R}_i\text{: } \mathbf{If}\ x_1 \text{ is } A_{i1} \ \mathbf{and}\ \ldots\ \mathbf{and}\ x_p \text{ is } A_{ip}\ \mathbf{then}\ y \text{ is } B_i, \quad i = 1, 2, \ldots, K\,. \tag{12}$$

Note that the above model is a special case of (4), as the fuzzy set A_i in (4) is obtained as the Cartesian product of fuzzy sets A_{ij}: $A_i = A_{i1} \times A_{i2} \times \cdots \times A_{ip}$. Hence, the degree of fulfillment (step 1 of Algorithm 3.1) is given by:

$$\beta_i = \mu_{A_{i1}}(x_1) \wedge \mu_{A_{i2}}(x_2) \wedge \cdots \wedge \mu_{A_{ip}}(x_p), \quad i = 1, 2, \ldots, K\,. \tag{13}$$

Other conjunction operators, such as the product, can be used. A set of rules in the conjunctive antecedent form divide the input domain into a lattice of fuzzy hyperboxes, parallel with the axes. Each of the hyperboxes is a Cartesian product-space intersection of the corresponding univariate fuzzy sets. This is shown in Fig. 6a. The number of rules in the conjunctive form, needed to cover the entire domain, is given by:

$$K = \Pi_{i=1}^{p} N_i, \tag{14}$$

where p is the dimension of the input space and N_i is the number of linguistic terms of the ith antecedent variable.

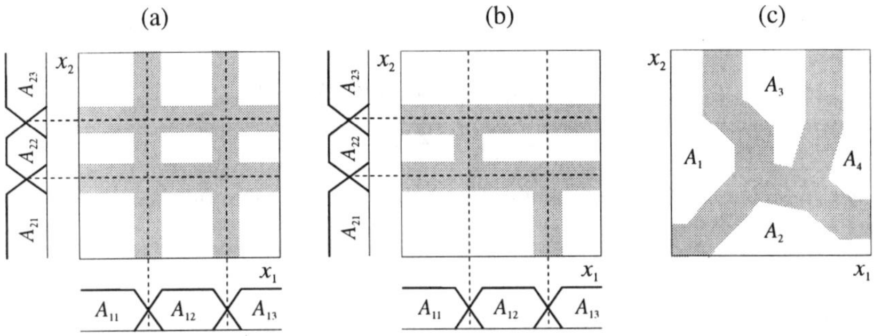

Figure 6. Different partitions of the antecedent space. Gray areas denote the overlapping regions of the fuzzy sets.

By combining conjunctions, disjunctions and negations, various partitions of the antecedent space can be obtained. The boundaries remain restricted to the rectangular grid defined by the fuzzy sets of the individual variables, see Fig. 6b. As an example consider the rule antecedent covering the lower left corner of the antecedent space in this figure:

If x_1 is **not** A_{13} **and** x_2 is A_{21} **then** ...

The degree of fulfillment of this rule is computed using the complement and intersection operators as:

$$\beta = [1 - \mu_{A_{13}}(x_1)] \wedge \mu_{A_{21}}(x_2)\,. \tag{15}$$

The antecedent form with multivariate membership functions (4) is the most general one, as there is no restriction as to the shape of the fuzzy regions. The boundaries between these regions can be arbitrarily curved and opaque to the axes, as depicted in Fig. 6c. Also the number of fuzzy sets needed to cover the antecedent space may be much smaller than in the previous cases. Hence, for complex multivariable systems, this partition may provide the most effective representation. Note that the fuzzy sets A_1 to A_4 in Fig. 6c still can be projected onto x_1 and x_2 to obtain an approximate linguistic interpretation of the regions described.

3.4 Defuzzification

In many applications, a crisp output $\mathbf{y}$ is desired. To obtain a crisp value, the output fuzzy set must be *defuzzified.* With the Mamdani inference scheme, the *center of gravity* (COG) defuzzification method is often used. This methods computes the y coordinate of the center of gravity of the area under the fuzzy set B':

$$y' = \text{cog}(B') = \frac{\sum_{j=1}^{F} \mu_{B'}(y_j) y_j}{\sum_{j=1}^{F} \mu_{B'}(y_j)}, \tag{16}$$

where F is the number of elements y_j in Y. A continuous domain Y thus must be discretized.

3.5 Singleton Model

A special case of the linguistic fuzzy model is obtained when the consequent fuzzy sets B_i are singleton fuzzy sets (see Fig. 3). These sets can be represented simply as real numbers b_i, yielding the following rules:

$$R_i\text{: } \textbf{If } \tilde{\mathbf{x}} \text{ is } A_i \textbf{ then } y = b_i, \quad i = 1, 2, \ldots, K\,. \tag{17}$$

This model is called the *singleton model.* A simplified version of (16), the *fuzzy-mean* defuzzification method is usually used with this model:

$$y = \frac{\sum_{i=1}^{K} \beta_i \, b_i}{\sum_{i=1}^{K} \beta_i}\,. \tag{18}$$

The singleton fuzzy model belongs to a general class of general function approximators, called the basis functions expansion [18] taking the form:

$$y = \sum_{i=1}^{K} \phi_i(\mathbf{x}) b_i\,. \tag{19}$$

Most structures used in nonlinear system identification, such as artificial neural networks, radial basis function networks, or splines, belong to this class of systems. Connections between these types of models have been investigated [19,20]. In the singleton model, the basis functions $\phi_i(\mathbf{x})$ are given by the normalized degrees of fulfillment of the rule antecedents, and the constants b_i are the consequents.

4 Fuzzy Relational Model

A fuzzy relational model [12,21] can be seen as an extension of the linguistic model, where the mapping between the input and output fuzzy sets is represented by a fuzzy relation.

Denote $\mathcal{A}_j$ and $\mathcal{B}$ the sets of linguistic terms defined for the antecedent and consequent variables, respectively:

$$\mathcal{A}_j = \{A_{jk} | k = 1, 2, \ldots, N_j\}, \quad j = 1, 2, \ldots, p,$$
$$\mathcal{B} = \{B_k | k = 1, 2, \ldots, M\}.$$

The rule base (12) can be represented as a crisp relation R between the linguistic terms in the antecedents and in the consequent:

$$R: \mathcal{A}_1 \times \mathcal{A}_2 \times \cdots \times \mathcal{A}_n \times \mathcal{B} \rightarrow \{0, 1\}. \tag{20}$$

Further, by denoting $\mathcal{A} = \mathcal{A}_1 \times \mathcal{A}_2 \times \cdots \times \mathcal{A}_n$ the Cartesian space of the antecedent linguistic terms, eq. (20) can be simplified to $R: \mathcal{A} \times \mathcal{B} \rightarrow \{0, 1\}$. A fuzzy relational model is obtained by generalizing R to a fuzzy relation:

$$R: \mathcal{A} \times \mathcal{B} \rightarrow [0, 1]. \tag{21}$$

In this model, each rule contains all the possible consequent terms, each with a different weighting factor, given by the respective elements of the fuzzy relation. With this weighting, one can more easily fine-tune the model, e.g., to fit some data. A single-input, single-output fuzzy relational model is depicted in Fig. 7.

It should be stressed that the relation R in (21) is different from the relation (6) encoding linguistic if–then rules. The latter relation is a multidimensional membership function defined in the product space of the input and output domains. Each element of this relation represents the degree of association between the individual *crisp* elements in the antecedent and consequent domains. In fuzzy relational models, however, the fuzzy relation represents associations between the individual linguistic terms.

The main advantage of the relational model is that the input–output mapping can be fine-tuned without changing the consequent fuzzy sets. In the linguistic model, the outcomes of the individual rules are restricted to the grid given by the centroids of the output fuzzy sets, which is not the case in the relational model.

The inference in the relational model proceeds in the following three steps:

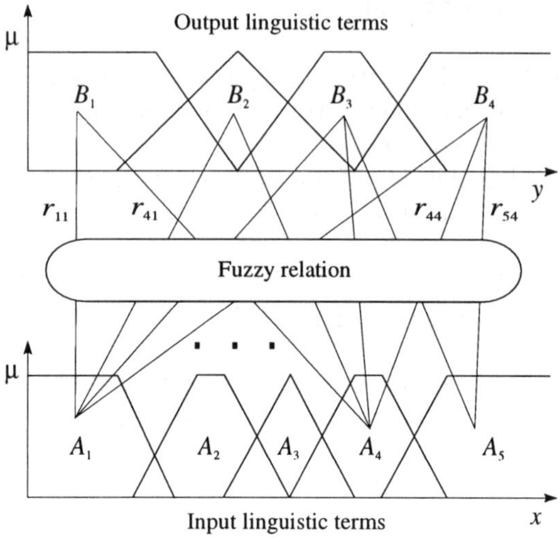

Figure 7. Fuzzy relational model. Reproduced from, [22] ©1996 Elsevier Science Ltd.

Algorithm 4.1 *Inference in fuzzy relational model.*

1. Compute the degree of fulfillment by:

$$\beta_i = \mu_{A_{i1}}(x_1) \wedge \cdots \wedge \mu_{A_{ip}}(x_p),\ i = 1, 2, \ldots, K\,.$$

2. Apply the relational composition $\boldsymbol{\omega} = \boldsymbol{\beta} \circ R$, given by:

$$\omega_j = \max_{1 \le i \le K} \big(\beta_i \wedge r_{ij}\big), \quad j = 1, 2, \ldots, M\,.$$

3. Defuzzify the consequent fuzzy set by:

$$y = \sum_{l=1}^{M} \omega_l\, b_l \Big/ \sum_{l=1}^{M} \omega_l,$$

where $b_l = \mathrm{cog}(\mathcal{B}_l)$.

5 Takagi–Sugeno Model

The Takagi–Sugeno (TS) fuzzy model, [23] involves crisp functions in the consequents. It can be seen as a combination of linguistic modeling and mathematical regression, in the sense that the antecedents describe fuzzy regions in the input space in which consequent functions are valid. The TS rules have the following form:

$$\mathcal{R}_i\text{: } \textbf{If } \mathbf{x} \text{ is } A_i \textbf{ then } \mathbf{y}_i = \mathbf{f}_i(\mathbf{x}), \qquad i = 1, 2, \ldots, K\,. \tag{22}$$

Contrary to the linguistic model, the input $\mathbf{x}$ is a crisp variable. The functions $\mathbf{f}_i$ are typically of the same structure, only the parameters in each rule are different. Generally, $\mathbf{f}_i$ is a vector-valued function, but for the ease of notation we will consider a scalar f_i in the sequel. A simple and practically useful parameterization is the affine (linear in parameters) form, yielding the rules:

$$\mathcal{R}_i\text{: } \textbf{If } \mathbf{x} \text{ is } A_i \textbf{ then } y_i = \mathbf{a}_i^T\mathbf{x} + b_i, \qquad i = 1, 2, \ldots, K, \tag{23}$$

where $\mathbf{a}_i$ is a parameter vector and b_i is a scalar offset. This model is called an *affine TS model.* Note that if $\mathbf{a}_i = 0$ for each i, the singleton model (17) is obtained.

5.1 *Inference Mechanism*

The inference formula of the TS model is a straightforward extension of the singleton model inference (18):

$$y = \frac{\sum_{i=1}^K \beta_i y_i}{\sum_{i=1}^K \beta_i} = \frac{\sum_{i=1}^K \beta_i(\mathbf{a}_i^T\mathbf{x} + b_i)}{\sum_{i=1}^K \beta_i}\,. \tag{24}$$

When the antecedent fuzzy sets define distinct but overlapping regions in the antecedent space and the parameters $\mathbf{a}_i$ and b_i correspond to a local linearization of a nonlinear function, the TS model can be regarded as a smoothed piece-wise linear approximation of that function, see Fig. 8.

5.2 *TS Model as a Quasi-Linear System*

The affine TS model can be regarded as a quasi-linear system (i.e., a linear system with input-dependent parameters). To see this, denote the normalized degree of fulfillment by

$$\gamma_i(\mathbf{x}) = \beta_i(\mathbf{x}) / \sum_{j=1}^K \beta_j(\mathbf{x})\,. \tag{25}$$

Here we write $\beta_i(\mathbf{x})$ explicitly as a function $\mathbf{x}$ to stress that the TS model is a quasi-linear model of the following form:

$$y = \left(\sum_{i=1}^K \gamma_i(\mathbf{x})\mathbf{a}_i^T\right)\mathbf{x} + \sum_{i=1}^K \gamma_i(\mathbf{x})b_i = \mathbf{a}^T(\mathbf{x})\mathbf{x} + b(\mathbf{x})\,. \tag{26}$$

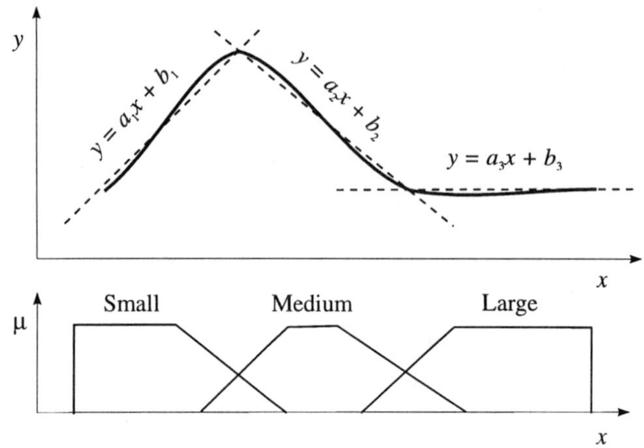

Figure 8. Takagi–Sugeno fuzzy model as a smoothed piece-wise linear approximation of a nonlinear function.

The 'parameters' $\mathbf{a}(\mathbf{x})$, $b(\mathbf{x})$ are convex linear combinations of the consequent parameters $\mathbf{a}_i$ and b_i, i.e.:

$$\mathbf{a}(\mathbf{x}) = \sum_{i=1}^{K} \gamma_i(\mathbf{x})\mathbf{a}_i, \quad b(\mathbf{x}) = \sum_{i=1}^{K} \gamma_i(\mathbf{x})b_i\,. \tag{27}$$

In this sense, a TS model can be regarded as a mapping from the antecedent (input) space to a convex region (polytope) in the space of the parameters of a quasi-linear system, as schematically depicted in Fig. 9.

This property facilitates the analysis of TS models in a framework similar to that of linear systems. Methods have been developed to design controllers with desired closed loop characteristics and to analyze their stability [24,25,26].

6 Modeling of Dynamic Systems

Before discussing dynamic fuzzy models, let us recall that time-invariant dynamic systems are in general modelled by static functions, by using the concept of the system's *state*. Given the state of a system and given its input, we can determine what the next state will be. In the discrete-time setting we can write

$$\mathbf{x}(k+1) = \mathbf{f}(\mathbf{x}(k), \mathbf{u}(k)), \tag{28}$$

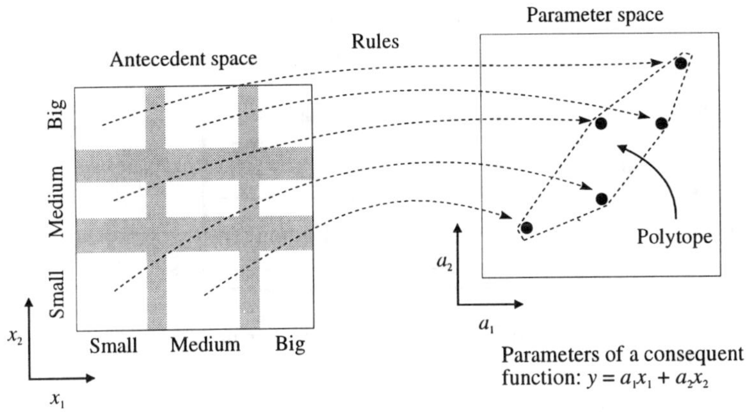

Figure 9. A TS model with affine consequents can be regarded as a mapping from the antecedent space to the space of the consequent parameters.

where $\mathbf{x}(k)$ and $\mathbf{u}(k)$ are the state and the input at time k, respectively, and $\mathbf{f}$ is a static function, called the *state-transition function*. Fuzzy models of different types can be used to approximate the state-transition function. As the state of a process is often not measured, *input-output* modeling is often applied. The most common is the NARX (Nonlinear AutoRegressive with eXogenous input) model:

$$y(k+1) = f(y(k), y(k-1), \ldots, y(k-n_y+1), u(k), u(k-1), \ldots, u(k-n_u+1)) \,. \quad (29)$$

Here $y(k), \ldots, y(k-n_y+1)$ and $u(k), \ldots, u(k-n_u+1)$ denote the past model outputs and inputs respectively and n_y, n_u are integers related to the model order. For example, a singleton fuzzy model of a dynamic system may consist of rules of the following form:

$$\begin{aligned} R_i\colon\ &\textbf{If } y(k) \text{ is } A_{i1} \textbf{ and } y(k-1) \text{ is } A_{i2} \textbf{ and} \ldots y(k-n+1) \text{ is } A_{in} \\ &\textbf{and } u(k) \text{ is } B_{i1} \textbf{ and } u(k-1) \text{ is } B_{i2} \textbf{ and} \ldots u(k-m+1) \text{ is } B_{im} \\ &\textbf{then } y(k+1) \text{ is } c_i \,. \end{aligned} \quad (30)$$

In this sense, we can say that the dynamic behavior is taken care of by external dynamic filters added to the fuzzy system, see Fig. 10. In (30), the input dynamic filter is a simple generator of the lagged inputs and outputs, and no output filter is used.

Since fuzzy models can approximate any smooth function to any degree of accuracy, [27] models of type (30) can approximate any observable and controllable modes of a large class of discrete-time nonlinear systems [28].

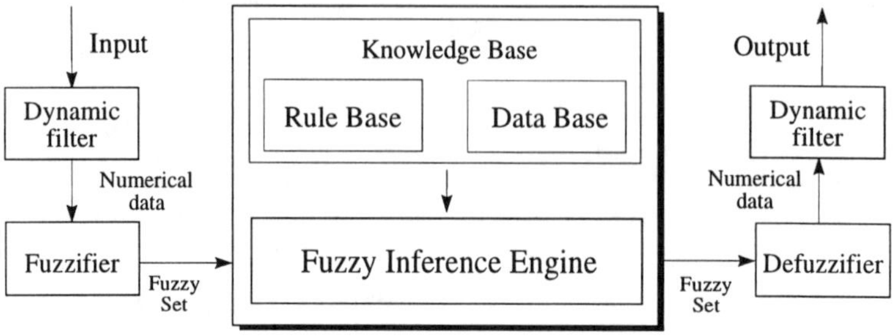

Figure 10. A generic fuzzy system with fuzzification and defuzzification units and external dynamic filters. Reproduced from, [22] ©1996 Elsevier Science Ltd.

7 Building Fuzzy Models

Two common sources of information for building fuzzy models are the *prior knowledge* and *data* (measurements). Prior knowledge can be of a rather approximate nature (qualitative knowledge, heuristics), which usually originates from "experts", i.e., process designers, operators, etc. In this sense, fuzzy models can be regarded as simple *fuzzy expert systems* [29].

For many processes, data are available as records of the process operation or special identification experiments can be designed to obtain the relevant data. Building fuzzy models from data involves methods based on fuzzy logic and approximate reasoning, but also ideas originating from the field of neural networks, data analysis and conventional systems identification. The acquisition or tuning of fuzzy models by means of data is usually termed *fuzzy identification.*

Two main approaches to the integration of knowledge and data in a fuzzy model can be distinguished:

1. The expert knowledge expressed in a verbal form is translated into a collection of if–then rules. In this way, a certain model structure is created. Parameters in this structure (membership functions, consequent singletons or parameters) can be fine-tuned using input-output data. The particular tuning algorithms exploit the fact that at the computational level, a fuzzy model can be seen as a layered structure (network), similar to artificial neural networks, to which standard learning algorithms can be applied. This approach is usually termed *neuro-fuzzy modeling* [30].

2. No prior knowledge about the system under study is initially used to formulate the rules, and a fuzzy model is constructed from data. It is

expected that the extracted rules and membership functions can provide an a posteriori interpretation of the system's behavior. An expert can confront this information with his own knowledge, can modify the rules, or supply new ones, and can design additional experiments in order to obtain more informative data. This approach can be termed *rule extraction*. Fuzzy clustering is one of the techniques that are often applied [31,32,33].

These techniques, of course, can be combined, depending on the particular application. In the sequel, we describe the main steps and choices in the knowledge-based construction of fuzzy models, and the main techniques to extract or fine-tune fuzzy models by means of data.

7.1 Structure and Parameters

With regard to the design of fuzzy (and also other) models, two basic items are distinguished: the structure and the parameters of the model. The structure determines the flexibility of the model in the approximation of (unknown) mappings. The parameters are then tuned (estimated) to fit the data at hand. A model with a rich structure is able to approximate more complicated functions, but, at the same time, has worse *generalization* properties. Good generalization means that a model fitted to one data set will also perform well on another data set from the same process. In fuzzy models, structure selection involves the following choices:

- *Input and output variables.* With complex systems, it is not always clear which variables should be used as inputs to the model. In the case of dynamic systems, one also must estimate the order of the system. For the input-output NARX model (29) this means to define the number of input and output lags n_y and n_u, respectively. Prior knowledge, insight in the process behavior and the purpose of modeling are the typical sources of information for this choice. Sometimes, automatic data-driven selection can be used to compare different choices in terms of some performance criteria.

- *Structure of the rules.* This choice involves the model type (linguistic, singleton, relational, Takagi-Sugeno) and the antecedent form (refer to Section 3.3). Important aspects are the purpose of modeling and the type of available knowledge.

- *Number and type of membership functions for each variable.* This choice determines the level of detail (granularity) of the model. Again, the purpose of modeling and the detail of available knowledge, will influence this choice. Automated, data-driven methods can be used to add or remove membership functions from the model.

- *Type of the inference mechanism, connective operators, defuzzification method.* These choices are restricted by the type of fuzzy model (Mamdani, TS). Within these restrictions, however, some freedom remains, e.g., as to the choice of the conjunction operators, etc. To facilitate data-driven optimization of fuzzy models (learning), differentiable operators (product, sum) are often preferred to the standard min and max operators.

After the structure is fixed, the performance of a fuzzy model can be fine-tuned by adjusting its parameters. Tunable parameters of linguistic models are the parameters of antecedent and consequent membership functions (determine their shape and position) and the rules (determine the mapping between the antecedent and consequent fuzzy regions). In fuzzy relational model, this mapping is encoded in the fuzzy relation. Takagi-Sugeno models have parameters in antecedent membership functions and in the consequent functions ($\mathbf{a}$ and b for the affine TS model).

7.2 Data-Driven Acquisition and Tuning of Fuzzy Models

In this section, we assume that a set of N input-output data pairs $\{(\mathbf{x}_i, y_i)|i = 1, 2, \ldots, N\}$ is available. Recall that $\mathbf{x}_i \in \mathbb{R}^p$ are input vectors and y_i are output scalars. Denote $\mathbf{X} \in \mathbb{R}^{N\times p}$ a matrix having the vectors $\mathbf{x}_k^T$ in its rows, and $\mathbf{y} \in \mathbb{R}^N$ a vector containing the outputs y_k:

$$\mathbf{X} = [\mathbf{x}_1, \ldots, \mathbf{x}_N]^T, \quad \mathbf{y} = [y_1, \ldots, y_N]^T. \tag{31}$$

7.3 Least-Squares Estimation of Consequents

Note that the defuzzification formulas of the singleton and TS models are linear in the consequent parameters, $\mathbf{a}_i$, b_i (see equations (18) and (24), respectively). Hence, these parameters can be estimated from the available data by least-squares techniques. Denote $\mathbf{\Gamma}_i \in \mathbb{R}^{N\times N}$ the diagonal matrix having the normalized membership degree $\gamma_i(\mathbf{x}_k)$ of (25) as its kth diagonal element. By appending a unitary column to $\mathbf{X}$, the extended matrix $\mathbf{X}_e = [\mathbf{X}, \mathbf{1}]$ is created. Further, denote $\mathbf{X}'$ the matrix in $\mathbb{R}^{N\times KN}$ composed of the products of matrices $\mathbf{\Gamma}_i$ and $\mathbf{X}_e$

$$\mathbf{X}' = [\mathbf{\Gamma}_1\mathbf{X}_e,\ \mathbf{\Gamma}_2\mathbf{X}_e,\ \ldots,\ \mathbf{\Gamma}_K\mathbf{X}_e]\,. \tag{32}$$

The consequent parameters $\mathbf{a}_i$ and b_i are lumped into a single parameter vector $\theta \in \mathbb{R}^{K(p+1)}$:

$$\theta = \left[\mathbf{a}_1^T, b_1, \mathbf{a}_2^T, b_2, \ldots, \mathbf{a}_K^T, b_K\right]^T . \tag{33}$$

Given the data $\mathbf{X}$, $\mathbf{y}$, eq. (24) now can be written in a matrix form, $\mathbf{y} = \mathbf{X}'\theta + \epsilon$. It is well known that this set of equations can be solved for the parameter θ by:

$$\theta = \left[(\mathbf{X}')^T\mathbf{X}'\right]^{-1}(\mathbf{X}')^T\mathbf{y} . \tag{34}$$

This is an optimal least-squares solution which gives the minimal prediction error, and as such is suitable for prediction models. At the same time, however, it may bias the estimates of the consequent parameters as parameters of local models. If an accurate estimate of local model parameters is desired, a weighted least-squares approach applied per rule may be used:

$$[\mathbf{a}_i^T, b_i]^T = \left[\mathbf{X}_e^T\mathbf{\Gamma}_i\mathbf{X}_e\right]^{-1}\mathbf{X}_e^T\mathbf{\Gamma}_i\mathbf{y} . \tag{35}$$

In this case, the consequent parameters of individual rules are estimated independently of each other, and therefore are not "biased" by the interactions of the rules. By omitting $\mathbf{a}_i$ for all $1 \le i \le K$, and by setting $\mathbf{X}_e = \mathbf{1}$, equations (34) and (35) directly apply to the singleton model (17).

7.4 Neuro-Fuzzy Modeling

We have seen that parameters that are linearly related to the output can be (optimally) estimated by least-squares methods. In order to optimize also the parameters which are related to the output in a nonlinear way, training algorithms known from the area of neural networks and nonlinear optimization can be employed. These techniques exploit the fact that, at the computational level, a fuzzy model can be seen as a layered structure (network), similar to artificial neural networks. Hence, this approach is usually referred to as neuro-fuzzy modeling [19,20,30,34]. Figure 11 gives an example of a singleton fuzzy model with two rules represented as a network. The rules are:

If x_1 is A_{11} **and** x_2 is A_{21} **then** $y = b_1$.

If x_1 is A_{12} **and** x_2 is A_{22} **then** $y = b_2$.

The nodes in the first layer compute the membership degree of the inputs in the antecedent fuzzy sets. The product nodes Π in the second layer represent the antecedent conjunction operator. The normalization node N and the summation node Σ realize the fuzzy-mean operator (24).

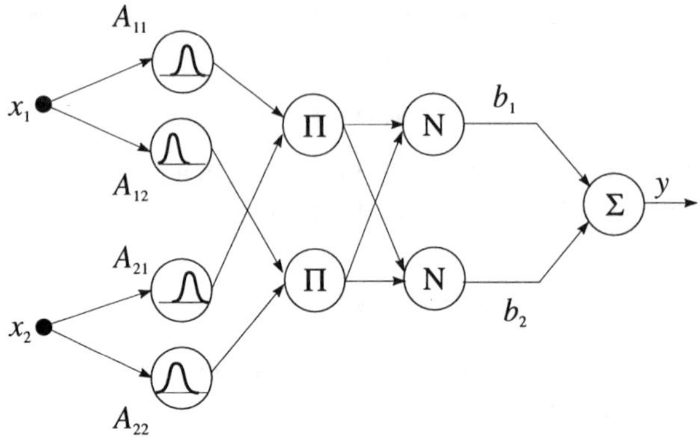

Figure 11. An example of a singleton fuzzy model with two rules represented as a (neuro-fuzzy) network.

By using smooth antecedent membership functions, such as the Gaussian functions:

$$\mu_{A_{ij}}(x_j; c_{ij}, \sigma_{ij}) = \exp\left(-(\frac{x_j - c_{ij}}{2\sigma_{ij}})^2\right), \tag{36}$$

the c_{ij} and σ_{ij} parameters can be adjusted by gradient-descent learning algorithms, such as back-propagation [30]. This allows for a fine-tuning of the fuzzy model to the available data in order to optimize its prediction accuracy.

7.5 Fuzzy Clustering

Identification methods based on fuzzy clustering originate from data analysis and pattern recognition, where the concept of graded membership is used to represent the degree to which a given object, represented as a vector of features, is similar to some prototypical object. The degree of similarity can be calculated using a suitable distance measure. Based on the similarity, feature vectors can be clustered such that the vectors within a cluster are as similar (close) as possible, and vectors from different clusters are as dissimilar as possible.

This idea is depicted in Fig. 12a, where the data is clustered into two groups with prototypes $\mathbf{v}_1$ and $\mathbf{v}_2$, using the Euclidean distance measure. The partitioning of the data is expressed in the *fuzzy partition matrix* whose elements μ_{ij} are degrees of membership of the data points $[x_i, y_i]$ in a fuzzy cluster with prototypes v_j.

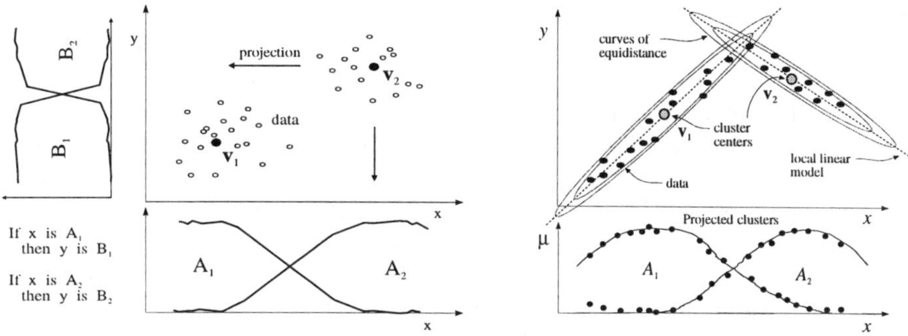

(a) Rule-based interpretation of fuzzy clusters.

(b) Hyperellipsoidal fuzzy clusters.

Figure 12. Identification by fuzzy clustering. Reproduced from, [35] ©1997 Taylor & Francis Ltd.

Fuzzy if-then rules can be extracted by projecting the clusters onto the axes. Figure 12a shows a data set with two apparent clusters and two associated fuzzy rules. The concept of similarity of data to a given prototype leaves enough space for the choice of an appropriate distance measure and of the character of the prototype itself. For example, the prototypes can be defined as linear subspaces, [36] or the clusters can be ellipsoids with adaptively determined elliptical shape, [37] see Fig. 12b. From such clusters, the antecedent membership functions and the consequent parameters of the Takagi–Sugeno model can be extracted: [38]

$$\begin{aligned} &\textbf{If } x \text{ is } A_1 \textbf{ then } y = a_1 x + b_1 , \\ &\textbf{If } x \text{ is } A_2 \textbf{ then } y = a_2 x + b_2 . \end{aligned}$$

Each obtained cluster is represented by one rule in the Takagi–Sugeno model. The membership functions for fuzzy sets A_1 and A_2 are generated by point-wise projection of the partition matrix onto the antecedent variables. These point-wise defined fuzzy sets are then approximated by a suitable parametric function. The consequent parameters for each rule are obtained as least-squares estimates (34) or (35).

8 Control Based on Fuzzy Models

Traditional knowledge-based fuzzy control [11] can be viewed as a heuristic, or model-free, control approach. The controller is designed and tuned without any explicit model knowledge about the process. However, fuzzy control can also be used when an explicit process model is available. In the academic world there is a large amount of work being done on model-based fuzzy control. The key issues here are analysis and synthesis methods for fuzzy control systems. The predominating standpoint is to view fuzzy control as special case of nonlinear control where the fuzzy logic is used for parameterizing the nonlinearities. Alternatively, fuzzy systems are viewed as function approximation schemes that can be used for nonlinear system identification.

8.1 Inverse Control

The simplest approach to design a controller for a nonlinear process when a process model is available is to use inverse control. This approach can also be used in fuzzy control, in which case the controller is based on the inverse of the fuzzy process model.

The inversion is here only explained for SISO fuzzy models without any delay from the input to the output. In this case the fuzzy model can be written as

$$y(k+1) = \mathcal{F}(x(k), u(k)),$$

where $x(k)$ is the state vector consisting of past process outputs and process inputs, i.e., $x(k) = [y(k), \ldots, y(k-n_y+1), u(k-1), \ldots, u(k-n_u+1)]$, $u(k)$ is the current process input, and $\mathcal{F}(\cdot,\cdot)$ is a fuzzy mapping. This model predicts the system's output at the next sample time. The objective of the inverse control method is to compute the input $u(k)$, such that the system output at the next sampling instant is equal to the desired (reference) output $r(k+1)$. This can be achieved if the process model can be inverted according to:

$$u(k) = \mathcal{F}^{-1}(x(k), r(k+1))$$

If it is possible to obtain the inverse model, an open-loop (feedforward) control scheme can be used according to Figure 13.

As with any open-loop control scheme, fuzzy inverse control only works if there is no disturbances or modeling errors. Inversion based approaches can also only be applied to stable systems with a minimum phase behavior (systems whose inverted dynamics are stable).

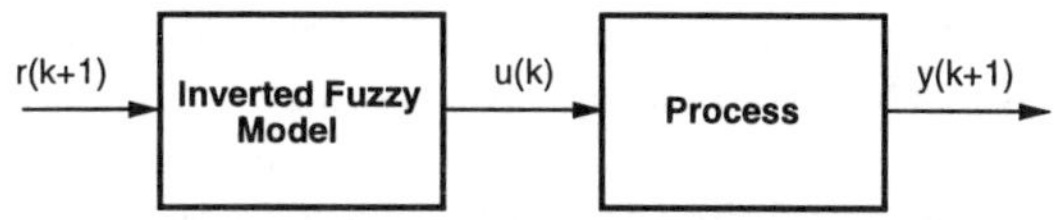

Figure 13. Fuzzy inverse control.

Fuzzy inverse control has been proposed by a number of authors, e.g., [39,12,16,40,41]. In most cases only an approximate inverse is used. In [7,42] exact fuzzy inverse schemes are given for the case of fuzzy singleton models.

8.2 Internal Model Control

Disturbances and model errors cause problems for open-loop schemes. The internal model control (IMC) scheme is one way of compensating for this. A fuzzy IMC controller scheme consists of of three parts: the controller based on an inverse of the process model, the fuzzy model, and a feedback filter, according to Figure 14.

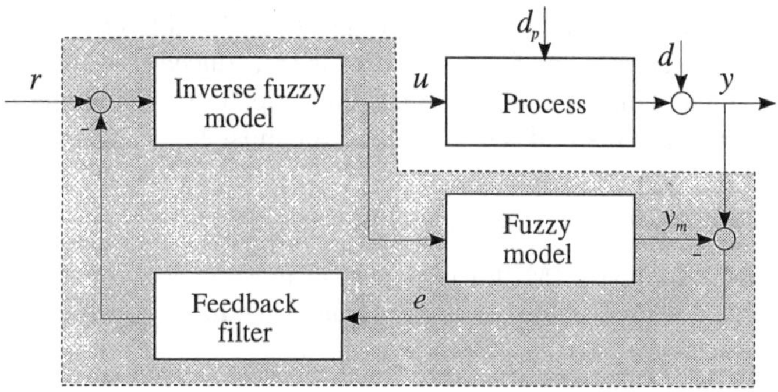

Figure 14. Fuzzy internal model control scheme.

The purpose of the fuzzy model working in parallel with the process is to subtract the effect of the control action from the process output. If the predicted and the measured process outputs are equal, the error e is zero and the controller works in the feedforward configuration. If a disturbance d acts on the process output, the feedback signal e is equal to the influence of the disturbance and is not affected by the effects of the control action. This signal is simply subtracted from the reference and the controller works in the open-loop feedback structure. With a perfect process model, the IMC scheme is hence able to cancel the effect of unmeasured output-additive disturbances and does not suffer from the disadvantages of feedforward controllers. However, the same is not true for input-additive disturbances, e.g., load disturbances.

Two basic properties of the ideal IMC are inherent stability and perfect control. Inherent stability means that if the controller and the process are input–output stable and a perfect model of the process is available, the closed loop system is input–output stable. If the system is not input–output stable, but it can be stabilized by feedback, IMC still can be applied. Perfect control means that if the controller is an exact inverse of the model, and the closed-loop system is stable, then the control is error-free, i.e. $y(k) = r(k), \forall k$. Control without steady-state offsets is attained for asymptotically constant references.

However, in practice, the model is never an exact representation of the process. The feedback signal then contains both the effect of unmeasured disturbances and the effects of modeling errors, and it becomes a true feedback. For large modeling errors it deteriorates the performance of the control system and may introduce stability problems. An important difference between linear and nonlinear systems is that for linear systems, disturbances can be assumed to act additively at the output, while for nonlinear systems, unmeasured disturbances d_p acting on the process generally lead to differences between the model and the process. The feedback filter is introduced in order to filter out the measurement noise and to stabilize the loop by reducing the loop gain. With nonlinear systems and models, the filter must be designed empirically.

8.3 Model Predictive Control

Model predictive control (MPC) is a general methodology of solving control problems in the time domain. It is based on three main concepts:

1. Explicit use of a model to predict the process output at future discrete time instants, over a prediction horizon.

2. Computation of a sequence of future control actions over a control horizon by minimizing a given objective function, such that the predicted process output is as close as possible to a desired reference signal.

3. Receding horizon strategy, so that only the first control action in the sequence is applied, the horizons are moved towards the future and optimization is repeated.

The future process outputs are predicted over the *prediction horizon* H_p using a model of the process. The predicted output values, denoted $\hat{y}(k+i)$ for

$i = 1, \ldots, H_p$, depend on the state of the process at the current time k (for input–output models, for instance, represented by a collection of past inputs and outputs) and on the future control signals $u(k+i)$ for $i = 0, \ldots, H_c - 1$, where H_c is the *control horizon*. If H_c is chosen such that $H_c < H_p$, the control signal is manipulated only within the control horizon and remains constant afterwards, i.e., $u(k+i) = u(k+H_c-1)$ for $i = H_c, \ldots, H_p - 1$, see Figure 15.

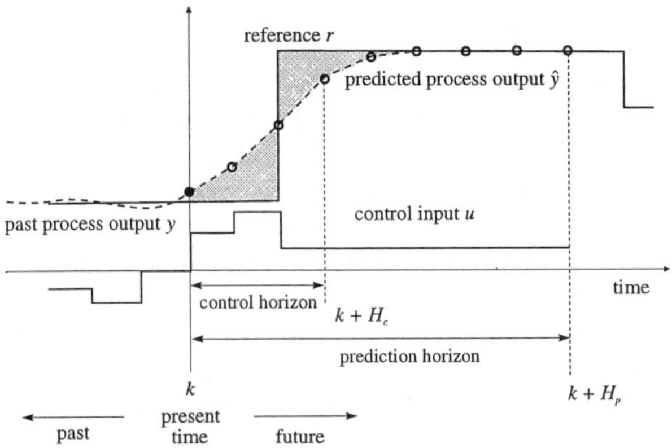

Figure 15. The basic principle of model predictive control.

Objective function

The sequence of future control signals $u(k+i)$ for $i = 0, \ldots, H_c - 1$ is computed by optimizing a given objective (cost) function, in order to bring and keep the process output as close as possible to the given reference trajectory r, which can be the set-point itself or, more often, some filtered version of it. Most often used objective functions are modifications of the following quadratic function [43]:

$$J = \sum_{i=1}^{H_p} \alpha_i \left(r(k+i) - \hat{y}(k+i)\right)^2 + \sum_{i=1}^{H_c} \beta_i \Delta u(k+i-1)^2 \tag{37}$$

The first term accounts for minimizing the variance of the process output from the reference, while the second term represents a penalty on the control effort (related for instance to energy). The latter term can also be expressed by using u itself or other filtered forms of u, depending on the problem [44]. The vectors α and β define the weighting of the output error and the control effort with respect to each other and with respect to the prediction step. Constraints, e.g., level and rate constraints of the control input or other process variables can be specified as a part of the optimization problem. Generally, any other suitable cost function can be used, but for a quadratic cost function, a linear, time-invariant model, and in the absence of constraints, an explicit analytic solution of the above optimization problem can be obtained. Otherwise, numerical (usually iterative) optimization methods must be used.

Receding horizon principle

Only the control signal $u(k)$ is applied to the process. At the next sampling instant, the process output $y(k+1)$ is available and the optimization and prediction can be repeated with the updated values. This is called the receding horizon principle. The control action $u(k+1)$ computed at time step $k+1$ will be generally different from the one calculated at time step k, since more up-to-date information about the process is available.

Predictive control can be regarded as a generalization of the inverse model control approach. Without constraints and without penalizing the control action, the one-step-ahead predictive control strategy is equivalent to the inverse model control strategy, where the inverse is computed numerically, by means of function minimization. Extending the prediction and control horizons, adding the control signal to the objective function and including constraints can be regarded as generalization of model inverse control.

Because of the optimization approach and the explicit use of the process model, MPC can realize multivariable optimal control, deal with nonlinear processes, and can efficiently handle constraints. Since a model of the process is a part of the control scheme, this model can be adapted on-line in order to minimize the difference between the expected and real process outputs. Fuzzy sets can be applied in several ways in the context of MPC, e.g., at the modeling level [45,46,47,48,49,50], in optimization [51], and in the specification of the control objectives [52].

Fuzzy Models in MPC

A fuzzy model acting as a numerical predictor of the process' output can be directly integrated in the MPC scheme shown in Fig. 16. The IMC scheme is usually employed to compensate for the disturbances and modeling errors.

Since fuzzy models are in general nonlinear models, the associated optimization problem in MPC is inherently nonlinear and generally non-convex. The most straightforward way of minimizing (37) is to use some numerical

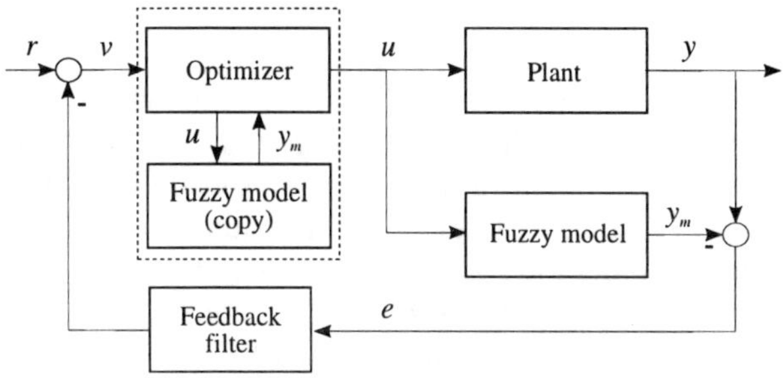

Figure 16. Fuzzy model in the MPC scheme with an internal model and a feedback to compensate for disturbances and modeling errors.

optimization technique, such as the Nelder-Mead method [53] or sequential quadratic programming [54]. These algorithms, however, require a significant computing power which may be a serious obstacle in a real-time implementation. Moreover, convergence to a global minimum is not guaranteed and the algorithm can be trapped in a local minimum, which may result in undesired control actions and poor performance. Also the number of iterations needed to reach a solution can drastically differ from sample to sample. Premature termination of the iteration process is often necessary to fulfill the real-time constraints, which results in inferior solutions and hence a decreased control performance. Alternative optimization schemes are sought which can solve the optimization problem satisfactorily, i.e., find the global minimum and fulfill the real-time requirements.

One possibility is to look for nonlinear models of a specific structure, e.g., nonlinear affine models. For instance, an input–output TS model which does not contain the inputs in the antecedent part is an affine nonlinear form. For affine models, input–output linearization techniques can be applied to obtain a globally linearized process model [55]. The main idea of this technique is that, if the process is minimum-phase, it is possible to linearize it by means of a nonlinear static feedback law. Additionally, if the nonlinear process model is an affine input–output model, and assuming no modeling errors, the linearization scheme is exact, i.e., the resulting linear system is valid over the complete operating range. This is a main difference from the classical linearization through Taylor expansions, where resulting linear model is valid only in the neighborhood of the nominal operating point. In the absence of constraints, a linear predictive control law can be applied to the linearized process model.

9 Conclusions

Fuzzy modeling is a framework in which different modeling and identification methods are combined, providing, on the one hand, a transparent interface with the designer or the operator and, on the other hand, a flexible tool for nonlinear system modeling and control. The rule-based character of fuzzy models allows for a model interpretation in a way that is similar to the one humans use to describe reality. Conventional methods for statistical validation based on numerical data can be complemented by the human expertise, that often involves heuristic knowledge and intuition.

Fuzzy models can be involved in model-based control schemes such as inverse control, internal model control and model predictive control. In this way, a powerful control design methodology is available that allows us to develop controllers for nonlinear systems in a consistent and systematic manner.

References

1. S. Yasunobu and S. Miyamoto. Automatic train operation system by predictive fuzzy control. In M. Sugeno, editor, *Industrial Applications of Fuzzy Control*, pages 1–18. North-Holland, 1985.
2. J.J. Østergaard. FUZZY II – the new generation of high level kiln control. *Zement-Kalk-Gips*, (11), 1990.
3. Kandel and Langholz, editors. *Fuzzy Control Systems.* CRC Press, Boca-Raton, Fl., 1994.
4. E.H. Mamdani. Applications of fuzzy algorithms for control of simple dynamic plant. In *Proceedings IEE*, number 121, pages 1585–1588, 1974.
5. R. Palm, D. Driankov, and H. Hellendoorn. *Model Based Fuzzy Control.* Springer, Berlin, 1997.
6. D. Driankov and R. Palm, editors. *Advances in Fuzzy Control.* Springer, Heidelberg, Germany, 1998.
7. R. Babuška. *Fuzzy Modeling for Control.* Kluwer Academic Publishers, Boston, 1998.
8. B. Kosko. Fuzzy systems as universal approximators. *IEEE Trans. Computers*, 43:1329–1333, 1994.
9. L.-X. Wang. *Adaptive Fuzzy Systems and Control, Design and Stability Analysis.* Prentice Hall, New Jersey, 1994.
10. X.J. Zeng and M.G. Singh. Approximation theory of fuzzy systems – MIMO case. *IEEE Trans. Fuzzy Systems*, 3(2):219–235, 1995.
11. K.-E. Arzén and M. Johansson. Fuzzy control: From heuristic PID to optimization-based nonlinear control. In *this volume.*

12. W. Pedrycz. *Fuzzy Control and Fuzzy Systems (second, extended,edition).* John Willey and Sons, New York, 1993.
13. L.A. Zadeh. Fuzzy sets. *Information and Control*, 8:338–353, 1965.
14. L.A. Zadeh. Outline of a new approach to the analysis of complex systems and decision processes. *IEEE Trans. Systems, Man, and Cybernetics*, 1:28–44, 1973.
15. E.H. Mamdani. Application of fuzzy logic to approximate reasoning using linguistic systems. *Fuzzy Sets and Systems*, 26:1182–1191, 1977.
16. D. Driankov, H. Hellendoorn, and M. Reinfrank. *An Introduction to Fuzzy Control.* Springer, Berlin, 1993.
17. G.J. Klir and B. Yuan. *Fuzzy sets and fuzzy logic; theory and applications.* Prentice Hall, 1995.
18. J.H. Friedman. Multivariate adaptive regression splines. *The Annals of Statistics*, 19(1):1–141, 1991.
19. J.-S.R. Jang and C.-T. Sun. Functional equivalence between radial basis function networks and fuzzy inference systems. *IEEE Transactions on Neural Networks*, 4(1):156–159, January 1993.
20. M. Brown and C. Harris. *Neurofuzzy Adaptive Modelling and Control.* Prentice Hall, New York, 1994.
21. W. Pedrycz. An identification algorithm in fuzzy relational systems. *Fuzzy Sets and Systems*, 13:153–167, 1984.
22. R. Babuška and H.B. Verbruggen. An overview of fuzzy modeling for control. *Control Engineering Practice*, 4(11):1593–1606, 1996.
23. T. Takagi and M. Sugeno. Fuzzy identification of systems and its application to modeling and control. *IEEE Trans. Systems, Man and Cybernetics*, 15(1):116–132, 1985.
24. K. Tanaka and M. Sugeno. Stability analysis and design of fuzzy control systems. *Fuzzy Sets and Systems*, 45(2):135–156, 1992.
25. J. Zhao. *Fuzzy logic in modeling and control.* PhD dissertation, CESAME, Louvain la Neuve, Belgium, 1995.
26. K. Tanaka, T. Ikeda, and H.O. Wang. Robust stabilization of a class of uncertain nonlinear systems via fuzzy control: Quadratic stability, H_∞ control theory and linear matrix inequalities. *IEEE Transactions on Fuzzy Systems*, 4(1):1–13, 1996.
27. L.-X. Wang. Fuzzy systems are universal approximators. In *Proc. IEEE Int. Conf. on Fuzzy Systems 1992*, pages 1163–1170, San Diego, USA, March 1992.
28. I.J. Leonaritis and S.A. Billings. Input-output parametric models for non-linear systems. *International Journal of Control*, 41:303–344, 1985.
29. H.-J. Zimmermann. *Fuzzy Sets, Decision Making and Expert Systems.*

Kluwer Academic Publishers, Boston, 1987.
30. J.-S.R. Jang, C.-T. Sun, and E. Mizutani. *Neuro-Fuzzy and Soft Computing; a Computational Approach to Learning and Machine Intelligence.* Prentice-Hall, Upper Sadle River, 1997.
31. Y. Yoshinari, W. Pedrycz, and K. Hirota. Construction of fuzzy models through clustering techniques. *Fuzzy Sets and Systems*, 54:157–165, 1993.
32. Y. Nakamori and M. Ryoke. Identification of fuzzy prediction models through hyperellipsoidal clustering. *IEEE Trans. Systems, Man and Cybernetics*, 24(8):1153–73, 1994.
33. R. Babuška, J.M. Sousa, and H.B. Verbruggen. Inverse fuzzy model based predictive control. In D. Driankov and R. Palm, editors, *Advances in Fuzzy Control*, pages 129–154. Springer, Heidelberg, Germany, 1998.
34. J.-S.R. Jang. ANFIS: Adaptive-network-based fuzzy inference systems. *IEEE Transactions on Systems, Man & Cybernetics*, 23(3):665–685, May 1993.
35. R. Babuška and H.B. Verbruggen. Fuzzy set methods for local modeling and identification. In R. Murray-Smith and T. A. Johansen, editors, *Multiple Model Approaches to Nonlinear Modeling and Control*, pages 75–100. Taylor & Francis, London, UK, 1997.
36. J.C. Bezdek. *Pattern Recognition with Fuzzy Objective Function.* Plenum Press, New York, 1981.
37. D.E. Gustafson and W.C. Kessel. Fuzzy clustering with a fuzzy covariance matrix. In *Proc. IEEE CDC*, pages 761–766, San Diego, CA, USA, 1979.
38. R. Babuška and H.B. Verbruggen. Identification of composite linear models via fuzzy clustering. In *Proceedings European Control Conference*, pages 1207–1212, Rome, Italy, September 1995.
39. M. Braae and D.A. Rutherford. Theoretical and linguistic aspects of the fuzzy logic controller. *Automatica*, 15:553–577, 1979.
40. C.J. Harris, C.G. Moore, and M. Brown. *Intelligent Control, Aspects of Fuzzy Logic and Neural Nets.* World Scientific, Singapore, 1993.
41. C. Raymond, S. Boverie, and A Titli. Fuzzy multivariable control design from the fuzzy system model. In *Proceedings Sixth IFSA World Congress*, Sao Paulo, Brazil, July 1995.
42. P. Baranyi, I.M. Bavelaar, R. Babuška, L.T. Kóczy, A. Titli, and H.B. Verbruggen. A method to invert a linguistic fuzzy model. *International Journal of Systems Science*, 29(7):711–721, 1998.
43. D.W. Clarke, C. Mohtadi, and P.S. Tuffs. Generalised predictive control. part 1: The basic algorithm. part 2: Extensions and interpretations. *Automatica*, 23(2):137–160, 1987.

44. A.R.M. Soeterboek. *Predictive Control - A Unified Approach.* PhD dissertation, Delft University of Technology, Delft, The Netherlands, November 1990.
45. J.M. Sousa, R. Babuska, and H.B. Verbruggen. Fuzzy predictive control applied to an air-conditioning system. *To appear in Control Engineering Practice*, 1997.
46. Y. Nakamori. Fuzzy modeling for adaptive process control. In A. Kanel and G. Langholz, editors, *Fuzzy Control Systems.* CRC Press, 1994.
47. M. Pottmann and D.E. Seborg. A nonlinear predictive control strategy based on radial basis function models. *Comp. Chem. Engng.*, 21:965–980, 1997.
48. M. Fischer, M. Schmidt, and K. Kavsel-Biasizzo. Nonlinear predictive control based on the extraction of step response models from Takagi-Sugeno fuzzy systems. In *Proc. of the American Control Conference*, pages 1210–1216, 1997.
49. J. Valente de Oliveira and J.M. Lemos. Long-range predictive adaptive fuzzy relational control. *Fuzzy Sets and Systems*, 70:337–357, 1995.
50. J.A. Roubos, R. Babuska, P.M. Bruijn, and H.B. Verbruggen. Predictive control by local linearization of a Takagi-Sugeno fuzzy model. In *Proc. of the IEEE International Conference on Fuzzy Systems*, 1998.
51. Y.-Z. Lu, M. He, and C.-W. Xu. Fuzzy modeling and expert optimization control for industrial processes. *IEEE Trans. Control Systems Tech.*, 5:2–12, 1997.
52. U. Kaymak, J.M. Sousa, and H.B. Verbruggen. A comparative study of fuzzy and conventional criteria in model-based predictive control. In *Proc. of IEEE International Conference on Fuzzy Systems*, volume 2, pages 907–914, 1997.
53. G.R. Walsh. *Methods of optimization.* John Wiley & Sons, New York, USA, 1975.
54. P.E. Gill, W. Murray, and M.H. Wright. *Practical Optimization.* Academic Press, New York and London, 1981.
55. H. Nijmeijer and A.J. van der Schaft. *Nonlinear Dynamical Control Systems.* Springer-Verlag, New York, USA, 1990.

FUZZY CONTROL: FROM HEURISTIC PID TO OPTIMIZATION-BASED NONLINEAR CONTROL

K.-E. ÅRZÉN AND M. JOHANSSON
Department of Automatic Control, Lund Institute of Technology,
P.O. Box 118, 221 00 Lund, Sweden
Email: karlerik@control.lth.se

A tutorial on fuzzy control is presented. It is based on the viewpoint that fuzzy systems are a special parameterization form of nonlinear mappings and that fuzzy modeling and control, therefore, from many respects can be seen as a special class of nonlinear modeling and control. Emphasis is given to heuristic design of fuzzy controllers and to model-based fuzzy control, specially Lyapunov-based analysis and synthesis methods.

1 Introduction

Fuzzy sets, the foundation of fuzzy control, were introduced thirty years ago, [1], as a way of expressing non-probabilistic uncertainties. Since then, fuzzy set theory has developed and found applications in database management, operations analysis, decision support systems, signal processing, data classifications, computer vision, etc. The application area that has attracted most attention is, however, control. In 1974, the first successful application of fuzzy logic to control was reported [2]. Control of cement kilns was an early industrial application [3]. Since the first consumer product using fuzzy logic was marketed in 1987, the use of fuzzy control has increased substantially. A number of CAD environments for fuzzy control design have emerged together with VLSI hardware for fast execution. Fuzzy control is being applied industrially in an increasing number of cases, e.g., [4], [5], [6], [7], and [8].

The early work in fuzzy control was motivated by a desire to

- directly express the control actions of an experienced human operator in the controller, i.e., to mimic his behavior, and
- to obtain smooth interpolation between discrete controller outputs.

Since then the application range of fuzzy control has widened substantially. However, the two main motivations still persevere. The linguistic nature of fuzzy control makes it possible to express process knowledge concerning how the process should be controlled or how the process behaves. The interpolation aspect of fuzzy control has led to the viewpoint where fuzzy systems are seen as smooth function approximation schemes.

In most cases a fuzzy controller is used for direct feedback control. However, it can also be used on the supervisory level as, e.g., a self-tuning device in a conventional PID controller. Also, fuzzy control is no longer only used to directly express a priori process knowledge. For example, a fuzzy controller can be derived from a fuzzy model obtained through system identification. Therefore, it is difficult to define what a fuzzy controller is. A very general definition is:

> A *Fuzzy Controller* is a controller that contains an, often non-linear, mapping that has been defined using fuzzy logic-based rules.

The key issues in this definition are the non-linear mapping and the fuzzy logic-based rules.

Writing a tutorial on fuzzy control is not an easy task. The reason for this is the dual role that fuzzy systems play. They can be seen as rule-based systems that can be used to describe system behaviors in linguistic and qualitative terms. They can also be seen as nonlinear function approximation schemes. The second reason why it is difficult is the vast theoretical body that has been developed within fuzzy logic since it was introduced in the 1960s. However, luckily enough, the majority of the work in fuzzy control only makes use of a small subset of the general fuzzy logic theory. A third reason for problems is the close relationship between fuzzy systems and neural networks. In many cases there is a one-to-one mapping between fuzzy systems and neural networks. This is the reason for the use of the term 'neuro-fuzzy' systems. This implies that many of the results that have been developed in the neural network community can directly be mapped over to the fuzzy systems area. The approach taken in this tutorial is based on the viewpoint that fuzzy systems are a special parameterization form of nonlinear mappings and that fuzzy modeling and control, therefore, from many respects can be seen as a special class of nonlinear modeling and control.

Many important issues have been left out due to space limitations. Examples are neuro-fuzzy control and fuzzy adaptive control. For a more detailed treatment we refer to the recent survey [9]. The chapter also does not explain in detail how the fuzzy inference process is performed. This can, however, be found in a number of other texts on fuzzy control.

Chapter Outline

Section 2 discusses the basic motivation for using fuzzy control. In Section 3 the two most used types of fuzzy systems, Mamdani systems and Takagi-Sugeno systems, are described. The nonlinear mapping defined by fuzzy rules is just one part of a fuzzy controller. In Section 4, the overall structure of a fuzzy controller is discussed. Fuzzy control can be divided into two areas. In the heuristic, or model-free, fuzzy control, discussed in Section 5, the controller is designed primarily from heuristics. The controllers developed are usually based on the standard PID controller. In model-based fuzzy control the controller is designed from a fuzzy or nonfuzzy process model. Fuzzy versions have been developed for most of the conventional linear and nonlinear control design methods. Model-based fuzzy control is discussed in Section 6. Special emphasis is put on Lyapunov-based methods for design of stabilizing state feedback controllers.

2 Background

Increased industrial demands on quality and performance over a wide range of operating regions have led to an increased interest in nonlinear control methods during recent years. The advent of 'new' techniques such as fuzzy control, neural networks, wavelets, and hybrid systems has amplified the interest. Nonlinear control is of interest, e.g., when the process that should be controlled is nonlinear and/or when the performance specifications are nonlinear. Basically all real processes are nonlinear, either through nonlinear dynamics or through constraints on, e.g., states and inputs.

Nonlinear techniques may be used for process modeling. The derived process model can be used, e.g., as the basis for model-based control design. Here, the model may be used off-line during the design or on-line, as a part of the controller. Nonlinear techniques can also be used to design the controller directly, without any process model. Nonlinear elements can be used in the feedback or in the feedforward path. In practice the nonlinear elements are often combined with linear filters.

A variety of methods may be used to define nonlinearities, e.g, analytical equations, fuzzy systems, sigmoidal neural networks, splines, radial basis functions, wavelets, local models/controllers, discrete switching logic, lookup tables, etc. The methods can be viewed as different ways of parameterizing nonlinearities, see Fig. 1. Many of the methods have been shown to be universal function approximators for certain classes of functions. For example, fuzzy logic systems are universal approximators for continuous functions defined on compact sets [10]. The same holds for sigmoidal neural networks [11]. Wavelets are universal approximators for the more general Besov function class [12]. This means that many of the methods are equivalent with respect to which nonlinearities that they can generate. Hence, it is of little value to argue whether one of the methods is better than the others if

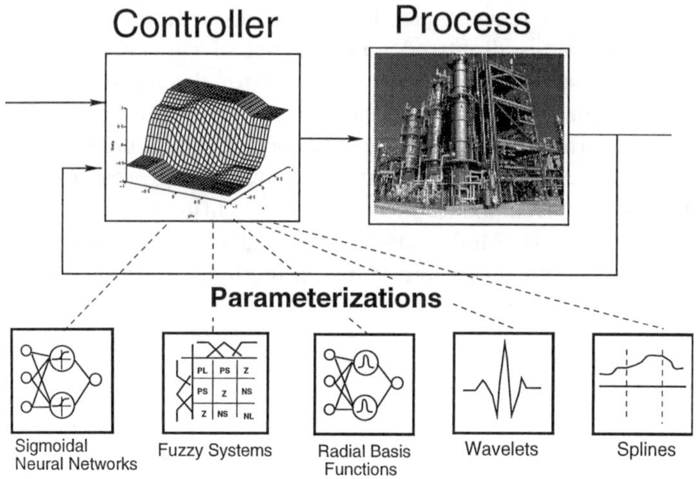

Figure 1. Different parameterizations of nonlinear controllers.

one considers only the closed loop control behavior. From the process' point of view it is the nonlinearity that matters and not how the nonlinearity is parameterized.

When comparing different nonlinear methods there are a number of important points to consider. As pointed out, one of them is the function approximation properties of the method. Another related issue is the efficiency of the approximation method in terms of the number of parameters needed to approximate a given function. The extreme case here is the table lookup method. Of great practical importance is whether the methods are local or global. Local methods allow local adjustments. Examples of local methods are radial basis functions, splines, and fuzzy systems. How well the methods support the generation of nonlinearities from input/output data, i.e., identification/learning/training, is also of large interest. The backpropagation method is well-known in the neural network case. Other important issues are how well theoretically grounded the methods are, i.e., to what extent analysis and synthesis methods are available; how transparent the methods are, i.e., how readable the methods are and how easy it is to express prior process knowledge; the computational efficiency of the method; the availability of computer tools; and finally, subjective preferences such as how comfortable the designer/operator is with the method, and the level of training needed to use and understand the method.

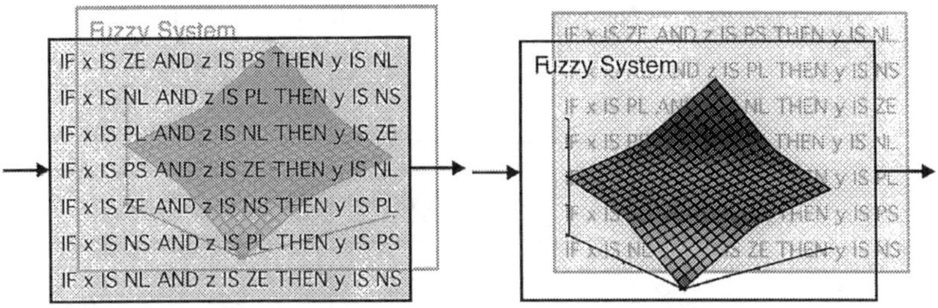

Figure 2. The views of fuzzy control

Fuzzy logic systems compare favorable with respect to most of these criteria. They are universal approximators and, if certain design choices are made, the approximation is reasonable efficient. Depending on how the membership functions are defined the method can be either global or local. It has similar estimation properties as, e.g., sigmoidal neural networks. Fuzzy logic systems can be very transparent and thereby they make it possible to express prior process knowledge well. A number of computer tools are available for fuzzy control implementation.

When working with fuzzy control it is important to always have in mind the two views of fuzzy control in mind. The first view consists of the fuzzy logic-based rules that are used to locally define the nonlinear mapping. This view can be seen as the user interface part of fuzzy systems. The second view consists of the nonlinear mapping that is generated from the rules and the inference process. The two views are shown in Fig. 2.

3 Fuzzy Control Systems

Fuzzy set theory uses fuzzy inferencing to reason about linguistic variables, i.e., variables described by fuzzy sets. A number of different inference systems have been developed. In fuzzy control it is, however, two inference systems that dominate: Mamdani fuzzy systems, also known as linguistic fuzzy systems, and Takagi-Sugeno fuzzy systems.

3.1 Mamdani Fuzzy Systems

A Mamdani Fuzzy System contains a knowledge base consisting of fuzzy IF-THEN rules and membership function definitions, together with an inference engine that applies the fuzzy rules to the fuzzy input variables, generating fuzzy output variables according to Fig. 3. The rules in a Mamdani fuzzy

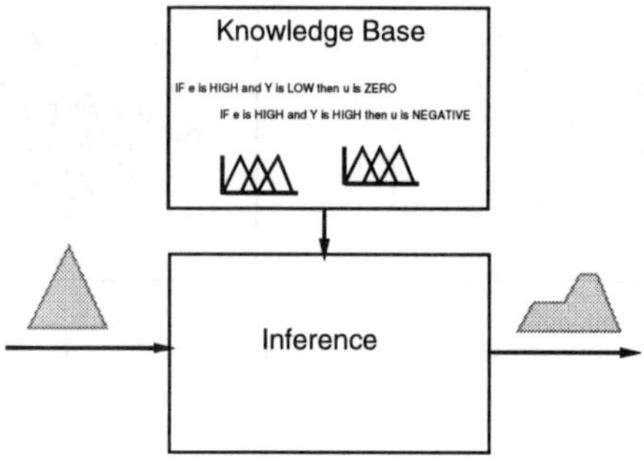

Figure 3. The knowledge base and inference engine of a Mamdani fuzzy system.

systems are on the form

$$\mathcal{R}^{(l)} : \text{ IF } x_1 \text{ IS } A_1^{(l)} \text{ AND } \ldots \text{ AND } x_n \text{ IS } A_n^{(l)} \text{ THEN } y \text{ IS } B^{(l)} \tag{1}$$

where x_i are the input variables and y is the output variable. Both the rule antecedent and consequent are defined using fuzzy sets. In the above example only the AND connective is used. Rule bases on this form are known as being written on conjunctive form. In the general case, also other logical connectives may be used, e.g., OR and NOT.

In fuzzy control the inputs and outputs are crisp rather than fuzzy. Therefore fuzzification and defuzzification are needed. Fuzzification transforms the numerical inputs into matching fuzzy sets, and defuzzification approximates the output fuzzy set with a single crisp number. The structure is extended according to Fig. 4.

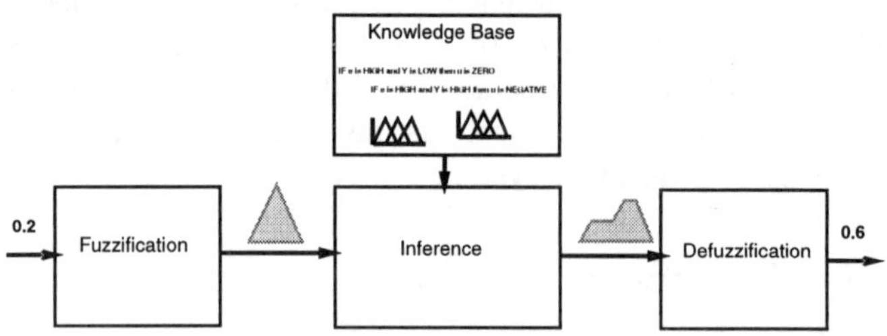

Figure 4. A Mamdani fuzzy system.

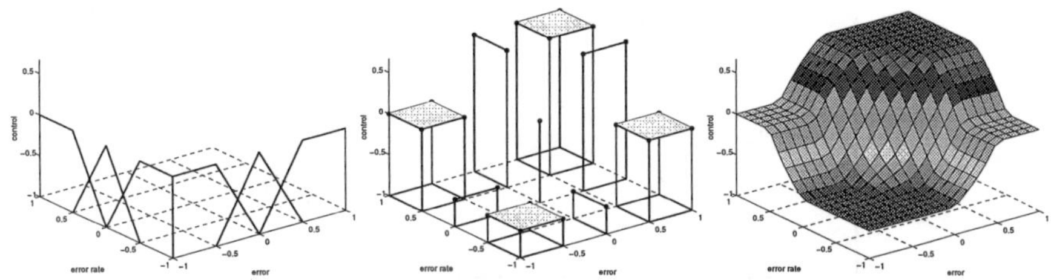

Figure 5. Left: The membership functions partition the input space. Middle: Each rule defines the output value for one point or area in the input space. Right: The fuzzy logic interpolates between the constant values.

In Mamdani fuzzy systems the antecedent and consequent fuzzy sets are often chosen to be triangular or gaussian. It is also common that the input membership functions overlap in such a way that the membership values of the rule antecedents always sum to one. In this case, and if the rule base is on conjunctive form, one can interpret each rule as defining the output value for one point in the input space. The input space point is the point obtained by taking the centers of the input fuzzy sets and the output value is the center of the output fuzzy set. The fuzzy logic is used to obtain smooth interpolation between the points in the input space. The situation is shown in Fig. 5. With this interpretation a Mamdani system can be viewed as defining a piecewise constant function with extensive interpolation. Depending on which inference methods that is used different interpolations are obtained. By proper choices it is even possible to obtain linear or multilinear interpolation.

Mamdani fuzzy systems are quite close in nature to manual control. The controller is defined by specifying what the output should be for a number of of different input signal combinations. Each input signal combination is represented as a rule.

3.2 Takagi-Sugeno Fuzzy Systems

A Takagi-Sugeno (TS) fuzzy system [13] uses rules on the form

$$R^{(l)} : \text{IF } x_1 \text{ IS } A_1^{(l)} \text{ AND } \ldots \text{ AND } x_n \; IS \; A_n^{(l)} \text{ THEN } y = h^{(l)}(x) \qquad (2)$$

The output functions $h^{(l)}(x)$ could, in principal, be arbitrary functions of the inputs. However, in most cases they are chosen to be a linear combination of the inputs, i.e.,

$$h^{(l)}(x) = l_1^{(l)} x_1 + \cdots + l_n^{(l)} x_n$$

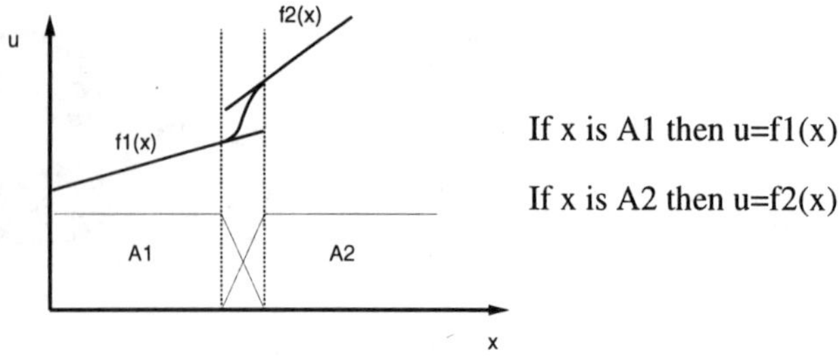

Figure 6. Affine TS fuzzy system in one dimension.

This is called a linear Takagi-Sugeno (TS) system. Sometimes also a constant term is added, i.e.,

$$h^{(l)}(x) = l_0^{(l)} + l_1^{(l)} x_1 + \cdots + l_n^{(l)} x_n.$$

This structure is known as an affine TS system.

In a TS fuzzy system, fuzzy logic is only used in the antecedent part of the rules. The consequent part is described using ordinary, crisp, numbers. The total output from a TS system is obtained as a weighted average of the crisp outputs of the individual rules with the weights taken as the degree of belief in the rules. Hence, the defuzzification is built-in to the inference system.

When TS fuzzy systems are used it is common that the input fuzzy sets are trapezoidal. Each fuzzy set determines a region in the input space where, in the linear case, the output is determined by a linear function of the inputs. Fuzzy logic is only used to interpolate in the cases where the regions in the input space overlap. The situation is illustrated for the one-dimensional case in Fig. 6. As seen in the figure, TS fuzzy systems can be viewed as piecewise linear (affine) functions with limited interpolation.

TS fuzzy systems are close in nature to gainscheduling. For each input region a linear controller is defined. The total controller is obtained by selecting one of the controllers based on the value of the inputs, or by interpolating between several of the linear controllers.

3.3 Fuzzy nonlinear mappings

Fuzzy systems generate static nonlinear input-output mappings. The nonlinear mappings can in general be written as basis function expansions, i.e., as a weighted sum of basis functions $g_i(x)$ according to,

$$f(x;\theta) = \sum_{i=1}^{M} \overbrace{g_i(x)}^{\text{IF-part}} \underbrace{w_i}_{\text{THEN-part}} \tag{3}$$

The exact nature of the basis functions is determined by which inference system that is used and how the fuzzy operations are defined. The basis function expansion can be viewed as canonical form that is common for several nonlinear function approximation methods, e.g., sigmoidal neural networks, radial basis function, splines, and wavelets. Under certain conditions it is possible to get an exact correspondence between certain fuzzy systems and other function approximation schemes, e.g., radial basis functions.

4 Fuzzy Controller Structure

A fuzzy system mapping is just one part of a fuzzy controller. Often, signal processing is required both before and after the fuzzy system evaluation. A general structure that captures most applications of fuzzy systems to control is illustrated in Fig. 7.

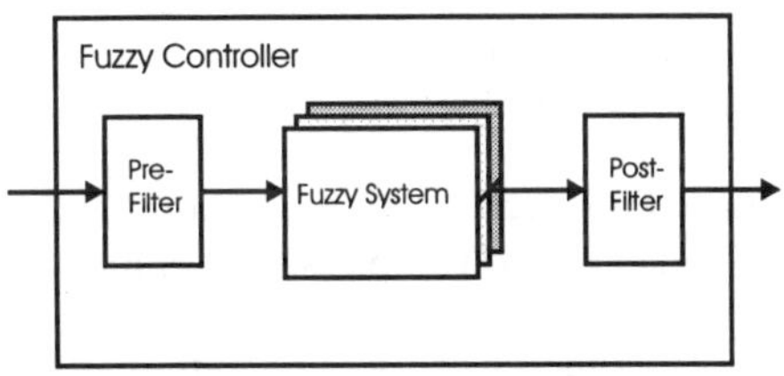

Figure 7. A general fuzzy controller structure, consisting of a prefiltering device, fuzzy system mappings and a postfiltering device.

The structure consists of three parts:

1. A Prefiltering Device – for computing the fuzzy system inputs

2. One (or several) Fuzzy System Mapping(s)

3. A Postfiltering Device – for computing the actual control signal

The prefiltering device represents the signal processing performed on the controller inputs in order to obtain the inputs of the fuzzy system. The prefiltering device may, for instance, perform some of the following operations on the input signals

Signal Conditioning. It is sometimes convenient to work with signals on a normalized domain, e.g., $[-1, 1]$. This is accomplished by the introduction of normalization gains. The normalization gain is a linear gain that scales the input into the normalized domain $[-1, 1]$. Values that fall outside the normalized domain are mapped onto the appropriate endpoint.

Dynamic Filtering. This includes both linear and nonlinear filters. In a fuzzy PID controller, for instance, linear filters are used to obtain the control error, the error derivative and the error integral. Nonlinear filters are found in nonlinear observers, and in adaptive fuzzy control where they are used to obtain the fuzzy system parameter estimates.

Feature Extraction. Through extraction of different features numeric transformations of the controller inputs is performed. These transformations may be Fourier- or Wavelet-based transformations, coordinate transformations or other basic operations performed on the fuzzy controller inputs.

The postfiltering device represents the signal processing performed on the fuzzy system output to obtain the actual control signal. Operations that the postfiltering device may perform include

Signal Conditioning. This can be a denormalization gain that scales the output of the fuzzy system to the physical domain of the actuator signal.

Dynamic Filtering. In some cases, the output of the fuzzy system is the control increments. The actual control signal is then obtained by integrating the control increments. Of course, other forms of smoothing devices and even nonlinear filters may be considered.

Fuzzy control can in principle be applied at all levels in the control system hierarchy from the sensor and actuator level up to the production planning level. However, in practice fuzzy control is mostly used at the direct control level or at the supervisory control level.

At the direct control level the inputs to the controller are the measured values and reference signals and the output from the controller is the control signal(s). Depending on which signals that are used as the input to the fuzzy mapping, different fuzzy controller are obtained that are structurally equivalent to various conventional linear control schemes. For example, a fuzzy controller with the controller error, e, and the error derivative as inputs and the control signal as the output is structurally equivalent to a conventional PD-controller. A controller with the process state vector, x, as the inputs and the control signal as output is structurally equivalent to a state feedback controller. If the nonlinear mapping is designed to be linear then exact equivalence is obtained between the fuzzy controller and the corresponding conventional control scheme.

At the supervisory control level the outputs from the controller are typically reference signals to underlying controllers. An example is the cascade control scheme.

It is also possible to use fuzzy control in combination with a conventional controller. One example is to use a fuzzy system as a gain scheduling device that adjusts the parameters in a conventional controller. Another example is to use a fuzzy controller in parallel with a conventional controller. The outputs of the two controller is added to form the total control signal. The fuzzy controller is designed to only contribute to the control signal in operating regions where the linear control action of the conventional controller is insufficient.

5 Heuristic Fuzzy Control

A heuristic fuzzy controller is designed directly by mimicking the control actions of some human expert, typically a process operator. The controllers developed are usually based on the structure of the standard PID controller. This is not a coincidence. Linear PID control is the most widely used control structure in industry. The reason why PID control has been so successful is that it is intuitive and close to the way humans manually control a process. The control action depends on the current controller error (the P-term), the time history of the error (the I-term), and a prediction of the future value of the error (the D-term). Furthermore, a generic empirical observation is that most industrial processes can be controlled reasonably well with PID control provided that the demands on performance are not too high. Heuristic tuning rules exists for PID controller that work reasonably well for a large class of processes. Hence, it is possible to use a PID controller also in the case when a detailed process model is not available.

Linear PID control is the most widely used control structure in industry.

The PID controller is usually presented as

$$u(t) = K\left(e(t) + \frac{1}{T_i}\int_{\tau=0}^{t} e(\tau)\, d\tau + T_d \frac{d}{dt}e(t)\right) \tag{4}$$

An alternative parameterization of the PID controller is the velocity (or incremental) form

$$\frac{du}{dt} = K\left(\frac{d}{dt}e(t) + \frac{1}{T_i}e(t) + T_d \frac{d^2}{dt^2}e(t)\right) \tag{5}$$

The latter form has several advantages, e.g., less problems with integrator windup and bumpless mode changes.

However, there are many cases when it is an advantage to use a nonlinear PID controller. If the plant is linear, we may for instance improve transient performance by crafting a controller nonlinearity that works as a time optimal controller when the control error is large and as a linear controller when the error is small. If the plant is nonlinear, the controller nonlinearity can be designed so as to compensate for plant nonlinearities. A deadzone is another example of a nonlinearity that is useful in connection with PID control. This reduces the wear on the actuators.

Using fuzzy PID control it is possible to combine the advantages of ordinary linear PID control with the possibility to introduce nonlinearities in the control law. Using fuzzy control is also possible to expand the basic PID scheme in different ways. It is possible to introduce extra input and output signals, thus creating multivariable controllers. It is also possible to use fuzzy logic to implement gain scheduling in different ways and to combine continuous control laws with logic and constraint handling.

The different forms of fuzzy PID control can be summarized as:

Fuzzy P-control:	$u = \mathcal{F}(e(t))$
Fuzzy PD-control:	$u = \mathcal{F}\left(e(t), \frac{d}{dt}e(t)\right)$
Fuzzy PI-control (absolute form):	$u = \mathcal{F}\left(e(t), \int e(\tau)\right)$
Fuzzy PI-control (incremental form):	$\frac{du}{dt} = \mathcal{F}\left(\frac{d}{dt}e(t), e(t)\right)$
Fuzzy PID-control (absolute form):	$u = \mathcal{F}\left(e(t), \int e(\tau), \frac{d}{dt}e(t)\right)$
Fuzzy PID-control (incremental form):	$\frac{du}{dt} = \mathcal{F}\left(\frac{d}{dt}e(t), e(t), \frac{d^2}{dt^2}e(t)\right)$

where $\mathcal{F}(\cdot)$ is a fuzzy system mapping. In fuzzy PID control the incremental forms are the most common. The reason for this is that it is more intuitive to describe the controller actions in terms of changes to the control signal then in terms of the sum of the control error.

The rulebase in a fuzzy PID controller builds up a nonlinearity that, in its basic form, is very close to being linear. A fuzzy PD controller could be obtained with the following nine rules:

IF e IS NL AND $\dot{e}$ IS NL THEN u IS NL
IF e IS ZE AND $\dot{e}$ IS NL THEN u IS NS
IF e IS PL AND $\dot{e}$ IS NL THEN u IS ZE
IF e IS NL AND $\dot{e}$ IS ZE THEN u IS NS
IF e IS ZE AND $\dot{e}$ IS ZE THEN u IS ZE
IF e IS PL AND $\dot{e}$ IS ZE THEN u IS PS
IF e IS NL AND $\dot{e}$ IS PL THEN u IS ZE
IF e IS ZE AND $\dot{e}$ IS PL THEN u IS PS
IF e IS PL AND $\dot{e}$ IS PL THEN u IS PL

where we have introduced the labels NL, NS, ZE, PS and PL to mean *Negative Large*, *Negative Small*, *Zero*, *Positive Small* and *Positive Large* respectively.

The readability is enhanced by presenting the rules in the table

		$\dot{e}$		
		NL	ZE	PL
	PL	ZE	PS	PL
e	ZE	NS	ZE	PS
	NL	NL	NS	ZE

The rule table has the typically stripe diagonal form. In fuzzy control a simple difference $\Delta e = e(k) - e(k-1)$ is often used as a (poor) approximation for the derivative. Starting from this, essentially linear, fuzzy PD controller the shape of the nonlinearity can be easily adjusted by modifying the membership functions or by introducing new rules.

There are also other reasons why fuzzy control has been successful in industry. Fuzzy control is a direct approach to nonlinear control design. The rule-based formalism is intuitive and easy to understand for noncontrol engineers, specially in comparison with modern nonlinear control theory that is all but intuitive and user-friendly. Each rule represents local process knowledge about how the control signal should be selected for certain input signals. The local nature of the rules makes it possible to build up a controller in a step-wise fashion. Another reason for the success is the way the technique is packaged. Commercial CAD environments for fuzzy controller development have user-friendly, graphical environments. They are available on industrially accepted hardware and they can automatically generate C code. This makes it easy for industry to apply fuzzy control. To start experimenting with nonlinear control using existing industrial control systems is considerably more difficult. They often only provide pre-packaged PID blocks and anything in addition to that can be very difficult for the end-users to implement.

6 Model-Based Fuzzy Control

The heuristic approach to fuzzy control dominates the industrial applications. However, in academia most research on fuzzy control is focused on model-based fuzzy control. A model-based control scheme is based on an explicit process model. The model could be used off-line during controller design and/or on-line as a part of the controller, on direct or inverse form. The model could be derived from first principles, from process data, or using a combination of the two approaches.

In fuzzy model-based control the underlying model could either be fuzzy or nonfuzzy. A fuzzy model could either be based on Mamdani inference or on Takagi-Sugeno inference. However, it is the latter case that dominates. An affine TS model on autonomous state space form and in continuous time has the following form

$$\begin{aligned} R_l :\ & \text{IF } x_1 \text{ IS } F_{l,1} \text{ AND } \ldots \text{ AND } x_n \text{ IS } F_{l,n} \\ & \text{THEN } \dot{x} = A_l x + a_l, \end{aligned}$$

For state-space systems with inputs and outputs the rule consequent is replaced by

$$\begin{aligned} \dot{x} &= A_l x + B_l u + a_l \\ y &= C_l x + D_l u + c_l \end{aligned}$$

In discrete time, $\dot{x}$ is replaced by $x(k+1)$. Similar models can be obtained for systems on input-output form. Here the consequents could have the following form in the SISO case, $y(k) = a_1 y(k-1) + a_2 y(k-2) + b_1 u(k-1) + b_2 u(k-2)$, depending on the number of time delays and the system order.

Since the fuzzy logic can be viewed as a parameterization of a general nonlinear mapping, it is not surprising that a number of fuzzy versions of of standard linear and nonlinear control design techniques have been developed. These includes inverse model control, internal model control (IMC), model-predictive control (MPC), feedback linearization, sliding mode control, and Lyapunov-based state feedback design. Inverse model control, IMC and MPC is described in the companion tutorial [14]. Here the focus will be feedback linearization, sliding mode control, and Lyapunov-based design. However, we will first consider the case where a fuzzy system is used to compensate for static nonlinearities.

6.1 Compensation of Static Nonlinearities and Static Scheduling

In some cases, the main nonlinearities of a process are static nonlinearities on the input and output of the system, whereas the system dynamics is linear. In these situations, we can use the approximation capabilities of fuzzy systems to compensate for these nonlinearities.

Example 6.1 (Input Nonlinearities) *Consider a SISO plant with linear dynamics and an input nonlinearity*

$$\dot{x} = Ax + b \cdot g(u)$$

One approach to control this systems is to momentarily disregard the input nonlinearity and design a linear controller

$$v = L(x)$$

for the linear plant. We can then approximate the inverse (which may, or may not exist) of the input nonlinearity using a fuzzy system

$$f(v) \approx g^{-1}(v)$$

and apply the control

$$u = f(L(x))$$

In some cases, it can be motivated to introduce static nonlinearities at the input of a linear system, as described by the next example.

Example 6.2 (Static Scheduling) *In climate control systems, the temperature dynamics of a room can be modeled as a linear system*

$$\dot{x} = Ax + bq$$

where q is the heat delivered by the air conditioner (AC). It is straight forward to design a linear controller for this system. Typically, this controller measures the room temperature and computes the heat that the AC must provide. However, we cannot control the heat of the delivered air stream directly. The heat contained in the air delivered by the AC is proportional to the product of the air mass flow, w, and the temperature, T, of the air stream, i.e.

$$q \propto w \cdot T$$

We may then use fuzzy logic rules to design a schedule

$$w = f_1(q; p)$$
$$T = f_2(q; p)$$

that for a desired heat q and auxiliary operating conditions p suggests the most comfortable combination of air mass flow and air temperature. This control structure is illustrated in Fig. 8.

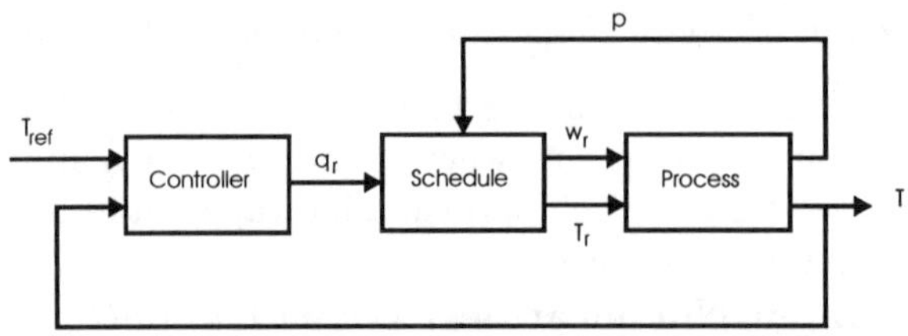

Figure 8. Static scheduling using fuzzy systems.

6.2 Fuzzy Feedback Linearization

Intuitively, a plant with significant nonlinear dynamics should be controlled using a nonlinear controller. However, control design for nonlinear systems is a hard problem, and at present date there are no general design methods available. Some promising progress have been made using techniques like feedback linearization and backstepping designs. Both methods can be applied to certain classes of systems. We choose to illustrate a special case of these design methods which is the nonlinear equivalent to pole placement.

Example 6.3 (Nonlinear Pole Placement) *Consider a nonlinear system on the form*

$$\begin{aligned} \dot{x}_1 &= x_2 \\ \dot{x}_2 &= x_3 \\ &\vdots \\ \dot{x}_{n-1} &= x_n \\ \dot{x}_n &= f(x) + g(x)u \end{aligned} \tag{6}$$

where $f(x)$ and $g(x)$ are nonlinear function of the process state vector. If we know $f(x)$ and $g(x) \neq 0$ perfectly, the feedback

$$u = \frac{1}{g(x)} \left(-f(x) + L^T x + r \right) \tag{7}$$

cancels the nonlinearity and gives the closed loop system the desired linear dynamics [15]

$$\dot{x} = \begin{bmatrix} 0 \; I_{n-1 \times n-1} \\ -L^T \end{bmatrix} x + \begin{bmatrix} 0_{n-1 \times 1} \\ 1 \end{bmatrix} r := A_m x + b_m r \tag{8}$$

If the functions $f(x)$ and $g(x)$ are not known, one approach is to obtain approximate models $\hat{f}(x)$ and $\hat{g}(x)$ from an identification experiment. The *approximations $\hat{f}(x)$ and $\hat{g}(x)$, can for instance be in terms of fuzzy systems whose parameters are optimized to fit the identification data.*

Using a certainty equivalence approach, we may then try the control

$$u = \frac{1}{\hat{g}(x)}\left(-\hat{f}(x) + L^T x + b_m r\right) \tag{9}$$

In this case the cancellation of the nonlinearity is only approximate.

Present approaches to model based fuzzy control in the above spirit simply assumes that the approximation of $f(x)$ and $g(x)$ is "sufficiently good". A more satisfactory approach would be to develop design algorithms that are robust with respect to bounded approximation errors.

6.3 Fuzzy Sliding Mode Control

A simple and highly robust control structure for uncertain nonlinear systems is the so called sliding mode controller [16] [17], [18]. The main idea behind sliding mode control is to transform the problem of stabilizing an nth order system (which is hard) into the problem of stabilizing a 1st order system (which is easier).

Sliding mode control can be demonstrated on the problem of globally stabilizing the system (6). Let $x(t)$ be the n-dimensional system state vector and consider the scalar function $s(x,t) = c^T x(t)$. This function defines a surface S of dimension n-1 in the system state space

$$S : s(x,t) = c_1 x_1 + c_2 x_2 + \cdots + c_n x_n = 0 \tag{10}$$

Substituting the n-1 first state equations of (6) into (10), we have

$$s(x,t) = c_1 x_1 + c_2 \dot{x}_1 + \cdots + c_n x^{(n-1)} = 0 \tag{11}$$

Consequently, S also defines an ordinary differential equation in x_1. If we define the surface S so that the equation (11) is exponentially stable, the system state is guaranteed to converge to the origin if it remains on the surface for all future times.

Thus, the n-dimensional problem of driving the state vector to origin has been transformed into the 1-dimensional problem to forcing the system state onto the surface S and keeping the state on the surface. Forcing the system state to the sliding surface is equivalent of forcing the scalar $s(x;t)$ to zero in finite time. This can be accomplished by finding a control u such that the Lyapunov-like "sliding condition"

$$\frac{1}{2}\frac{d}{dt}s^2(x;t) \leq -\eta |s(x;t)| \tag{12}$$

with $\eta > 0$ is fulfilled. Such a control forces the system state to the sliding surface S, and once on the surface the system state remains there, obeying the dynamics of (11). This is illustrated in Fig. 9.

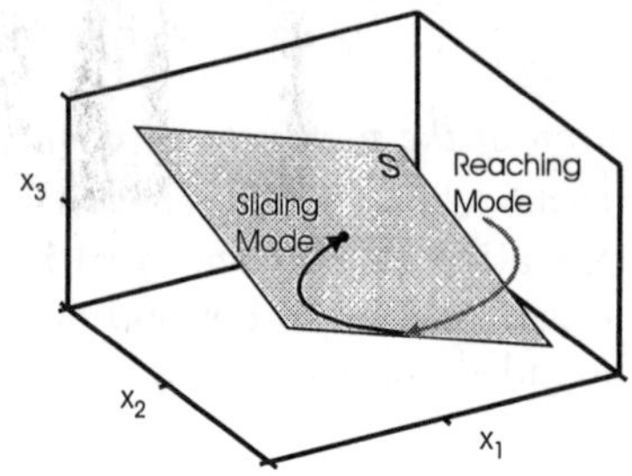

Figure 9. State trajectories for a system under sliding mode control

Next, we give an example of how sliding mode control can be used to guarantee stability for a nonlinear system in the presence of approximation errors.

Example 6.4 (Sliding Control using a Fuzzy Model) *Consider the problem of stabilizing the following nonlinear system*

$$\begin{aligned}\dot{x}_1 &= x_2\\ \dot{x}_2 &= f(x) + u\end{aligned}$$

Assume that a mathematical model of $f(x)$ is unknown, but that we have some heuristic knowledge that we can formulate into a fuzzy model $\hat{f}(x)$. We cannot expect this approximate model to be perfect, but assume that we can bound the approximation error by a function $F(x)$

$$|f(x) - \hat{f}(x)| \leq F(x)$$

A sliding mode controller can be designed for this system in the following steps. Define the sliding surface to be

$$S : s(x,t) = \begin{bmatrix}\lambda \ 1\end{bmatrix} x = \lambda x_1 + \dot{x}_1 = 0$$

with $\lambda > 0$. If the system state is on the sliding surface, it remains on the surface provided that $\dot{s} = 0$, i.e.

$$\dot{s} = (\lambda \dot{x}_1 + \dot{x}_2) = (\lambda x_2 + f(x) + u) = 0$$

Since we do not know $f(x)$, we use the control

$$u(x) = \underbrace{-\hat{f}(x) - \lambda x_2}_{\hat{u}} - \underbrace{k\,\mathrm{sgn}(s)}_{u^*} \tag{13}$$

The first part, $\hat{u}$, of the control is a certainty equivalence control (compare with (9)). To ensure that the state vector is forced onto the surface, we have added a second term, u^*, discontinuous over the sliding surface. If we set

$$k = F(x) + \eta$$

we can verify that this control satisfies the sliding condition (12):

$$\frac{1}{2}\frac{d}{dt}s^2 = \dot{s}s = \left(f(x) - \hat{f}(x)\right)s - k\,\mathrm{sgn}(s)s \leq -\eta|s|$$

From (13), we can see that the system state is directed towards the sliding surface "from both sides", by switching the control whenever the state trajectory crosses the sliding surface. In presence of noise or unmodelled dynamics, this results in a "chattering" control, inhibiting the direct implementation of sliding mode control. In practice, chattering can be eliminated by approximating the switching $\mathrm{sgn}(s)$ by some smooth function, resulting in a "boundary layer" around the sliding surface.

It has been suggested to design fuzzy controllers by approximating the function (13) by a fuzzy system, e.g., [19] and [20]. It is not always clear, however, why one would like to approximate the compact formula (13) by a fuzzy system, increasing the number of system parameters and the computational requirements.

7 Lyapunov-based analysis and synthesis

Using fuzzy systems it is possible to define very general nonlinearities, e.g., that contain discontinuities and jumps. In order to be able to derive any useful analytical results it is necessary to constrain the classes of nonlinearities that one consider. The class of systems that has achieved most attention is linear and affine Takagi-Sugeno systems on state-space form. For these systems both stability and synthesis results are available based on Lyapunov theory. The results are based on the fact that, by the appropriate restriction on inference parameters, see [21,22], the dynamics of this system can be described by the expression

$$\dot{x} = \sum_{l=1}^{L} \mu_l(x)(A_l x + a_l) \tag{14}$$

where $\mu_l(x)$ are normalized membership functions of the rule antecedents satisfying $0 \leq \mu_l(x) \leq 1$, $\sum_l \mu_l(x) = 1$. This also known as a polytopic linear differential inclusion (LDI).

For affine polytopic LDIs ($a_l = 0, \ \forall l$), a sufficient stability condition can be formulated in the following way.

Theorem 7.1 (Quadratic stability) *The equilibrium $x = 0$ of a polytopic LDI system is asymptotically stable if there exists a common matrix P such that*

$$P > 0$$
$$A_i^T P + P A_i < 0, \ i = 1, \cdots, L$$

This is equivalent to saying that one must find a single Lyapunov function,$V(x) = x^T P x$, that holds for all the linear dynamic systems A_i involved.

The quadratic stability problem for polytopic LDIs has no analytic solution. However, it can be conveniently solved using convex optimization [23]. The conditions in Theorem 7.1 are linear matrix inequalities (LMIs) in P. Efficient software is available for verifying LMI conditions [24]. Current software is capable of treating several hundreds of variables in a matter of seconds.

Quadratic Lyapunov functions can also be used to obtain stabilizing state feedback controllers. The polytopic LDI

$$\dot{x} = A(t)x + B(t)u, \ [A(t)B(t)] \in \mathbf{Co}\{[A_1 \, B_1], \cdots, [A_N \, B_N]\}$$

is quadratically stabilizable via linear state feedback if there exists a state-feedback control law $u = Kx$ such that the closed loop system $\dot{x} = (A(t) + B(t)K)x$ is quadratically stable.

This can be expressed as LMIs in the following way. Let us first assume that we have a linear time-invariant system $\dot{x} = Ax + Bu$. This system is quadratically stabilizable if and only if there exists $P > 0$ such that

$$(A + BK)^T P + P(A + BK) < 0$$

or, equivalently, there exists $Q > 0$ such that

$$Q(A + BK)^T + (A + BK)Q < 0$$

This is not an LMI. However, if make the change of variables $Y = KQ$ we obtain an equivalent condition that is an LMI, i.e.,

$$AQ + QA^T + BY + Y^T B^T < 0. \tag{15}$$

The original system is quadratically stabilizable if and only if there exist $Q > 0$ and Y such that (15) holds. The corresponding state feedback law is given by $u = YQ^{-1}x$. In the polytopic LDI case, condition (15) is replaced by

$$QA_i^T + A_iQ + B_iY + Y^TB_i^T < 0$$

The feedback law $u = Kx$ is nonfuzzy. However, similar techniques can also be used to find a stabilizing fuzzy state feedback controller. The controller can then be expressed as

$$u = \sum_{i=1}^{M} g_i(x)K_i$$

where $g_i(\cdot)$ are the basis functions for the fuzzy process model, i.e., the same fuzzy partition must be used in the model and in the controller. The closed loop system now becomes

$$\dot{x} = \sum_{i=1}^{M} g_i(x)\{A_i + B_i \sum_{j=1}^{M} g_j(x)K_j\}x.$$

From this it possible to state the search for the stabilizing K_i matrices as a LMI problem, [25],[21],[26].

In control synthesis, meeting performance specifications needs to be considered in addition to stability. This can in some cases also be stated as LMIs. For example, it is possible to state the search for a fuzzy state feedback that minimizes the exponential decay rate with which the states approach the origin, as an LMI problem [25], [27]. It is also possible to obtain results for other performance specifications, e.g., rejection of external disturbances and limitations on the control variables and process outputs [25,28].

It is also possible to develop fuzzy versions of LQG control. In [29] a design problem is defined that minimizes the upper bound on a quadratic performance criterion. In [28] it is shown how the design of a state observer can be posed as a quadratic LMI problem.

Fuzzy models are often used to describe complex nonlinear systems where a certain amount of uncertainties is involved. Therefore, it is important that the stabilizing controller that is designed is robust against model variations, i.e., robust stability must be ensured for a nominal model. This problem can also be approached with LMI techniques. In [25], it is shown how fuzzy and nonfuzzy state feedback controllers can be designed for fuzzy models with uncertainties both in the rule premises, i.e., in the shape and position of the membership functions, and in the rule consequents, i.e., in the linear dynamics.

Quadratic Lyapunov functions are very powerful if they can be found. However, in many cases it is very difficult to find a common global Lyapunov function, i.e., the approach is quite conservative. The conservatism of the global approaches to quadratic stability is twofold; no information of the region partition is taken into account and the Lyapunov function is restricted to be globally quadratic. A less conservative result has recently been suggested in [30]. The idea is to use a piecewise quadratic Lyapunov function that is tailored to fit the cell partition of the system. The search for a piecewise quadratic Lyapunov function can also be formulated as an LMI-problem. The ideas can be extended to more general uncertain and nonlinear systems [31], and similar techniques can also be used for performance analysis, such as $\mathcal{L}_2$ gain analysis of nonlinear systems [32]. Formulation of the stability results for affine fuzzy Sugeno systems can be found in [33] and [34]. A related result which also is based on piecewise Lyapunov functions is described in [35].

The optimization based analysis and synthesis of fuzzy control systems is a currently a very active area, and one may expect that the number of results in this area will increase substantially in the near future.

8 Summary

Fuzzy control is becoming an increasingly used method for implementing nonlinear control systems. Industrial control systems are dominated by linear PID controllers. In many cases it is an advantage to instead add various nonlinearities to the basic PID scheme. This is one reason why fuzzy PID control has been successful in industry. Another reason is the user-friendly interface and computer-based design environments available.

Also in the academic world a large amount of work is being done on fuzzy control. Here, the focus is on analysis and synthesis methods. For certain classes of fuzzy systems, e.g., linear Takagi-Sugeno systems, a substantial amount of analysis and synthesis results have been developed.

References

1. L.A. Zadeh. Fuzzy sets. *Information and Control*, 8:338–353, 1965.
2. E.H. Mamdani. Application of fuzzy algorithm for control of simple dynamic plant. *Proc. IEE*, 121:1585–1588, 1974.
3. L.P. Holmblad and J.J. Ostergaard. Control of a cement kiln by fuzzy logic. In M.M. Gupta and E. Sanchez, editors, *Fuzzy Information and Decision Processes*. North-Holland, Amsterdam, 1982.

4. Thomas Froese. Applying of fuzzy control and neuronal networks to modern process control systems. In *Proceedings of the EUFIT '93*, volume II, pages 559–568, Aachen, September 1993.
5. H. Hellendoorn. Design and development of fuzzy systems at siemens r&d. In *Proc. of the IEEE International Conference on Fuzzy Systems*, pages 1365–1370, 1993.
6. Piero P. Bonissone. Fuzzy logic controllers: an industrial reality. In Jacek M. Zurada, Robert J. Marks II, and Charles J. Robinson, editors, *Computational Intelligence: imitating life*, pages 316–327. IEEE Press, Piscataway, NJ, 1994.
7. K. Hirota, editor. *Industrial Applications of Fuzzy Technology.* Springer, Tokyo, 1993.
8. T. Terano, K. Asai, and M. Sugeno. *Applied Fuzzy Systems.* Academic Press, Inc., Boston, 1994.
9. K.-E. Årzén, M. Johansson, and R. Babuska. A survey on fuzzy control. Famimo deliverable d1.1b, Lund Institute of Technology, 1997.
10. J.L. Castro. Fuzzy logic controllers are universal approximators. *IEEE Transactions on Systems, Man and Cybernetics*, pages 629–635, April 1995.
11. R. Zbikowski, K.J. Hunt, A. Dzieliński, R. Murray-Smith, and P.J. Gawthrop. A Review of Advances in Neural Adaptive Control Systems. Technical report, Daimler-Benz AG and University of Glasgow, June 1994.
12. A. Juditsky, H. Hjalmarsson, A. Benveniste, B. Delyon, L. Ljung, J. Sjöberg, and Q. Zhang. Nonlinear black-box models in system identification: Mathematical foundations. *Automatica*, 31(12):1725–1750, 1995.
13. T. Takagi and M. Sugeno. Fuzzy identification of systems and its applications to modeling and control. *IEEE Transactions on Systems, Man, and Cybernetics*, 15:116–132, 1985.
14. R. Babuska. An overview of fuzzy modeling and model-based control. In *this volume.*
15. A. Isidori. *Nonlinear Control Systems: an Introduction.* Springer-Verlag, 1989.
16. V. I. Utkin. Variable structure systems with sliding modes: a survey. *IEEE Transactions on Automatic Control*, 22:212–222, 1977.
17. Jean-Jaqcues Slotine and Winpeg Li. *Applied Nonlinear Control.* Prentice Hall International, 1991.
18. John Y. Hung, Weibing Gao, and James C. Hung. Variable structure control: A survey. *IEEE Transactions on Industrial Electronics*, pages 2–21, February 1993.

19. R. Palm. Sliding mode fuzzy control. In *Proc. of the IEEE International Conference on Fuzzy Systems*, pages 519–526, 1992.
20. R. Palm and D. Driankov. Stability of fuzzy gain-schedulers: Sliding-mode based analysis. In *Proc. of the IEEE International Conference on Fuzzy Systems*, pages 177–183, 1997.
21. Kazuo Tanaka, Takayuki Ikeda, and Hua O. Wang. Robust stabilization of a class of uncertain nonlinear systems via fuzzy control: Quadratic stabilizability, H_∞ control theory and linear matrix inequalities. *IEEE Transactions on Fuzzy Systems*, 4(1):1–13, February 1996.
22. Li-Xin Wang. *Adaptive Fuzzy Systems and Control: Design and Stability Analysis.* Prentice Hall, 1994.
23. S. Boyd, L. El-Ghaoui, E. Feron, and V. Balakrishnan. *Linear Matrix Inequalities in System and Control Theory.* Studies in Applied Mathematics. SIAM, 1994.
24. Pascal Gahinet, Arkadi Nemirovski, Alan J. Laub, and Mahamoud Chilali. *LMI Control Toolbox for use with Matlab.* The Mathworks Inc., May 1995.
25. Jun Zhao. *System Modeling, Identification and Control using Fuzzy Logic.* PhD thesis, UCL, Université Catholique de Louvain, 1995.
26. K. Tanaka and M. Sano. A robust stabilization problem of fuzzy systems and its application to backing up control of a truck-trailer. *IEEE Trans. Fuzzy Systems*, 2(2):119–134, 1994.
27. K. Tanaka, T. Ikeda, and H.Q Wang. An LMI approach to fuzzy controller designs based on relaxed stability conditions. In *Proc. of the IEEE International Conference on Fuzzy Systems*, pages 171–176, 1997.
28. K. Tanaka, T. Ikeda, and H.O. Wang. Fuzzy regulators and fuzzy observerss: A linear matrix inequality approach. In *36th IEEE Conference on Decision and Control*, 1997.
29. K. Tanaka, T. Taniguchi, and H.O. Wang. A mixed control problem of robust fuzzy control and optimal fuzzy control and its solution via linear matrix inequalities. *IEEE Transaction on Fuzzy Systems (to appear)*, 1999.
30. Mikael Johansson and Anders Rantzer. Computation of piecewise quadratic Lyapunov functions for hybrid systems. Technical report, Department of Automatic Control, June 1996. Also available at `http://www.control.lth.se/~rantzer`.
31. M. Johansson and A. Rantzer. Computation of piecewise quadratic Lyapunov functions for hybrid systems. In *European Control Conference, ECC97*, 1997.
32. A. Rantzer and M. Johansson. Piecewise linear quadratic control. In

American Control Conference, ACC'97, 1997.
33. M. Johansson, J. Malmborg, A. Rantzer, B. Bernhardsson, and K.-E. Årzén. Modeling and control of fuzzy, heterogeneous and hybrid systems. In *Proc. of SICICA 97, Annecy, France*, 1997.
34. M. Johansson, A. Rantzer, and K.-E. Årzén. Piecewise quadratic stability for affine sugeno systems. In *Proc. of FUZZ-IEEE'98, Anchorage*, 1998.
35. S.G. Cao, N.W. Rees, and G. Feng. Analysis and design for a class of complex control systems part II: Fuzzy controller design. *Automatica*, 33(6):1017–1028, 1997.

PART II: SURVEYS ON ADVANCED METHODOLOGIES

NEURO-FUZZY METHODS

D. NAUCK
University of Magdeburg
Faculty of Computer Science (FIN-IWS)
Universitaetsplatz 2, D-39106 Magdeburg, Germany
E-mail: detlef.nauck@cs.uni-magdeburg.de,
http://fuzzy.cs.uni-magdeburg.de/~nauck

The manual construction of fuzzy systems can be a time-consuming and error-prone task. To support this process learning methods that can create fuzzy systems from training data are of interest. This chapter gives an introductory overview on *neuro-fuzzy systems*, which combine methods from neural network theory with fuzzy systems. Usually neuro-fuzzy approaches do not actually employ neural networks, even though they are very often depicted in form of some kind of neural network structure. However, all approaches display some kind of learning capability, as it is known from neural networks. This means, they use algorithms which enable them to determine their parameters from training data in an iterative process. This chapter reviews some neuro-fuzzy approaches in the areas of function approximation, control and classification. We illustrate our view of neuro-fuzzy techniques and our understanding on how these approaches should be used. From our point of view *neuro-fuzzy* means using heuristic learning strategies derived from the domain of neural network theory to support the development of a fuzzy system.

1 Introduction

Ever since fuzzy systems were applied in industrial applications, developers know that the construction of a well performing fuzzy system is not always easy. The problem of finding appropriate membership functions and fuzzy rules is often a tiring process of trial and error. Therefore the idea of applying learning algorithms to fuzzy systems was considered early. Approaches to so-called *adaptive* or *self-organizing fuzzy controllers* can be found e.g. in [32,33,34]. An overview of this area is presented in [9]. These kind of adaptive models usually use knowledge based methods. However, another possibility of learning parameters of fuzzy systems is given by neural networks.

The learning capabilities of neural networks made them a prime target for a combination with fuzzy systems in order to automate or support the process of developing a fuzzy system for a given task. The first so-called neuro-fuzzy approaches were considered mainly in the domain of (neuro-) fuzzy control, but today the approach is more general. Neuro-fuzzy systems are applied in various domains, e.g. control, data analysis, decision support, etc.

Modern neuro-fuzzy-systems are usually represented as a multilayer feedforward neural network[2,7,8,11,17,35], but fuzzifications of other neural network architectures are also considered, for example self-organizing feature maps[5,36]. In neuro-fuzzy models, connection weights and propagation and activation functions differ from common neural networks. Although there are a lot of different approaches[7,17], we want to restrict the term "neuro-fuzzy" to systems which display the following properties:

1. A neuro-fuzzy system is a fuzzy system that is trained by a learning algorithm (usually) derived from neural network theory. The (heuristic) learning procedure operates on local information, and causes only local modifications in the underlying fuzzy system. The learning process is not knowledge based, but data driven.

2. A neuro-fuzzy system can be viewed as a special 3-layer feedforward neural network. The units in this network use t-norms or t-conorms instead of the activation functions usually used in neural networks. The first layer represents input variables, the middle (hidden) layer represents fuzzy rules and the third layer represents output variables. Fuzzy sets are encoded as (fuzzy) connection weights.

   Some neuro-fuzzy models use more than 3 layers, and encode fuzzy sets as activation functions. In this case, it is usually possible to transform them into a 3-layer architecture.

   This view of a fuzzy system illustrates the data flow within the system and its parallel nature. However this neural network view is not a prerequisite for applying a learning procedure, it is merely a convenience.

3. A neuro-fuzzy system can always (i.e. before, during and after learning) be interpreted as a system of fuzzy rules. It is both possible to create the system out of training data from scratch, and it is possible to initialize it by prior knowledge in form of fuzzy rules.

4. The learning procedure of a neuro-fuzzy system takes the semantical properties of the underlying fuzzy system into account. This results in constraints on the possible modifications of the system's parameters.

5. A neuro-fuzzy system approximates an n-dimensional (unknown) function that is partially given by the training data. The fuzzy rules encoded within the system represent vague samples, and can be viewed as vague prototypes of the training data. A neuro-fuzzy system should not be seen as a kind of (fuzzy) expert system, and it has nothing to do with fuzzy logic in the narrow sense[15].

In this chapter *neuro-fuzzy* has to be understood as stated by the five points above. Therefore we consider *neuro-fuzzy* as a technique to derive a fuzzy system from data, or to enhance it by learning from examples. The exact implementation of the *neuro-fuzzy model* does not matter. It is possible to use a neural network to learn certain parameters of a fuzzy system, like using a self-organizing feature map to find fuzzy rules [31] (cooperative models), or to view a fuzzy system as a special neural network and to apply a learning algorithm directly [19] (hybrid models).

Approaches, where neural networks are used to provide inputs for a fuzzy system, or to change the output of a fuzzy system, we prefer to call *neural (network)/fuzzy (system) combinations* or *concurrent neural/fuzzy models* to stress the difference that in these approaches parameters of a fuzzy system are not changed by a learning process. If the creation of a neural network is the main target, it is possible to apply fuzzy techniques to speed up the learning process, or to fuzzify a neural network by the extension principle to be able to process fuzzy inputs. These approaches could be called *fuzzy neural networks* to stress that fuzzy techniques are used to create or enhance neural networks [6,10].

The following sections give an introduction to some neuro-fuzzy approaches. We refrained from presenting a lot of technical details in order to provide a readable overview on the ideas of neuro-fuzzy systems. For a comprehensive technical treatment of neuro-fuzzy methods the reader is referred to [17,25,26]. In Section 2 we develop a framework that allows us to represent a fuzzy system as a neural network. Two neuro-fuzzy methods in function approximation are presented in Section 3. Fuzzy control is discussed in Section 4. Section 5 considers a neuro-fuzzy method for classification.

2 A Fuzzy System in a Neural Network Structure

A lot of neuro-fuzzy approaches represent their models as a neural network. This is of course not necessary to apply a learning algorithm to a fuzzy system. However, it can be convenient because it visualizes the data flow through the system, as well for the input data as for the error signals that are used to update the system parameters. An additional benefit is that different models can easily be compared, and structural differences are clearly visible.

There can be also some applicational advantages. If a fuzzy system is represented in form of a network, and a neural network development tool is available that is flexible enough to let us define special activation and propagation functions, then it may be possible to use it. Fuzzy system development tools usually implement only very restrictive learning capabilities like, for instance, rule weights.

To describe the network structure of a neuro-fuzzy model in general, it is useful to have a generic model. It is easily possible to describe a *generic 3-layer fuzzy perceptron* that can be used as a foundation to specific neuro-fuzzy systems [17,25,26]. The name "fuzzy perceptron" refers to the structure of the model, that is similar to the perceptrons as they are known from the domain of neural networks. The term *fuzzy (multi-layer) perceptron* has also been used by other authors for their approaches [14,16,30]. We use our interpretation of this notion here to describe the structure of our generic model. Other definitions of the term "fuzzy perceptron" are also possible, of course.

By using a generic fuzzy perceptron to derive neuro-fuzzy systems for special domains, it would be possible to evaluate these different neuro-fuzzy approaches by means of the same underlying model. The fuzzy perceptron was used to derive some of the models which we discuss in the following sections.

A generic fuzzy perceptron has the architecture of a usual multilayer perceptron, but the weights are modeled as fuzzy sets and the activation, output, and propagation functions are changed accordingly to implement a common fuzzy inference path. The intention of this model is to provide a framework for learning algorithms, to be interpretable as a system of linguistic rules, and to be able to use prior rule based knowledge, so that the learning need not start from scratch.

A generic *3-layer fuzzy perceptron* is a 3-layer feedforward neural network $(U, W, \text{NET}, A, O, \text{ex})$ with the following specifications:

1. $U = U_1 \cup U_2 \cup U_3$ is a non-empty set of units (neurons). For all $i, j \in \{1,2,3\}, U_i \neq \emptyset$, and $U_i \cap U_j = \emptyset$ with $i \neq j$ holds. U_1 is called input layer, U_2 rule layer (hidden layer), and U_3 output layer.

2. The structure of the network (connections) is defined as

$$W : U \times U \to \mathcal{F}(\mathbb{R}),$$

such that there are only connections $W(u, v)$ with $u \in U_i$, $v \in U_{i+1}$ $(i \in \{1,2\})$. $\mathcal{F}(\mathbb{R})$ is the set of all fuzzy subsets of $\mathbb{R}$.

3. By A an activation function A_u for each $u \in U$ is given to calculate the activation a_u

   (a) for input and rule units $u \in U_1 \cup U_2$:

$$A_u : \mathbb{R} \to \mathbb{R}, \quad a_u = A_u(\text{net}_u) = \text{net}_u,$$

(b) for output units $u \in U_3$:

$$\begin{aligned} A_u &: \mathcal{F}(\mathbb{R}) \to \mathcal{F}(\mathbb{R}), \\ a_u &= A_u(\text{net}_u) = \text{net}_u. \end{aligned}$$

4. O defines for each $u \in U$ an output function O_u to calculate the output o_u

   (a) for input and rule units $u \in U_1 \cup U_2$:

$$O_u : \mathbb{R} \to \mathbb{R}, \quad o_u = O_u(a_u) = a_u,$$

   (b) for output units $u \in U_3$:

$$\begin{aligned} O_u &: \mathcal{F}(\mathbb{R}) \to \mathbb{R}, \\ o_u &= O_u(a_u) = \text{DEFUZZ}_u(a_u), \end{aligned}$$

   where DEFUZZ_u is a suitable defuzzification function.

5. NET defines for each unit $u \in U$ a propagation function NET_u to calculate the net input net_u

   (a) for input units $u \in U_1$:

$$\text{NET}_u : \mathbb{R} \to \mathbb{R}, \quad \text{net}_u = ex_u,$$

   (b) for rule units $u \in U_2$:

$$\begin{aligned} \text{NET}_u &: (\mathbb{R} \times \mathcal{F}(\mathbb{R}))^{U_1} \to [0,1], \\ \text{net}_u &= \mathop{\top}_{u' \in U_1} \{W(u',u)(o_{u'})\}, \end{aligned}$$

   where $\top$ is a t-norm,

   (c) for output units $u \in U_3$:

$$\begin{aligned} \text{NET}_u &: ([0,1] \times \mathcal{F}(\mathbb{R}))^{U_2} \to \mathcal{F}(\mathbb{R}), \\ \text{net}_u &: \mathbb{R} \to [0,1], \\ \text{net}_u(x) &= \mathop{\perp}_{u' \in U_2} \{\top(o_{u'}, W(u',u)(x))\}, \end{aligned}$$

   where $\perp$ is a t-conorm.

6. $\text{ex} : U_1 \to \mathbb{R}$, defines for each input unit $u \in U_1$ its external input $\text{ex}(u) = \text{ex}_u$. For all other units ex is not defined.

A fuzzy perceptron can be viewed as a fuzzy system represented as a graph or as a usual 3-layer perceptron that is *fuzzified to a certain extent*. Only the weights, the net inputs, and the activations of the output units are modeled as fuzzy sets. A fuzzy perceptron is like a usual perceptron used for function approximation. The advantage is the interpretation of its structure in the form of linguistic rules, because the fuzzy weights can be associated with linguistic terms. The network can also be created partly, or in the whole, out of linguistic (fuzzy if-then) rules.

In the following three sections we consider neuro-fuzzy models in the domain of function approximation. This means we want to describe a function by means of a fuzzy rule base. First we consider general function approximation, where a function is given by (noisy) data samples. In this case, we can use plain supervised learning. Then we discuss two special cases of function approximation by learning: fuzzy control and fuzzy classification.

We encounter a special case of function approximation in fuzzy control [1]. Here, learning is usually done indirectly by reinforcement, because the outputs (control actions) for given input values (system states) are unknown, if no other controller exists for the considered problem. The sought-after (control) function is not given by data samples in this case, but by the behavior of the process to be controlled and by an external performance or reinforcement signal that guides the learning process.

In neuro-fuzzy classification plain supervised learning is used again, because we use labelled training data. In this case, we are not looking for a continuous function like in the other two cases, but for a discrete one by which we can classify input vectors.

3 Function Approximation

In this section we consider the problem of approximating an unknown continuous function by a fuzzy system, where the function is partly specified by a set of data samples. This is a supervised learning problem, because the error of the approximation is defined by the difference between the actual output of the fuzzy system and the target output given in the training data.

3.1 ANFIS

ANFIS (adaptive network-based fuzzy inference system) [12] is a neuro-fuzzy method to determine the parameters of a Sugeno-type fuzzy model which is represented as a special feedforward network (see Figure 1). It encodes fuzzy rules of the form

$$R_r\text{: } \mathbf{If}\ x_1\ \mathbf{is}\ A^{(1)}_{j_1} \wedge \ldots \wedge x_n\ \mathbf{is}\ A^{(n)}_{j_n}\ \mathbf{then}\ y = \alpha^{(r)}_0 + \alpha^{(r)}_1 x_1 + \ldots + \alpha^{(r)}_n x_n.$$

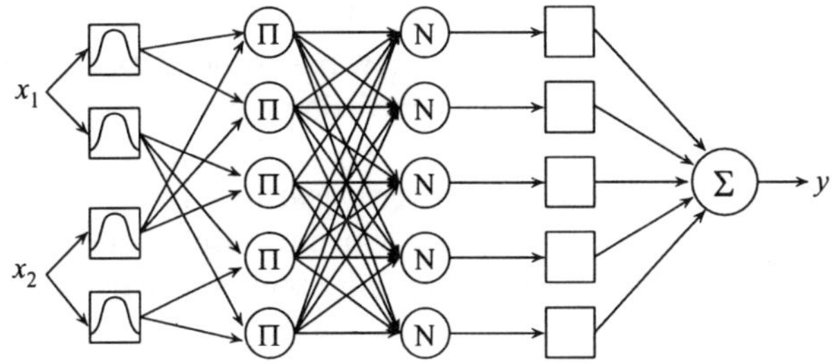

Figure 1: ANFIS encodes a Sugeno-type fuzzy model in a feedforward network structure [12]

Each of the k fuzzy rules owns an individual set of consequent parameters $\alpha_i^{(r)}$, and for each of the n input variables $x_i \in \boldsymbol{X}_i$ there are p_i fuzzy sets A_j^i. A fuzzy set A is represented by a bell-shaped membership function using three parameters $a, b, c \in \mathbb{R}$:

$$A(x) = \frac{1}{1+\left(\left(\frac{x-c}{a}\right)^2\right)^b}.$$

Each node of the first layer is connected to exactly one of the n input variables and stores the three parameters of a membership function. The k nodes in the second layer represent the antecedents of the fuzzy rules. They compute the degree of fulfillment by multiplying the degrees of membership. The k nodes of the third layer compute the relative degree of fulfillment for each rule. The output values of the rules are computed by the k nodes of layer 4. They store the consequent parameters. Each node in this layer is connected to one node of layer 3 and to all input variables (these connections not drawn in Figure 1). The output node in layer 5 computes the overall output value. If the model must compute $m > 1$ output values, then there are m output nodes and mk nodes in layer 4.

ANFIS uses only differentiable functions, and therefore it is easy to apply standard gradient descent learning procedures from neural network theory. For ANFIS a mixture of backpropagation (BP) and least mean square estimation (LSE) is suggested by Jang [12]. BP is used to compute the updates of the antecedent parameters, i.e. the parameters of the fuzzy sets, and LSE is used to determine the updates for the consequent parameters, i.e. the coefficients of the linear combinations in the consequents.

A step in the training procedure has two parts: In the first part the input patterns are propagated, and the optimal consequent parameters are estimated by an iterative LSE procedure, while the antecedent parameters are assumed to be fixed for the current cycle through the training set. In the second part the patterns are propagated again, and now backpropagation is used to modify the antecedent parameters, while the conclusion parameters remain fixed. This procedure is then iterated.

Jang suggests also to use an adaptive learning rate. If the error is decreased for four consecutive steps, the learning rate should be increased. If the error increases and decreases within four steps, the learning rate should be reduced [12].

3.2 NEFPROX

The NEFPROX model [24,28] (neuro fuzzy function approximation) is a neuro-fuzzy approach based on a Mamdani-type of fuzzy system. It can be derived from the generic fuzzy perceptron presented above, and implements fuzzy rules like :

$$R: \textbf{if } x_1 \text{ is } A_{j_1^{(1)}} \textbf{ and } \ldots \textbf{ and } x_n \text{ is } A_{j_n^{(n)}} \textbf{ then } y_1 \text{ is } B_{k_1}^{(1)}, \ldots, y_m \text{ is } B_{k_m}^{(m)}.$$

A NEFPROX system (see Fig. 2) is a special 3-layer fuzzy perceptron with the following specifications:

1. The input units are denoted as $x_1, \ldots, x_n$, the hidden rule units are denoted as $R_1, \ldots, R_k$, and the output units are denoted as $y_1, \ldots, y_m$.

2. Each connection is weighted with a fuzzy set, and is labelled with a linguistic term.

3. Connections that come from the same input unit and have identical labels, bear the same fuzzy weight at all times. These connections are called **linked connections**, and their weight is called a **shared weight**. An analogous condition holds for the connections that lead to the same output unit.

4. Let $L_{x,R}$ denote the label of the connection between an input unit x and a rule unit R. For all rule units R, R' $(\forall x\ L_{x,R} = L_{x,R'}) \Longrightarrow R = R'$ holds.

This definition makes it possible to interpret a NEFPROX system in terms of a fuzzy system; each hidden unit represents a fuzzy if-then rule. Condition 3 specifies that there have to be *shared* or *linked weights*. If this feature is missing, it would be possible for fuzzy weights representing identical linguistic terms to evolve differently during the learning process. If this is allowed to happen, the architecture of the NEFPROX system cannot be understood as a

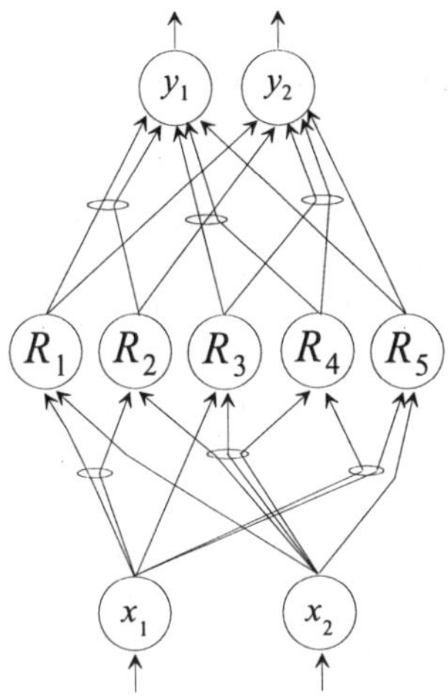

Figure 2: The structure of the NEFPROX model: Some of the connections are linked – they always have the same (fuzzy) weight

fuzzy rule base. Shared weights make sure that for each linguistic value (e.g. "x_1 is *positive big*") there is only one representation as a fuzzy set, i.e. the linguistic value has only one interpretation for all rule units (e.g. R_1 and R_2 in Fig. 2). It cannot happen that two fuzzy sets that are identical at the beginning of the learning process develop differently, and so the semantics of the rule base encoded in the structure of the network is not affected [17]. Connections that share a weight always come from the same input unit or lead to the same output unit. Condition 4 determines that there are no rules with identical antecedents.

In a function approximation problem we can use plain supervised learning, because we know for each given input vector the correct output vector (fixed learning problem). If we use a system of fuzzy rules to approximate the function, we can use prior knowledge. This means, if we already know suitable rules for certain areas, we can initialize the neuro-fuzzy system with them. The remaining rules have to be found by learning. If there is no prior knowledge we start with a NEFPROX system without hidden units and incrementally learn all rules.

The fuzzy set learning algorithm for NEFPROX is a simple, computationally inexpensive heuristic procedure, and not a gradient descent method, which would not be applicable, because the functions (min and max) used for evalutating the fuzzy rules are not differentiable. Based on the error measure at the output layer the fuzzy sets of the conclusions are shifted to higher or lower values, and the width of their support is modified. Then the error is propagated back to the rule nodes. Each rule node computes its individual error value and uses it to correct the spread and position of the antecedent membership functions. It is easy to define constraints for the learning procedure, e.g. that fuzzy sets must not pass each other, or that they must intersect at 0.5, etc. As a stopping criterion usually the error on an additional validation set is observed. Training is continued until the error on the validation set does not further decrease. This technique is well known from neural network learning, and is used to avoid over-fitting to the training data.

To start the learning process, we must specify initial fuzzy partitions for each input variable. This is not necessary for output variables. For them, fuzzy sets can be created during learning, by creating a fuzzy set of a given shape at the current output value, if there is no suitable fuzzy set so far.

The structure (rule) learning algorithm selects fuzzy rules based on a predefined partitioning of the input space given by the initial fuzzy sets (see also the description of the NEFCLASS rule learning algorithm in Sect. 5 and Fig. 6). If the algorithm creates too many rules, it is possible to evaluate them by determining individual rule errors and to keep only the best rules.

In this case, however, the approximation performance will suffer. Each rule represents a number of crisp samples of the (unknown) function by a fuzzy sample. If rules are deleted, some samples are not considered anymore. If parameter learning cannot compensate for this, then the approximation performance must decrease. For classification problems as they are handled by NEFCLASS (Sect. 5) rule pruning is not such a problem. This is due to the winner-takes-all interpretation which is not much influenced by small changes in the output units. In contrast, the output of NEFPROX is taken as a function result such that changes in the output units have a stronger influence.

We often encounter a trade-off in neuro-fuzzy approaches [3,4]. To obtain a high performance, we need complex training algorithms based on gradient descent, algorithms which demand Sugeno-type fuzzy systems. If we use a Mamdani-type fuzzy system, which is easier to interpret, we can use fast heuristics for training, but usually achieve a lower performance. ANFIS [12], for example, on the one hand sometimes provides better approximation than NEFPROX, which on the other hand learns much faster [24,28].

Another problem in neuro-fuzzy systems is rule learning. Either no rule learning procedure is defined for a given neuro-fuzzy model (e.g. ANFIS), or simple heuristics are used. However, those simple strategies are not always powerful enough to yield good (i.e. small and interpretable) rule bases. In this case, it can be useful to consider pruning techniques from neural networks to reduce the number of rules and variables in a neuro-fuzzy system [37]. It is also possible to use e.g. fuzzy clustering methods to find fuzzy rules, and initialize a neuro-fuzzy system with them.

4 Fuzzy Control

Learning fuzzy rules is especially hard in fuzzy control. In fuzzy control we encounter a special case of function approximation, where we want to approximate a control function (we restrict ourselves to static aspects). We assume that we want to find a fuzzy controller for some process by learning. We further assume that we do not have any previously recorded training data, i.e. we cannot use plain supervised learning. A solution is to use reinforcement learning, if either a model of the considered process is available, or training can be done online using the real process.

NEFCON[18] is a model for neural fuzzy controllers developed by our group, and it is also based on the architecture of the generic fuzzy perceptron described above. A NEFCON system is in fact a NEFPROX system with just one output variable, and it is trained by reinforcement learning.

Learning in fuzzy control is a difficult problem, because the correct output of the controller is (usually) not known. Instead we can only observe the behavior of the controlled system and judge its current state as (more or less) acceptable or unacceptable. We can express our level of acceptance by a so-called *external reinforcement signal*, which can be used by reinforcement learning[13]. Usually the reinforcement signal is simply binary and a signal different from zero indicates a system failure. This is not much information for a learning algorithm, and therefore usually an internal adaptive critic is used – a special neural network that learns to predict the external reinforcement signal. Its prediction is called *internal reinforcement signal* and is also used in training the controller. GARIC is a neuro-fuzzy sytem that uses an adaptive critic[2].

The NEFCON learning algorithm uses a rule based fuzzy error measure as reinforcement signal. The user sets up a simple fuzzy rule base that describes the desired system state (control goal). Such a description can usually easily be given. Note that this fuzzy error descriptions does not specify any actions of the controller. The output of the fuzzy error rulebase is used as a continuous reinforcement signal that contains a lot more information that a simple binary signal. Thus it is possible to define a reinforcement type learning algorithm without using an adaptive critic element.

Most neuro-fuzzy control approaches depend on the user to specify the rule base of the controller. In the learning process they only adapt the membership functions used by the rules. Rule learning in a reinforcement learning situation

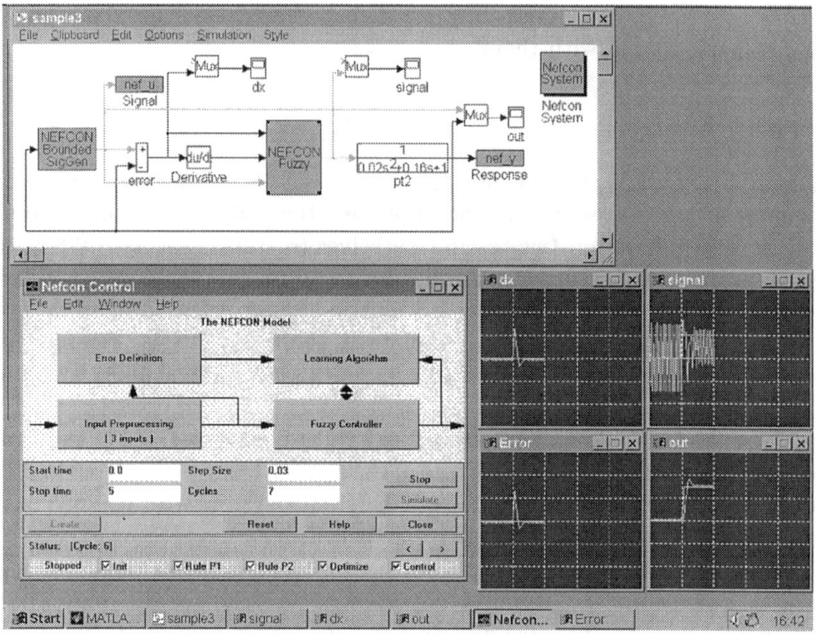

Figure 3: NEFCON for MATLAB/SIMULINK

is extremly difficult, as rules must be created on the fly. There is no training data that can be used to set up a rule base.

The learning algorithm of NEFCON is able to learn fuzzy sets as well as fuzzy rules. Learning a rule base is done by deleting (decremental rule learning) or inserting rules (incremental rule learning). Hence the learning process can work online and does not need previously recorded sample data.

One possibility to learn rules is to start with a NEFCON system that contains all fuzzy rules that can be defined due to the partitioning of the variables. Thus the system begins with an inconsistent initial rule base, which must be made consistent by learning. The idea of decremental rule learning is to try out existing rules and to evaluate them. Rule units that do not pass this test are eliminated from the network. This rule learning algorithm becomes very expensive, if there are a lot of fuzzy sets defined for a lot of variables. For this reason one should always try to use partial knowledge to avoid that all possible rules have to be created. If there are no known rules for certain input states, then only for these particular states all possible rules have to be created. This way the number of initial rule units can be reduced.

Incremental rule learning goes the opposite way, and creates a rule base from scratch by adding rule by rule. It does this by first classifying an input vector, i.e. finding that membership function for each variable that yields the highest membership value for the respective input value. By this a rule antecedent is formed. Then the algorithm tries to guess the output value by deriving it from the current fuzzy error. In a second phase the rule base is optimised by changing the conclusion to an adjacent membership function if necessary.

This idea provides a rule learning algorithm that is less expensive than the one previously described. It is not necessary to handle all possible rules at once, something that soon becomes impossible, especially if there are a lot of variables, or a lot of membership functions. After the rule base has been learned, the learning procedure for the fuzzy sets is invoked to tune the membership functions.

NEFCON has been implemented in several software tools under Unix, Windows and MATLAB/SIMULINK. A description of the MATLAB version can be found in [29].

5 Classification

In this section we discuss classification as another special case of function approximation. An input vector $\mathbf{x} = (x_1, \ldots, x_n) \in \mathbb{R}^n$ is mapped to class indicator C which represents a crisp subset of $\mathbb{R}^n$. We assume the intersection of two different classes to be empty. The training data consists of a set of labelled data, i.e. for each training sample its correct class is known. A classification problem can be represented by a function

$$\varphi : \mathbb{R}^n \to \{0,1\}^m,$$

where $\varphi(\mathbf{x}) = \mathbf{c} = (c_1, \ldots, c_m)$ such that $c_i = 1$ and $c_j = 0$ ($j \in \{1, \ldots, m\}$, $j \neq i$), i.e. $\mathbf{x}$ belongs to class C_i. This way, the class information is given by a 1-of-n code. If a fuzzy system is used to perform this task, the fuzzy rules look like this:

R : **if** x_1 is A_1 **and** ... **and** x_n is A_n
then pattern $(x_1, x_2, \ldots, x_n)$ belongs to class C,

where $A_1, \ldots, A_n$ are fuzzy sets which describe the patterns' features.

Because of the mathematics involved in the rule evaluation process, the rule base actually does not approximate the above mentioned function φ but the function $\varphi' : \mathbb{R}^n \to [0,1]^m$. We will get $\varphi(\mathbf{x})$ by $\varphi(\mathbf{x}) = \psi(\varphi'(\mathbf{x}))$, where ψ reflects the interpretation of the classification result obtained from the fuzzy system. Usually, we will map the highest component of each vector $\mathbf{c}$ to 1 and its other components to 0 (winner takes all).

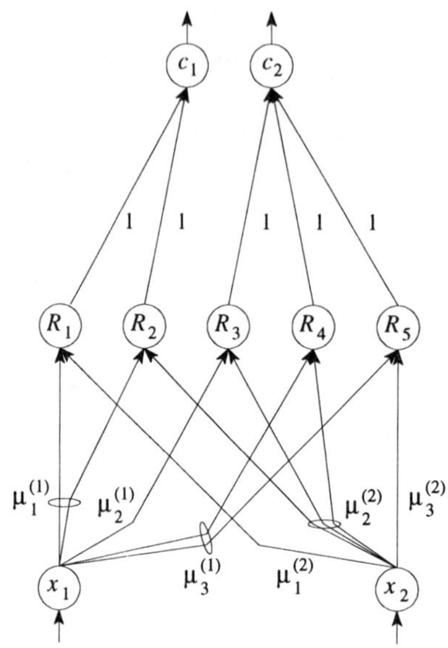

Figure 4: A NEFCLASS system with two inputs, five rules and two output classes

NEFCLASS (neuro fuzzy classification) [20,23] is a neuro-fuzzy model that is derived from the generic fuzzy perceptron presented above. An example for a NEFCLASS system is presented in Fig. 4. It shows a NEFCLASS system that classifies input patterns with two features into two distinct classes by using five linguistic rules. There are again shared weights on the connections from the input to the hidden layer. Compared to NEFCON and NEFPROX, the connections to the output layer are quite different, however. Each hidden (rule) unit is connected to exactly one output (class) unit. The weights on the connections from rule units to output units are fixed at 1 for semantical reasons (to avoid weighted rules).

The output value for an output unit (class), can either be computed as the mean of the activation values of all rule units it is connected to, or as the maximum of those values. The activation of a rule is the minimum of the membership values from its antecedent.

A NEFCLASS system can be built from partial knowledge about the patterns, and can then be refined by learning, or it can be created from scratch by learning. A user has to define a number of initial fuzzy sets partitioning the domains of the input features, and must specify the maximum number of rule nodes that may be created in the hidden layer.

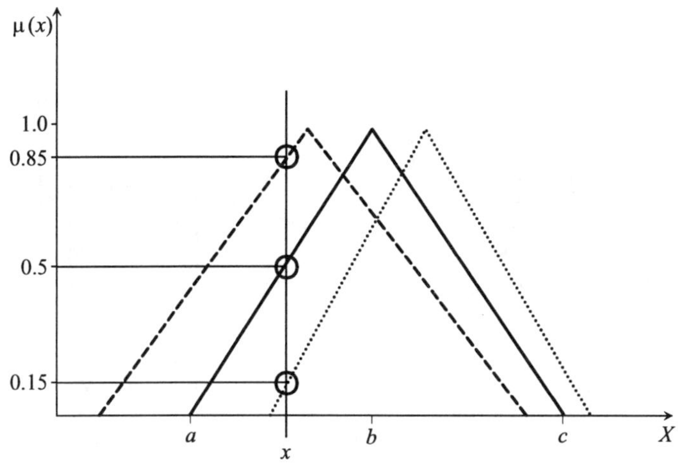

Figure 5: In NEFCLASS the adaption of fuzzy sets (right) is carried out by simply changing the parameters of its membership function in a way that the membership degree for the current feature value is increased or decreased (middle fuzzy set: initial situation, left fuzzy set: increase situation, right fuzzy set: decrease situation)

The idea of the learning algorithm is to create a rulebase first, and then to refine it by modifying the initially given membership functions (usually fuzzy partitions where the membership degrees of each value add up to 1.0). The rulebase will be created by finding for each pattern in the training set a rule that best classifies it. If a rule with an identical antecedent is not already in the rule base, it will be added. After all patterns are processed once, the rule base is complete. It is then possible to evaluate the performance of each rule, and delete some of them (if there are too many) to keep only the best rules.

The learning algorithm of the membership functions uses the output error that tells, whether the degree of fulfillment of a rule has to be higher or lower. This information is used to change the input fuzzy sets by shifting the membership functions, and making their supports larger or smaller (see Fig. 5). By changing only the fuzzy set that delivered the smallest membership degree for the current pattern, the changes are kept as small as possible. It is easy to define constraints for the learning procedure, e.g. that fuzzy sets must not pass each other, or that they must intersect at 0.5, etc. Constraints like these help to obtain an interpretable rule base, but may cause a loss of performance in classification.

The learning process is visualized in Fig. 6. The left part shows the situation after the rule learning algorithm has terminated. The predefined fuzzy partitioning on both input variables defines a partitioning of the input space,

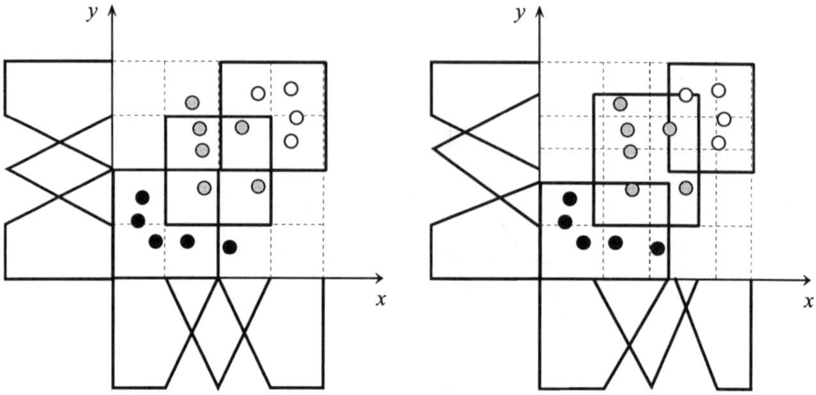

Figure 6: Visualization of a possible NEFCLASS learning process after rule creation (left), and after fuzzy set tuning (right)

created by overlapping hyper-boxes, where each hyper-box is formed by the Cartesian product of the support of n fuzzy sets. Each hyper-box represents the support of an n-dimensional fuzzy set, i.e. the antecedent of a fuzzy rule. During rule learning, hyper-boxes are selected due to the distribution of the patterns. Each hyper-box is mapped to the class of the pattern which caused its selection (i.e. the rule conclusion is determined). After all patterns are processed, the mapping of hyper-boxes to classes is re-evaluated and changed, where necessary. After this only the "best" hyper-boxes (fuzzy rules) are kept.

After rule learning there are usually some patterns which are not classified, because their hyper-box (rule) was not included in the set of $k_{\max}$ best rules. There are usually also some misclassifications. It is the task of the fuzzy set learning algorithm to improve this situation. By modifying the membership functions, the predefined partitioning is distorted. This results in the situation shown in the right part of Fig. 6. Because the learning algorithm for the fuzzy sets is constrained (e.g. a fuzzy set must not pass a neighbor, fuzzy sets must always overlap, etc.), it is possible, that some changes to the form of the hyper-boxes are not applicable. This is one reason that some classification errors can remain in more difficult problems than the one shown in Fig. 6. There will also be classification errors, if the user does not allow NEFCLASS to create enough rules to cover all patterns after fuzzy set learning. Too few rules can also lead to undesired forms of membership functions (e.g. too much overlapping). Classification errors will, of course, also occur, if there is substantial overlapping between classes, i.e. if the classes are indistinguishable in certain areas of the data space.

From the viewpoint of the NEFCLASS architecture and the flow of data, the fuzzy sets are trained by a backpropagation-like algorithm: the error is propagated from the output units towards the input units and is used to change the membership function parameters, but there is no gradient information involved. The adaptivity of a NEFCLASS system is restricted, because of the initially given input fuzzy partions, which define the form and maximal number of clusters, and by the constraints that do not admit certain changes in the fuzzy sets.

To interpret a fuzzy rule base it is important that there are as few rules as possible, and that superfluous variables are not used in the rules. With the pruning techniques of NEFCLASS[21] a rule base can be further improved. The rule editor of NEFCLASS supports the user with the rule simplification process. It checks the rule base for consistency after each pruning step, and tells the user which inconsistencies have occured. The user can decide what actions have to be taken in case of an inconsistency, for example which rule to delete, if there are two rules with identical antecedents, but different conclusions.

NEFCLASS offers five pruning strategies:

- Evaluate rules: The rule learning procedure of NEFCLASS is invoked again, and only the best k rules are kept. The user can specify a value for k. This strategy is useful, if the learning process was started with a large number of rules, and a many of them have a poor performance measure.

- Delete inputs: NEFCLASS computes the correlations of the input variables with the class information, and offers the possibility to delete input variables that have a low correlation to the output. The user can specify the absolute value for the minimum correlation. The deletion of inputs can lead to inconsistencies in the rule base. The rule editor will notify the user of this and support her or him in solving such problems.

- Delete rules: Rules that are only responsible for a few number of classifications can be deleted from the rule base, if they are not needed to cover assumptions in the data. The user can specify a percentage value that has to be reached by each rule to remain in the rule base.

- Delete antecedents: For each rule the antecedents that are not identical to the rule's degree of fulfillment in at least z percent, are deleted from the rule. The user can specify a value for z. This method can lead to an inconsistent rule base, which has to be repaired in the rule editor.

- Delete fuzzy sets: If a variable is partitioned by more than two fuzzy sets, sometimes the support for one or more of them can become quite large during learning. This can be seen as evidence, that such a fuzzy set is superfluous. The user can specify a percentage value, and all fuzzy sets that cover more percent of the domain are deleted, which leads to a reduction of variables in the antecedents. This procedure can also lead to an inconsistent rule base, which will be noticed by the rule editor.

Each of the five above mentioned pruning strategies is applied interactively by the user. It is possible to view the evaluations of the pruning strategy which shall be applied prior to using it. By this the user can decide on the appropriate parameter for the pruning method. After the pruning method was applied the rule editor is invoked and displays the new rule base. If pruning resulted in inconsistencies or other problems like identical rules, the user can solve these problems step by step. If the new rule base is not desirable, the pruning step can be cancelled. If the classification performance decreases after a pruning step, the learning process can be invoked to improve it again, if possible. By saving and restoring NEFCLASS systems the user can try different strategies to improve the rule base until a set of rules is found that yields acceptable performance and interpretability.

An example of applying NEFCLASS to a medical data set and a comparision to other classification approaches can be found in [27].

6 Conclusions

Neuro-fuzzy approaches like NEFPROX, NEFCON and NEFCLASS are are derived from the same generic model and follow the same principle: Keep the learning algorithms simple and do not touch the semantics of the underlying fuzzy systems. Other researchers may prefer other models, for example more sophisticated learning algorithms, e.g. ANFIS that uses a combination of backpropagation and least mean square estimation. However, more powerful learning strategies can be more difficult to handle. From our point of view neuro-fuzzy techniques should be used as tools, and not as automatic solution generators. Therefore we think it is important that the models and learning algorithms are easy to handle and that a user can easily interpret them.

From an applicational point of view, one could say: why bother with interpretability and semantics? It is important that the system does its job. It is of course possible to omit all constraints from the learning procedures of a neuro-fuzzy system, to consider it only as a convenient tool that can be initialized by prior knowledge and trained with sample data, and never to analyse the final system, as long as it performs in an acceptable way. But interpretability and clear semantics provide us with advantages like simply checking the system for plausibility and to maintain it during its life cycle.

It is also possible to consider even simpler learning mechanisms like weighted fuzzy rules, as they can be found in several commercial fuzzy shells. However, they give rise to some semantical problems, which we addressed in [22,25]. Allowing the weights to be selected from $[0, 1]$ could be interpreted as something like a degree of support for a rule, i.e. a value less than 1 would then indicate an ill-defined rule, that supports a class only to some extent. Some approaches allow the weights to assume any value in $\mathbb{R}$, but this leaves the semantics of fuzzy rules behind. It is not clear how rules weighted by absolute values greater than 1 or by negative values should be interpreted (for rules with negative weights sometimes an interpretation as *if not* rules is suggested [11]).

Rule weights can always be replaced by changes in the fuzzy sets of a rule. However these changes can lead to non-normal fuzzy sets and to situations in which identical linguistic values are represented differently in different rules. Rule weights can destroy the interpretation of a fuzzy system completely. Therefore we always refrain from learning weights in our approaches.

Our view of neuro-fuzzy approaches as heuristics to determine parameters of fuzzy systems by processing training data with a learning algorithm, is expressed by the list of five points at the end of Section 1. We think that neuro-fuzzy systems should be seen as development tools that can help to construct a fuzzy system. They are not automatic "fuzzy system generators". The user should always supervise the learning process and try to interpret its results. We also have to keep in mind that – like in neural networks – the success of the learning process is not guaranteed. The same guidelines for selecting and preprocessing training data that are known from neural networks apply to neuro-fuzzy systems. But if the applicaton of neuro-fuzzy methods is well-considered, they can be a powerful tool in the development process of fuzzy systems.

The approaches discussed in this chapter are implemented in software tools which can be obtained via the Internet at http://fuzzy.cs.uni-magdeburg.de.

References

1. Robert Babuska. *Fuzzy Modeling for Control.* Kluwer Academic Publishers, Boston, 1998.
2. Hamid R. Berenji and Pratap Khedkar. Learning and tuning fuzzy logic controllers through reinforcements. *IEEE Trans. Neural Networks*, 3:724–740, September 1992.
3. Hugues Bersini and Gianluca Bontempi. Now comes the time to defuzzify neuro-fuzzy models. *Fuzzy Sets and Systems*, 90:161–170, 1997.

4. Hugues Bersini, Gianluca Bontempi, and Mauro Birattari. Is readability compatible with accuracy? from neuro-fuzzy to lazy learning. In *Fuzzy-Neuro Systems '98 – Computational Intelligence. Proc. 5th Int. Workshop Fuzzy-Neuro-Systems '98 (FNS'98) in Munich, Germany*, volume 7 of *Proceedings in Artifical Intelligence*, pages 10–25, Sankt Augustin, 1998. infix.
5. James C. Bezdek, Eric Chen-Kuo Tsao, and Nikhil R. Pal. Fuzzy Kohonen clustering networks. In *Proc. IEEE Int. Conf. on Fuzzy Systems 1992*, pages 1035–1043, San Diego, CA, 1992.
6. James J. Buckley and Esfandiar Eslami. Fuzzy neural networks: Capabilities. In Witold Pedrycz, editor, *Fuzzy Modelling: Paradigms and Practice*, pages 167–183. Kluwer, Boston, 1996.
7. James J. Buckley and Yoichi Hayashi. Fuzzy neural networks: A survey. *Fuzzy Sets and Systems*, 66:1–13, 1994.
8. James J. Buckley and Yoichi Hayashi. Neural networks for fuzzy systems. *Fuzzy Sets and Systems*, 71:265–276, 1995.
9. Dimiter Driankov, Hans Hellendoorn, and Michael Reinfrank. *An Introduction to Fuzzy Control.* Springer-Verlag, Berlin, 1993.
10. T. Feuring and W. M. Lippe. Learning in fuzzy neural networks. In *Proc. IEEE Int. Conf. on Neural Networks 1996 (ICNN'96)*, pages 1061–1066, Washington, 1996.
11. Saman K. Halgamuge and Manfred Glesner. Neural networks in designing fuzzy systems for real world applications. *Fuzzy Sets and Systems*, 65:1–12, 1994.
12. J. S. Roger Jang. ANFIS: Adaptive-network-based fuzzy inference systems. *IEEE Trans. Systems, Man & Cybernetics*, 23:665–685, 1993.
13. Leslie Pack Kaelbling, Michael H. Littman, and Andrew W. Moore. Reinforcement learning: A survey. *J. Artificial Intelligence Research*, 4:237–285, 1996.
14. James M. Keller and Hossein Tahani. Backpropagation neural networks for fuzzy logic. *Information Sciences*, 62:205–221, 1992.
15. Rudolf Kruse, Jörg Gebhardt, and Frank Klawonn. *Foundations of Fuzzy Systems.* Wiley, Chichester, 1994.
16. Sushmita Mitra and Ludmilla Kuncheva. Improving classification performance using fuzzy mlp and two-level selective partitioning of the feature space. *Fuzzy Sets and Systems*, 70:1–13, 1995.
17. Detlef Nauck, Frank Klawonn, and Rudolf Kruse. *Foundations of Neuro-Fuzzy Systems.* Wiley, Chichester, 1997.

18. Detlef Nauck and Rudolf Kruse. NEFCON-I: An X-Window based simulator for neural fuzzy controllers. In *Proc. IEEE Int. Conf. Neural Networks 1994 at IEEE WCCI'94*, pages 1638–1643, Orlando, FL, June 1994. .
19. Detlef Nauck and Rudolf Kruse. Designing neuro-fuzzy systems through backpropagation. In Witold Pedrycz, editor, *Fuzzy Modelling: Paradigms and Practice*, pages 203–228. Kluwer, Boston, 1996.
20. Detlef Nauck and Rudolf Kruse. A neuro-fuzzy method to learn fuzzy classification rules from data. *Fuzzy Sets and Systems*, 89:277–288, 1997.
21. Detlef Nauck and Rudolf Kruse. New learning strategies for NEFCLASS. In *Proc. Seventh International Fuzzy Systems Association World Congress IFSA'97*, volume IV, pages 50–55, Prague, 1997.
22. Detlef Nauck and Rudolf Kruse. How the learning of rule weights affects the interpretability of fuzzy systems. In *Proc. IEEE Int. Conf. on Fuzzy Systems 1998*, pages 1235–1240, Anchorage, May 1998.
23. Detlef Nauck and Rudolf Kruse. NEFCLASS-X – a soft computing tool to build readable fuzzy classifiers. *BT Technology Journal*, 16(3):180–190, 1998.
24. Detlef Nauck and Rudolf Kruse. A neuro-fuzzy approach to obtain interpretable fuzzy systems for function approximation. In *Proc. IEEE Int. Conf. on Fuzzy Systems 1998*, pages 1106–1111, Anchorage, May 1998.
25. Detlef Nauck and Rudolf Kruse. Neuro-fuzzy methods in fuzzy rule generation. In Didier Dudois and Henri Prade, editors, *Approximate Reasoning and Fuzzy Information Systems*, Handbook of Fuzzy Sets, chapter 5. Kluwer, Norwell, MA, 1998.
26. Detlef Nauck and Rudolf Kruse. Neuro-fuzzy systems. In E. Ruspini, P. Bonissone, and W. Pedrycz, editors, *Handbook of Fuzzy Computation*, chapter D.2. Institute of Physics Publishing Ltd., Philadelphia, PA, 1998.
27. Detlef Nauck and Rudolf Kruse. Obtaining interpretable fuzzy classification rules from medical data. *Artificial Intelligence in Medicine*, 1998. Accepted.
28. Detlef Nauck and Rudolf Kruse. Neuro–fuzzy systems for function approximation. *Fuzzy Sets and Systems*, 1999. Accepted.
29. Andreas Nürnberger, Detlef Nauck, and Rudolf Kruse. Neuro-fuzzy control based on the NEFCON model under MATLAB/SIMULINK. In Pravir Chawdry, Rajkumar Roy, and R. K. Pant, editors, *Proc. 2nd Online World Conference on Soft Computing (WSC2)*, London, June 1997. Springer-Verlag.
30. S. K. Pal and S. Mitra. Multi-layer perceptron, fuzzy sets and classifica-

tion. *IEEE Trans. Neural Networks*, 3:683–697, 1992.
31. Witold Pedrycz and H. C. Card. Linguistic interpretation of self-organizing maps. In *Proc. IEEE Int. Conf. on Fuzzy Systems 1992*, pages 371–378, San Diego, CA, 1992.
32. T. J. Procyk and E. H. Mamdani. A linguistic self-organizing process controller. *Automatica*, 15:15–30, 1979.
33. Wu Zhi Qiao, Wang Pei Zhuang, Teh Hoon Heng, and Song Shou Shan. A rule self-regulating fuzzy controller. *Fuzzy Sets and Systems*, 47:13–21, 1992.
34. Shihuang Shao. Fuzzy self-organizing controller and its application for dynamic processes. *Fuzzy Sets and Systems*, 26:151–164, 1988.
35. Nadine Tschichold-Gürman. The neural network model rulenet and its application to mobile robot navigation. *Fuzzy Sets and Systems*, 85:287–303, 1997.
36. Petri Vuorimaa. Fuzzy self-organizing map. *Fuzzy Sets and Systems*, 66:223–231, 1994.
37. Hans Georg Zimmermann, Ralph Neuneier, Hubert Dichtl, and Stefan Siekmann. Modeling the german stock index dax with neuro-fuzzy. In *Proc. Fourth European Congress on Intelligent Techniques and Soft Computing (EUFIT96)*, Aachen, September 1996.

COMPLEXITY REDUCTION METHODS FOR FUZZY SYSTEMS DESIGN

MAGNE SETNES

Control Laboratory
Faculty of Information Technology and Systems
Delft University of Technology
P.O.Box 5031, 2600GA Delft, The Netherlands

Many methods can be applied to obtain complexity reduction in fuzzy systems, focusing on various aspects like pruning of rules, merging of terms and rules, combination of inputs, and modular structuring of the rule base. This chapter presents a survey of some of these methods. Approaches are considered that are applicable before, during and after the construction of a fuzzy system. The performance of several data-driven methods are illustrated on a real-world MISO modeling problem.

1 Introduction

Combinatorial explosion in the number of fuzzy rules is a problem often encountered in the construction of fuzzy systems. Typically, an initial model structure is created from expert knowledge in the form of if–then rules. Parameters in this structure can be fine-tuned using input-output data. When no prior knowledge about the system under study is available, fuzzy models can be constructed entirely on the basis of numerical data. Most approaches, however, pay little attention to the complexity of the resulting models. Automated modeling techniques can also introduce unnecessary redundancy into the rule base as a result of their focus on numerical optimization [1].

In this chapter, a survey is given of some methods proposed in the literature for the purpose of complexity reduction in fuzzy systems. Section 2 discuss the pruning of insignificant rules and assumes that a rule base is already available, or that some initial structure has been chosen. In Section 3, the use of fuzzy clustering to obtain a suitable input space partition is discussed. The method discussed in Section 4 seeks to reduce of the number of linguistic values, or terms, used in the rule base and is generally applicable to existing rule based systems with fuzzy antecedents. If the rule base still has to be constructed, one can reduce the problem's dimension through fusion of the input variables or by introducing a decentralized or hierarchical structure. This approach is discussed in Section 5, while Section 6 summarizes the chapter.

2 Elimination and selection

We assume that a rule base has already been constructed or that some initial rule base structure (premise) has been obtained and is to be adapted by some supervised learning algorithm. In many cases, especially if a complete rule base has been constructed from combination of linguistic terms defined for the input variables, the rule base will have rules defined for regions of the input space that will not be visited by the physical system. The objective of methods discussed in this section is to reduce the complexity of the rule base. They address this problem by removing, or pruning, rules in the rule base. The methods addressed below can be applied as well after as during construction of a rule base.

2.1 Iterative parameter adaptation techniques

Many training methods for fuzzy systems makes use of the fact that a wide class of fuzzy inference systems can be interpreted as a feed-forward network. This enables the use of parameter adaptation techniques known form the artificial neural networks theory, like back-propagation, in order to learn the parameters of a fuzzy system from training data. However, most such techniques focus only on the quantitative approximation of the given training data, while paying little attention to the complexity of the resulting rule base. There have been some attempts to remedy this by reducing the number of generated rules, and maintaining a transparent rule structure [2,3,4,5]. Two such methods are described below.

Adaptive membership function fusion and annihilation
This method [2] was proposed for reducing the number of rules in a fuzzy Min-Max inference system without violating the completeness of the rule base. The method is applicable also during data-driven construction, where a (high) number of equidistant symmetrical membership functions are initially defined for all variables. These membership functions are then adapted using supervised learning applied to training data. Under the assumption that the inference surface is smooth, the process of adaptation can reveal overdetermination (i.e. to many rules) of the fuzzy system in two ways. First, if two membership functions comes sufficiently close to each other they can be fused into a single membership function. This closeness is determined the consistency measure:

$$s(A, B) = \max[\min(\mu_A(x), \mu_B(x)] \,, \tag{1}$$

which determine the maximum of the intersection of the two membership functions μ_A and μ_B. Second, if a membership function becomes to narrow with respect to an adjacent membership function, it can be annihilated. In both cases, the number of rules is reduced as shown in Fig. 1.

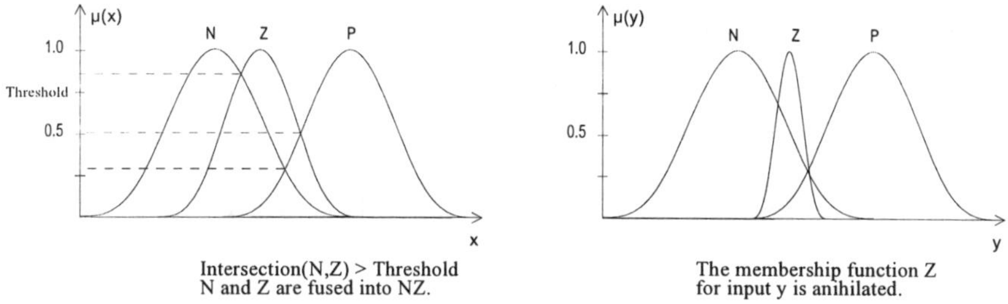

y \ x	N	Z	P
N	Z	P	P
Z	Z	Z	P
P	N	N	Z

→

y \ x	NZ	P
N	?	P
Z	Z	P
P	N	Z

→

y \ x	NZ	P
N	?	P
P	N	Z

Figure 1: Original rule table (left), after fusion (center) and after annihilation (right).

Structure learning

A similar method had been proposed as a part of the structure learning in fuzzy neural networks [5]. The structure learning method also applies a similarity measure to detect compatible fuzzy sets that are then fused. Besides applying a different similarity measure[a], the main difference is that the structure learning method does not remove narrow membership functions, but rather prunes individual rules without paying attention to completeness of the resulting rule base. This allows for pruning of rules defined for conditions that can not occur in the physical system.

The used similarity measure is the fuzzy Jaccard-index [5]. It measures the degree of overlap of two membership functions as the cardinality of their intersection divided by the cardinality of their union:

$$c(A,B) = \frac{|min(\mu_A(x), \mu_B(x))|}{|max(\mu_A(x), \mu_B(x))|}, \tag{2}$$

where min and max model the intersection and the union, respectively, and $|\cdot|$

[a] Many methods have been proposed to assess the similarity, or compatibility, of fuzzy concepts, and various analysis and overviews of such measures have been reported [6,7,8].

denotes the cardinality of a fuzzy set[b]. To detect irrelevant rules, the structure learning approach evaluate the cumulative relative contribution, CRC, of each rule. The CRC of a rule R^j is measured by taking the sum of its relative degree of firing for each training data sample k in the used training data set:

$$\gamma_j = \sum_{k=1}^{N} \frac{\beta_j(k)}{\sum_{j'=1}^{M} \beta_{j'}(k)} , \quad j = 1, \ldots, M , \tag{3}$$

where γ_j is the CRC of the jth rule. If the CRC of a rule is lower than a given threshold, $\gamma_j < \tau$, then this rule is considered irrelevant for the model output and can be deleted. The CRC threshold τ reflects the users willingness to trade accuracy for less complexity. Higher values of τ remove more rules.

Example 2.1 Results obtained by the two previous methods are shown in Fig. 2. The studied system concerns the pressure dynamics of a fermenter tank (Fig. 2a). Systems measurements are obtained by keeping the air inlet at the bottom of the tank constant while the outlet valve is controlled. A model of the pressure dynamics is sought of the form $y(k+1) = f(y(k), u(k))$ where $y(k)$ and $u(k)$ are the pressure and outlet valve position, respectively, at time sample k, and $f(\cdot)$ is a fuzzy system of the Takagi-Sugeno type[9]. The figures 2b – 2d show the training data projected onto the initial (complete) and final fuzzy premises of two rule-based models obtained by the two previous learning methods. □

2.2 Two pass OLS

Orthogonal least-squares (OLS) learning of a fuzzy basis function (FBF) expansion from data was introduced as a data-driven modeling technique[10]. This learning method was extended[11] in order to reduce the initial FBF expansion and the corresponding rule-based system by selecting only the membership functions of the most significant rules to be a part of the final FBF expansion. This method of reduction is applicable also to an existing fuzzy rule-based system, as it can be written as a FBF expansion.

Fuzzy logic system
Consider a fuzzy logic system (FLS) with rules in the following form:

$$R^j : \text{IF } x_1 \text{ is } F_1^j \text{ and } \cdots \text{ and } x_n \text{ is } F_n^j \text{ THEN } y \text{ is } \theta_j , \quad j = 1, 2, \cdots, M , \tag{4}$$

[b]The cardinality of a fuzzy set $A(x)$ is given by $|A| = \sum_x \mu_A(x)$. For continuous domains $\mathcal{X}$, the summation is replaced by an integration.

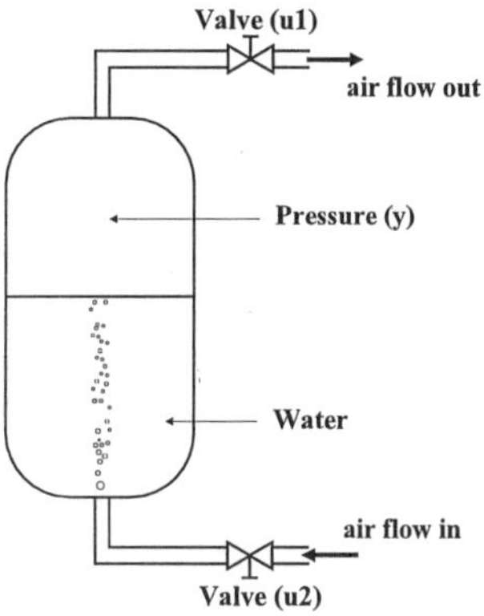

(a) Fermenter tank

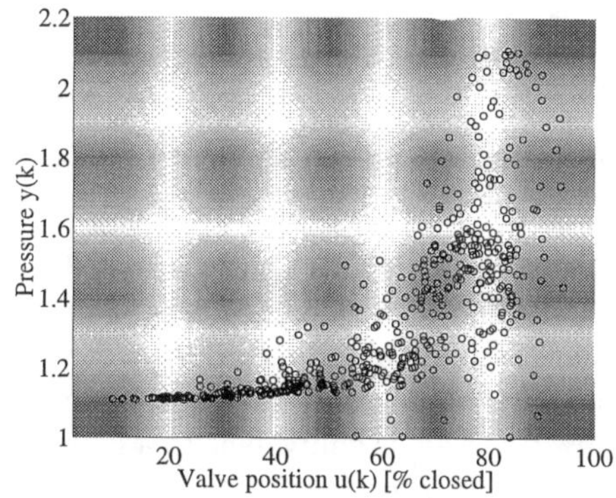

(b) Initial rule base premise

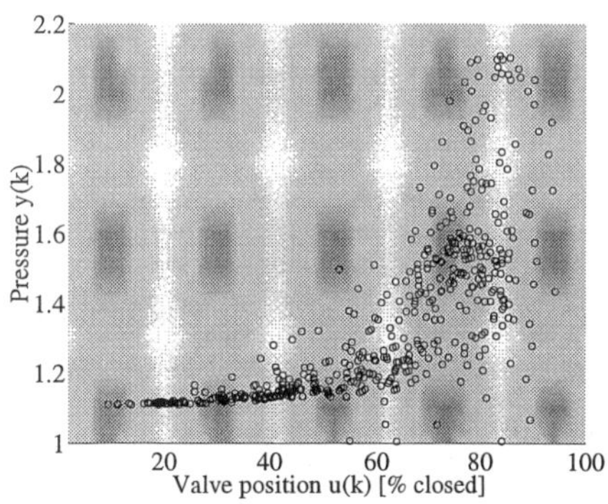

(c) Trained with fusion and annihilation

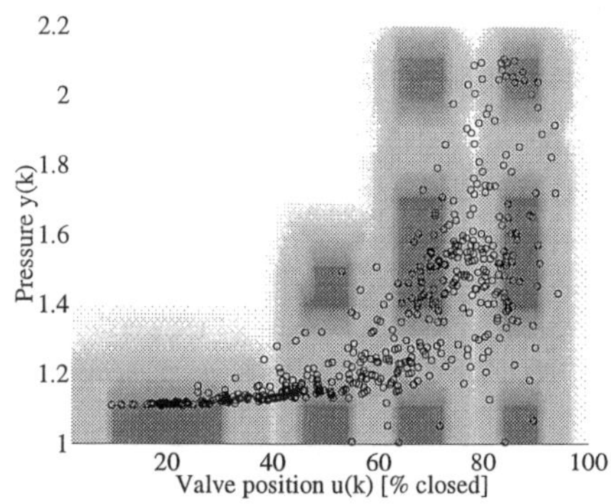

(d) Trained with structure learning

Figure 2: The fermenter tank (a) and the premise of the initial rule base (b), and the results obtained with training using fusion and annihilation (c) and the structure learning method (d). Note that the structure learning has pruned irrelevant rules in regions not populated by the training data (circles).

where F_i^j are fuzzy sets defined by their respective membership functions $\mu_{F_i^j}(x_i), i = 1, 2, \cdots, n$, and $\theta_j \in \mathbb{R}$ are singleton (crisp) rule consequents.

When a product and-operator and a product implication method are used together with the center of gravity defuzzification method, this leads to a FLS with the following form:

$$y = f(\boldsymbol{x}) = \frac{\sum_{j=1}^{M} (\prod_{i=1}^{n} \mu_{F_i^j}(x_i))\theta_j}{\sum_{j=1}^{M} \prod_{i=1}^{n} \mu_{F_i^j}(x_i)}, \tag{5}$$

which coincides with the "zero-order" Takagi-Sugeno (TS) model [9].

When all the parameters of the FLS in (5) are considered free, methods such as back-propagation learning can be applied. The idea behind the OLS learning [10] is to fix the premise parameters of the FLS such that the resulting fuzzy system is equivalent to a linear combination of nonlinear functions called *fuzzy basis functions* (FBFs) and then to apply the more efficient OLS learning algorithm to determine the most significant FBFs and the corresponding rule consequents θ.

Fuzzy basis function expansion

Consider the fuzzy logic system in (5). Define a fuzzy basis function (FBF) as

$$p_j(\boldsymbol{x}) = \frac{\prod_{i=1}^{n} \mu_{F_i^j}(x_i)}{\sum_{j=1}^{M} \prod_{i=1}^{n} \mu_{F_i^j}(x_i)}, \qquad j = 1, 2, \cdots, M, \tag{6}$$

where $\boldsymbol{x} = [x_1, \ldots, x_n]$. Now the FLS (5) is equivalent to a linear combination of FBFs

$$f(\boldsymbol{x}) = \sum_{j=1}^{M} p_j(\boldsymbol{x})\theta_j. \tag{7}$$

From (5) and (6) we see that each fuzzy IF-THEN rule has a corresponding FBF that gives the relative degree to which the rule fires. The numerator gives the degree to which a particular rule fires, while the denominator in the FBF gives the sum of the degrees of all rules. Due to the normalization of the denominator in the FBF, each FBF is a function of all the membership functions in the FLS (5). Figure 3 shows an example of FBFs in a one dimensional premise space.

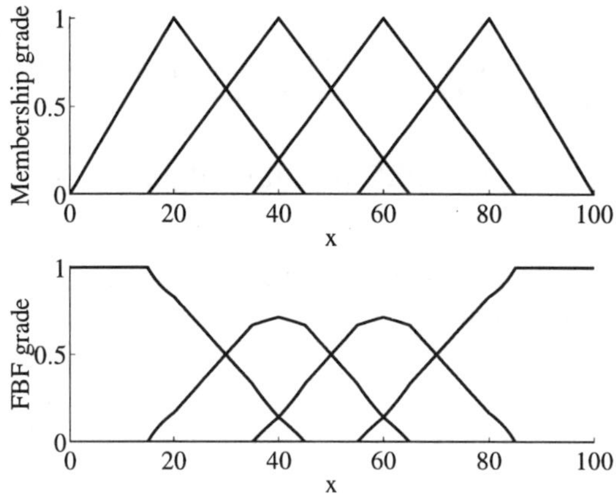

Figure 3: One dimensional premise space of four rules (top) and the corresponding fuzzy basis functions (bottom).

OLS learning
The FBF expansion in (7) is a special case of the linear regression model

$$d(k) = \sum_{j=1}^{M} p_j(k)\theta_j + e(k)\,, \tag{8}$$

where $d(k)$ is the system output, θ_j are real parameters, $p_j(k) = p_j(\boldsymbol{x}(k))$ are known regressors, and $e(k)$ is an error signal. Given N input-output pairs $[\boldsymbol{x}^0(k), d^0(k)]$, $k = 1, \cdots, N$, we seek a FBF expansion $f(\boldsymbol{x})$ such that $\sum_k (f(\boldsymbol{x}^0(k)) - d^0(k))^2$ is minimized.

The linear regression model (8) can be written in a matrix form as

$$\boldsymbol{d} = \mathrm{P}\boldsymbol{\theta} + \boldsymbol{e}\,, \tag{9}$$

where $\boldsymbol{d} = [d(1), \ldots, d(N)]^T, \mathrm{P} = [\boldsymbol{p}_1, \ldots, \boldsymbol{p}_M]$ with $\boldsymbol{p}_j = [p_j(1), \ldots, p_j(N)]^T$, $\boldsymbol{\theta} = [\theta_1, \ldots, \theta_N]^T$, and $\boldsymbol{e} = [e(1), \ldots, e(N)]$. The Gram-Schmidt OLS algorithm performs an orthogonal decomposition for P, that is, P = WA, where W is an orthogonal matrix ($\mathrm{W}^T\mathrm{W} = \mathrm{I}$), and A is an upper-triangular matrix with unity diagonal elements. Substituting P = WA into (9), we have $\boldsymbol{d} = \mathrm{WA}\boldsymbol{\theta} + \boldsymbol{e} = \mathrm{W}\boldsymbol{g} + \boldsymbol{e}$, where $\boldsymbol{g} = \mathrm{A}\boldsymbol{\theta}$. The solution $\boldsymbol{\theta}$ can then be found by $\boldsymbol{\theta} = \mathrm{A}^{-1}\boldsymbol{g}$ since the matrix A is invertible. The OLS algorithm presented below [11] does not decompose the complete matrix P, but selects some dominant columns of P.

Starting from an initial rule base (5), the parameters of the FBF expansion are derived following (6) and (7)[c]. By fixing the parameters in $p_j(\boldsymbol{x})$, the OLS algorithm picks the $M_S < M$ most significant FBF's.

Algorithm 2.1 (Two-pass OLS learning algorithm)

- Step 1: *Select the first basis vector* $\boldsymbol{w}_1$
  For $1 \le i \le M$, calculate

$$\boldsymbol{w}_1^{(i)} = \boldsymbol{p}_i \quad , \quad g_1^{(i)} = \frac{\left(\boldsymbol{w}_1^{(i)}\right)^T \boldsymbol{d}^0}{\left(\boldsymbol{w}_1^{(i)}\right)^T \boldsymbol{w}_1^{(i)}} \tag{10}$$

  and the error-reduction ratio

$$[err]_1^{(i)} = \frac{\left(g_1^{(i)}\right)^2 \left(\boldsymbol{w}_1^{(i)}\right)^T \boldsymbol{w}_1^{(i)}}{(\boldsymbol{d}^0)^T \boldsymbol{d}^0} \tag{11}$$

  where $\boldsymbol{p}_i = [p_i(\boldsymbol{x}^0(1)), \ldots, p_i(\boldsymbol{x}^0(N))]^T$ is given by the initial FBF expansion. Find the FBF with the largest error reduction ratio

$$[err]_1^{(i_1)} = \max_{1 \le i \le M} \left([err]_1^{(i)}\right) \tag{12}$$

  and select the first basis vector $\boldsymbol{w}_1$ and the first element g_1 of the OLS solution vector

$$\boldsymbol{w}_1 = \boldsymbol{w}_1^{(i_1)} = \boldsymbol{p}_{i_1} \quad , \quad g_1 = g_1^{(i_1)} \tag{13}$$

- Step 2: *Select the next basis vectors* $\boldsymbol{w}_k$
  Repeat for $2 \le k \le M_S$:
  For $1 \le i \le M$, $i \ne i_1, \cdots, i \ne i_{k-1}$, calculate

$$\alpha_{jk}^{(i)} = \frac{\boldsymbol{w}_j{}^T \boldsymbol{p}_i}{\boldsymbol{w}_j{}^T \boldsymbol{w}_j} \quad , \quad 1 \le j < k \tag{14}$$

$$\boldsymbol{w}_k^{(i)} = \boldsymbol{p}_i - \sum_{j=1}^{k-1} \alpha_{jk}^{(i)} \boldsymbol{w}_j \quad , \quad g_k^{(i)} = \frac{\left(\boldsymbol{w}_k^{(i)}\right)^T \boldsymbol{d}^0}{\left(\boldsymbol{w}_k^{(i)}\right)^T \boldsymbol{w}_k^{(i)}} \tag{15}$$

---

[c] OLS learning can also be used to identify an initial rule base from data [10].

$$[err]_k^{(i)} = \frac{\left(g_k^{(i)}\right)^2 \left(\boldsymbol{w}_k^{(i)}\right)^T \boldsymbol{w}_k^{(i)}}{(\boldsymbol{d}^0)^T \boldsymbol{d}^0} \tag{16}$$

Find the remaining FBF with the largest error reduction ratio

$$[err]_k^{(i_k)} = \max_{1 \le i \le M, i \ne i_1, \cdots, i \ne i_{k-1}} \left([err]_k^{(i)}\right) \tag{17}$$

and select the kth basis vector $\boldsymbol{w}_k$ and the kth element g_k of the OLS solution vector.

$$\boldsymbol{w}_k = \boldsymbol{w}_k^{(i_k)} \quad , \quad g_k = g_k^{(i_k)} \tag{18}$$

- Step 3: *Remove rules*
  Keep only the antecedent parameters of the M_S selected rules. Set $M = M_S$, to reduce the number of rules. Update the FBF expansion (the remaining FBFs are now normalized by $\sum_{j=1}^{M_S} \prod_{i=1}^{n} \mu_{F_i^j}(x_i)$). Only the $\boldsymbol{p}_i$ vectors for the M_S rules are needed, and the number of $\boldsymbol{p}_i$ vectors is thus reduced from M to M_S.

- Step 4: *Calculate rule consequents*
  Apply OLS again to obtain the $\boldsymbol{\theta}^{(M_S)}$ parameters. The objective is to orthogonalize the modified set of M_S FBFs by using equations (10), (14) and (15), and then to determine the $\boldsymbol{\theta}^{(M_S)}$ vector by solving

$$\mathrm{A}^{(M_S)} \boldsymbol{\theta}^{(M_S)} = \boldsymbol{g}^{(M_S)} \tag{19}$$

which is given by

$$\begin{bmatrix} 1 & \alpha_{1,2}^{i_2} & \alpha_{1,3}^{i_3} & \cdots & \alpha_{1,M_S}^{i_{M_S}} \\ & 1 & \alpha_{2,3}^{i_3} & \cdots & \alpha_{2,M_S}^{i_{M_S}} \\ & 0 & \ddots & & \vdots \\ & & & 1 & \alpha_{M_S-1,M_S}^{i_{M_S}} \\ & & & & 1 \end{bmatrix} = \begin{bmatrix} g_1^{(M_S)} \\ g_2^{(M_S)} \\ \vdots \\ g_{M_S}^{(M_S)} \end{bmatrix} \cdot \begin{bmatrix} \theta_1^{(M_S)} \\ \theta_2^{(M_S)} \\ \vdots \\ \theta_{M_S}^{(M_S)} \end{bmatrix} \tag{20}$$

The final FBF expansion is

$$f(\boldsymbol{x}) = \sum_{j=1}^{M_S} p_{i_j}(\boldsymbol{x}) \theta_j^{(M_S)} \tag{21}$$

where $p_{i_j}(\boldsymbol{x})$ are the subset of the initial FBFs with i_j determined by the preceding steps.

Remarks:

1. The $[err]_k^{(i)} = \frac{\left(g_k^{(i)}\right)^2 \left(\boldsymbol{w}_k^{(i)}\right)^T \boldsymbol{w}_k^{(i)}}{(\boldsymbol{d}^0)^T \boldsymbol{d}^0}$ represents the error-reduction ratio due to $\boldsymbol{w}_k^{(i)}$. Hence, the most significant FBFs are retained (based on their error-reduction ratio) in the final FBF expansion.

2. It is not necessary to prespecify M_S. Step 2 in the algorithm can be ended at the M_Sth repetition when $1 - \sum_{j=1}^{M_S} [err]_j < \rho$, where $0 < \rho < 1$ is a prespecified tolerance.

Example 2.2 Consider the fuzzy model of the fermenter pressure dynamics from Example 2.1. The premise space partition in the initial model is shown in Fig. 4a along with the rule labels. The OLS approach order the importance of the rules based on their error reduction ration as shown in Fig. 4b. Note that low importance is assigned to the rules located in the upper left quadrant of the premise space. □

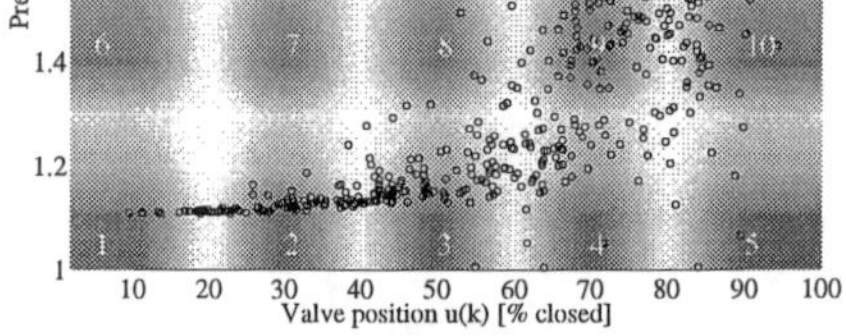

(a) The position of the rules in the premise and their corresponding number

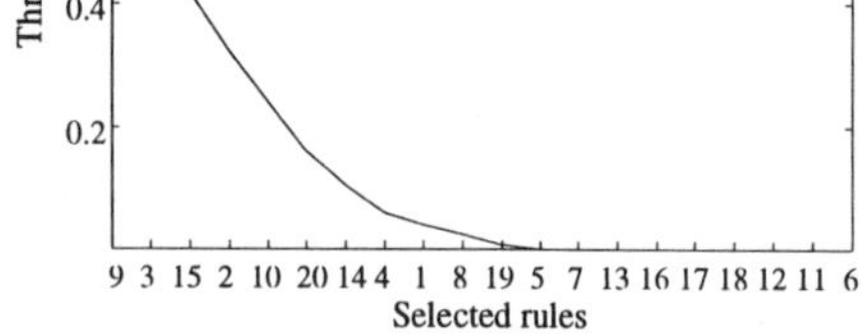

(b) Rule selection according to two-pass OLS learning

Figure 4: The rules in the premise of the pressure dynamics model with their corresponding labels (a), and the error reduction ratio following the two-pass OLS (b) showing which rules are selected depending on the used threshold.

3 Fuzzy Clustering

Fuzzy clustering methods partition a set of data into a number of overlapping clusters based on the distance in a metric space between the data points and the cluster prototypes. Two popular clustering algorithms are the fuzzy c-means (FCM) algorithm [12] and the Gustafson-Kessel (GK) algorithm [13]. The GK algorithm is an extension of the FCM that employs an adaptive distance norm in order to detect clusters of different geometric shapes in one data set. This makes the GK algorithm suitable for the identification of TS fuzzy models from sampled data [14].

Two issues of concern for fuzzy clustering methods are the determination of the relevant number of clusters in the data and the influence of the data distribution on the clustering results. When there is a high concentration of data points in a specific part of the data space, the fuzzy clustering algorithm tends to find more clusters in that region, even though those clusters essentially capture the same information. Typically external cluster validity techniques [15] are used to assess the goodness of a given partition with a specific number of clusters considering criteria like the compactness of the clusters and the distance between the clusters. A drawback of cluster validity methods is the need for repetitive clustering of the data using different number of clusters each time. Another approach to determining the number of clusters is based on starting with a larger number of clusters than required, and then iteratively merging compatible clusters [16]. This offers a more automated and computational less expensive way of determining the right partition.

An extension of the FCM has been introduced [17] that is less sensitive to the bias in the distribution of the data. The algorithm uses *volume cluster prototypes*. Through iterative merging based on *cluster similarity*, the algorithm also determines a relevant number of clusters for the studied data. It differs from other cluster merging methods in that the merging takes place during the optimization procedure for determining the fuzzy partition. Below these extensions are discussed and applied to the GK fuzzy clustering algorithm.

3.1 GK Fuzzy Clustering

Let $\{\boldsymbol{x}_1, \boldsymbol{x}_2, \ldots, \boldsymbol{x}_N\}$ be a set of N data objects represented by n-dimensional feature vectors $\boldsymbol{x}_k = [x_{1k}, \ldots, x_{nk}]^T \in \mathbb{R}^n$. A set of N feature vectors is represented as a $n \times N$ data matrix X. Fuzzy clustering partitions X into M fuzzy clusters. A fuzzy partition can be represented as a matrix U, whose elements $\mu_{jk} \in [0,1]$ represents the membership degree of $\boldsymbol{x}_k$ in cluster j. Hence, the jth row of U contains values of the jth *membership function* in the fuzzy partition. The sum of each column of U is constrained to 1, but the distribution of memberships among the M fuzzy subsets is not constrained. Also, there can be no empty clusters and no cluster may contain all the objects.

Each cluster has its own norm-inducing matrix A_j, giving the inner-product norm:

$$d^2_{jkA_j} = (\boldsymbol{x}_k - \boldsymbol{v}_j)^T A_j (\boldsymbol{x}_k - \boldsymbol{v}_j) \,. \tag{22}$$

The GK fuzzy clustering algorithm is based on minimization of

$$J(X; U, V, A) = \sum_{j=1}^{M} \sum_{k=1}^{N} (\mu_{jk})^m d^2_{jkA_j} \tag{23}$$

where $V = [\boldsymbol{v}_1, \boldsymbol{v}_2, \ldots, \boldsymbol{v}_M], \boldsymbol{v}_j \in \mathbb{R}^n$ is a vector of *cluster prototypes* which have to be determined, and $m \in (1, \infty)$ is a weighting exponent which determines the fuzziness of the clusters. The minimization of (23) represents a nonlinear optimization problem which is solved by the clustering algorithm in an iterative manner.

3.2 Clustering with volume prototypes

Usually, cluster prototypes are considered points in a metric space, and the clusters have membership 1 only in a single point. Often, however, there are a number of points that are close by, which can also be considered to belong fully to a cluster. This is the case especially when there are some clusters that are well separated from the others. It is then sensible to extend the core of a cluster from a single point to a region in the space. The volume prototypes [17] meet this purpose. A volume prototype $\boldsymbol{v} \in \mathbb{R}^n$ is a n-dimensional, convex and compact subspace of $X \in \mathbb{R}^n$.

The extended clustering algorithm measures the distance to the volume prototypes, and data points that are elements of a prototype have by definition a membership of 1.0 in that particular cluster. The size of the cluster prototypes can be obtained from the fuzzy variances of the clusters using

$$r_j^2 = \mathrm{AVR}_{i=1}^{n} \left(\frac{\Sigma_{k=1}^{N} \mu_{jk}^m (x_{ki} - v_{ji})^2}{\Sigma_{k=1}^{N} \mu_{jk}^m}, \right) \quad 1 \le j \le M, \tag{24}$$

where AVR represents an averaging operation over the directions along different variables. According to this definition, the points within a distance of average standard deviation are taken to belong fully to a single cluster. The influence of these points on the remaining clusters is removed and hence the tendency to localize clusters in regions with high concentrations of data points decreases.

3.3 Determining the number of clusters

The determination of the number of "natural" groups in the data is important for the successful application of fuzzy clustering methods. The extended clustering algorithm[17] initializes the fuzzy partition with an estimated upper limit on the number of clusters. Cluster similarity is evaluated during the two-step optimization and similar clusters are merged. The similarity of clusters is assessed by considering the multidimensional fuzzy clusters in the data space. If the similarity between clusters is higher than a given threshold λ, the clusters that are most similar are merged at each iteration of the two-step algorithm.

A fuzzy inclusion measure is used to assess cluster similarity. Consider two fuzzy clusters, $\mu_l(\boldsymbol{x}_k)$ and $\mu_m(\boldsymbol{x}_k)$, defined pointwise in X as rows l and m, respectively, of the partition matrix U. The fuzzy inclusion measure is defined as

$$S_{lm} = \left(\sum_{k=1}^{N} \min(\mu_{lk}, \mu_{mk}) \right) / \left(\sum_{k=1}^{N} \mu_{lk} \right) . \tag{25}$$

The inclusion measure is asymmetric, and it represents the ratio of the cardinality of the intersection of the two fuzzy sets divided by the cardinality of $\mu_l(\boldsymbol{x}_k)$. This is a set-theoretic similarity measure which is not influenced by the scaling of the clustered data[1].

The parameter λ is the threshold above which merging takes place. Since only the most similar pair of clusters are merged at each iteration step, the number of clusters decreases at most by one at every cycle of the iteration. The merging process stops when the similarity between all pairs of clusters is reduced below the given threshold. The algorithm then continues until the change in the elements of the partition matrix is below a defined threshold.

Example 3.1 Consider the pressure system from Example 2.1. The extended GK clustering algorithm is applied to the available systems measurements in the input space. The membership functions of the rule base derived from the resulting clusters[14] are shown in Fig. 5a, and the corresponding premise partition is depicted in Fig. 5b. The clustering algorithm was initialized with 10 random clusters, and a similarity threshold $\lambda = 1/3$ was used[17]. The resulting partition contains only three clusters, corresponding to a fuzzy rule base of three rules. □

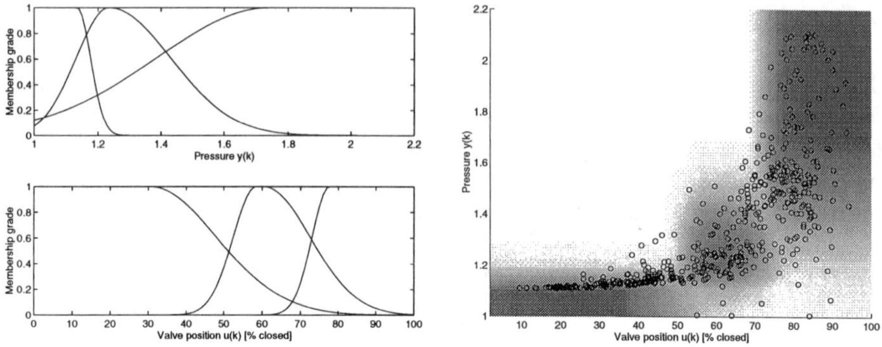

(a) *Membership functions*

(b) *Premise of rule base*

Figure 5: Membership functions (a) and the corresponding premise partition (b) of the rule base identified by the extended GK clustering algorithm.

4 Rule base simplification

Rule-based models obtained from data are often unnecessary complex as their construction is driven by the minimization of some cost function. The resulting rule base can contain unnecessary redundancy in the form of highly overlapping or compatible membership functions. This is often due to the fact that the rules defined in the multidimensional premise are overlapping in one or more dimensions. As a result, more membership functions will describe approximately the same concept, see Fig. 6. Rule base simplification methods, based on similarity between membership functions, have been proposed to remedy this problem [1,18].

4.1 Similarity driven simplification

A similarity relation is used to quantify the similarity between the fuzzy sets in the rule base. Similar fuzzy sets, representing compatible concepts, are aggregated (merged) in order to obtain a generalized concept represented by a common fuzzy set that can replace the similar ones in the rule base. This reduces the number of concepts (linguistic terms) used in the model. Fuzzy sets estimated from data can also possess high similarity to a universal set, that is, a set in which all elements of a domain have a membership close to 1. Such sets add no information to the model and can be removed. The process of aggregation is driven by the similarity among fuzzy sets defined on the domain

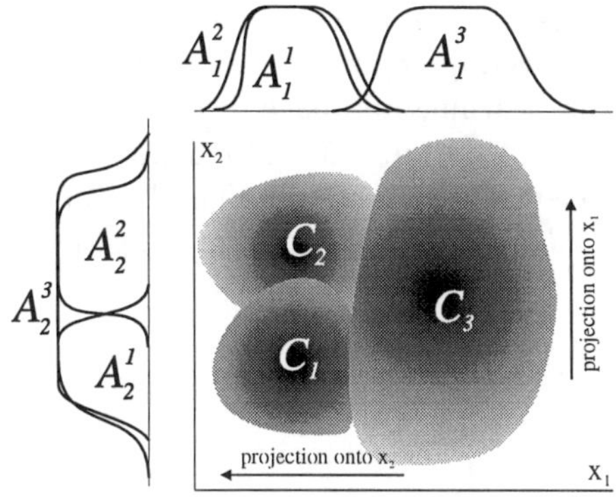

Figure 6: Example of redundancy in the fuzzy term set introduced during identification of a rule base by means of fuzzy clustering

of the same antecedent variable, and not in the product space of the inputs. As such, a reduction of the model's term set is made possible without necessarily any rules being merged. Reduction of the number of rules some times follows as a result of the aggregation of similar fuzzy sets when the antecedents of two or more rules become equal. Such rules are aggregated into a new, general rule, which replaces the rules with the equal premise.

Similarity analysis
For each antecedent variable, the similarity relation is obtained in two steps. First a compatibility relation is constructed from pair-wise compatibility assessment of the fuzzy sets by means of a similarity measure. To assess the compatibility of the fuzzy sets $F_i^j, i = 1, 2, \ldots, n,\ j = 1, 2, \ldots, M$, the fuzzy analog to the Jaccard index is applied:

$$c_{ilm} = \frac{\int (F_i^l \cap F_i^m) dx_i}{\int (F_i^l \cup F_i^m) dx_i}, \tag{26}$$

where $l, m = 1, 2, \ldots, M$. When the fuzzy sets F_i^l and F_i^m are defined on a discrete domain by the membership function $\mu_{F_i^l}(x_i)$ and $\mu_{F_i^m}(x_i)$, respectively, the fuzzy Jaccard index can be written as

$$c_{ilm} = \frac{\mid \min(\mu_{F_i^l}(x_i), \mu_{F_i^m}(x_i)) \mid}{\mid \max(\mu_{F_i^l}(x_i), \mu_{F_i^m}(x_i)) \mid}, \tag{27}$$

where the min and max operators model the intersection and the union, respectively, and $\mid \cdot \mid$ denotes the cardinality of a fuzzy set [19]. The compatibility measure takes on values $c_{ilm} \in [0,1]$, where $c_{ilm} = 1$ reflects equality, and $c_{ilm} = 0$ refers to non-overlapping fuzzy sets. This is the same measure that is used in the structure learning approach discussed in Section 2.1.

For each input $i = 1, 2, \ldots, n$, a $M \times M$ binary fuzzy relation $\mathrm{C}_i = [c_{ilm}]$ is calculated, whose elements c_{ilm} are obtained by (27). In order to obtain a transitive *similarity relation*, S_i, the max-min transitive closure, C_{Ti}, of C_i is calculated [19]:

1. $\mathrm{C}'_i = \max[\mathrm{C}_i, (\mathrm{C}_i \circ \mathrm{C}_i)]$.

2. If $\mathrm{C}'_i \neq \mathrm{C}_i$, set $\mathrm{C}_i = \mathrm{C}'_i$ and go to 1.

3. Stop: $\mathrm{C}_{Ti} = \mathrm{C}'_i$, set $\mathrm{S}_i = \mathrm{C}_{Ti}$.

Here $\circ$ is the max-min composition. The lmth element of the $M \times M$ binary fuzzy similarity relation $\mathrm{S}_i = [s_{ilm}]$ gives the similarity between the concepts represented by the fuzzy sets F_i^l and F_i^m.

Aggregation
For each antecedent variable, the aggregation of similar fuzzy sets takes place by applying a threshold $\lambda \in (0,1)$ to the similarity relation. The set of fuzzy sets having a similarity greater than the threshold λ are aggregated to produce a fuzzy set representing a generalization of the individual concepts represented by the similar fuzzy sets. After aggregation, the updated rule base is checked for fuzzy sets similar to the universal set. Such sets are removed from the premise of the rules in which they occur. Since the simplification does not require additional data acquisition or computationally expensive optimization, the user can investigate the effect of different thresholds. The approach is summarized in Algorithm 4.1.

Algorithm 4.1 (Aggregation algorithm)
Given a rule base $\boldsymbol{R} = \{R^j | j = 1, \ldots, M\}$, R^j: IF x_1 is F_1^j and ... and x_n is F_n^j THEN ..., where $F_i^j, i = 1, \ldots, n$, are fuzzy sets with membership functions $\mu_{F_i^j}$, select $\lambda \in (0,1)$.

Repeat for $i = 1, 2, \ldots, n$

- Step 1: *Calculate similarity relation:*

$$
\begin{aligned}
\mathrm{C}_i &= [c_{ilm}], \quad l, m = 1, 2, \ldots, M, \\
\mathrm{S}_i &= [s_{ilm}] = \mathrm{C}_{Ti}.
\end{aligned}
$$

- Step 2: *Aggregate similar fuzzy sets:*

for $l = 1, 2, \ldots, M$

$$\boldsymbol{F}_i^l = \{F_i^m \mid s_{ilm} > \lambda\}, \quad m \in \{1, 2, \ldots, M\}$$
$$F_{Gi}^l = \text{aggregate}\{\boldsymbol{F}_i^l\},$$

end

$$\text{set } F_i^l = F_{Gi}^l, \quad l = 1, 2, \ldots, M.$$

- Step 3: *Remove fuzzy sets similar to universal set*

$$c_i^j = \frac{\mid \min(F_i^j, U_i^j) \mid}{\mid \max(F_i^j, U_i^j) \mid}, \quad j = 1, 2, \ldots, M,$$

where $\mu_{U_i} = 1, \forall x_i$.
If $c_i^j > \lambda$, remove F_i^j from the antecedent of R^j.

Having identified a set $\boldsymbol{F}_i^l$ of fuzzy sets F_i^m with a similarity higher than the given threshold, $\boldsymbol{F}_i^l = \{F_i^m \mid s_{ilm} > \lambda\}$, the aggregation of these sets is accomplished by letting the support of the general fuzzy set F_{Gi}^l be the support of $\cup \boldsymbol{F}_i^m$. This guarantees preservation of the premise space. The kernel of F_{Gi}^l is obtained by averaging of the kernels of $\boldsymbol{F}_i^l$. If two or more rules end up with a common antecedent part as a result of the aggregation, they can be replaced by *one* general rule R^G. If training data is available, the consequent parameters of the reduced rule base should be reestimated.

Example 4.1 Consider the pressure system from Example 2.1. A fuzzy rule-based model has been obtained from systems measurements using the Gustafson-Kessel fuzzy clustering algorithm[13] in the product-space of the systems in- and outputs[14]. The membership functions of the rule base are shown in Fig. 7a, and the corresponding premise partition is depicted in Fig. 7b. After application of the similarity driven simplification, using $\lambda = 1/4$, a reduced model with four rules and six membership functions is obtained. The membership functions and the premise partition of the reduced rule base are shown in Fig. 7c and Fig. 7d, respectively. Note the similarity between these results and the results obtained with the extended GK clustering algorithm in Example 3.1 □

5 Structured systems

To facilitate the design of a multivariable FLC with a high number of variables, one can define a particular structure (decentralized, hierarchical) according

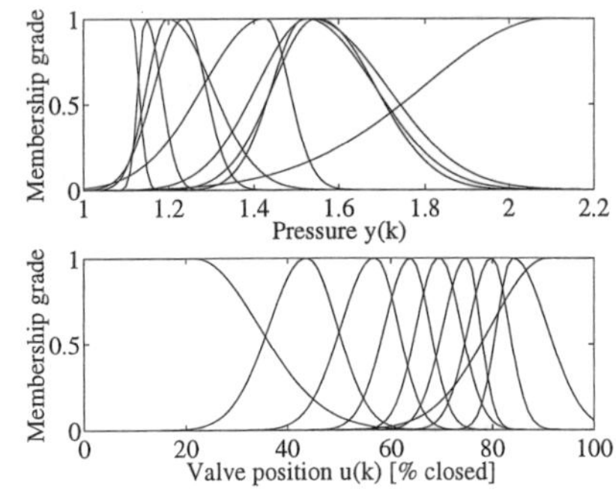

(a) Membership functions, initial RB

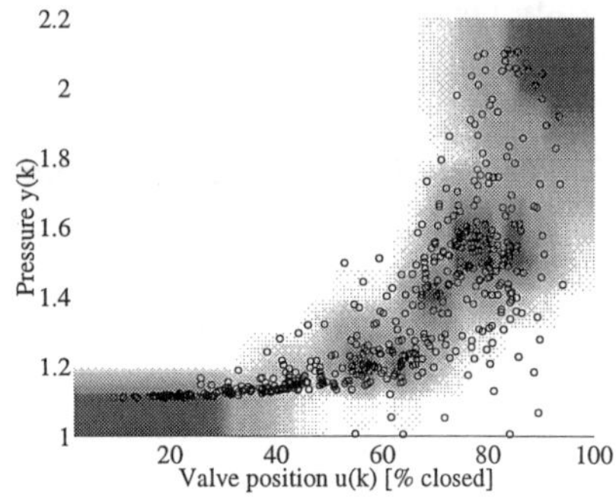

(b) Premise of initial rule base

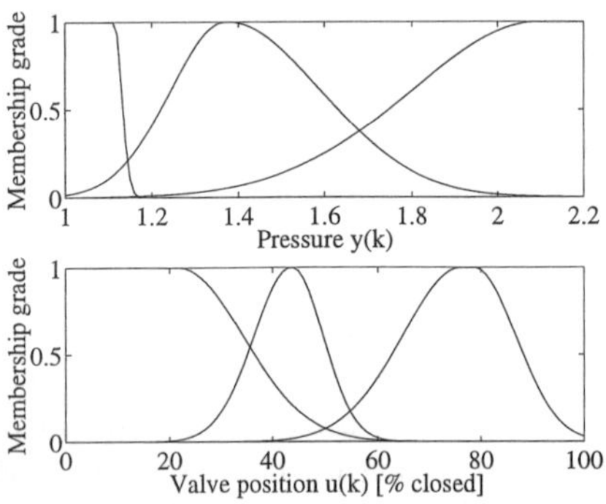

(c) Membership functions, reduced RB

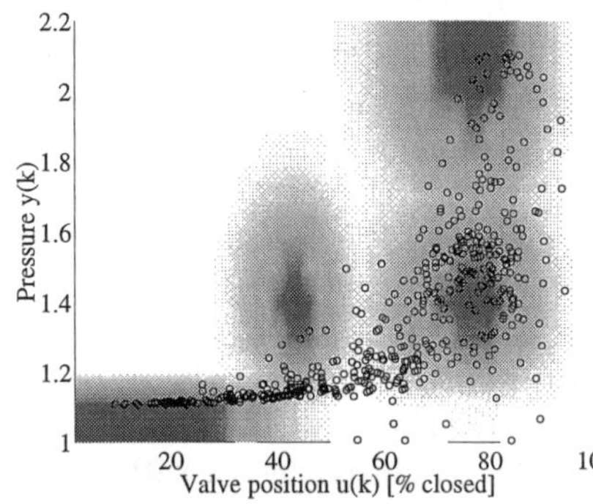

(d) Premise of reduce rule base

Figure 7: Membership functions (a) and the premise partition (b) of the rule base identified from clustering, and the membership functions (c) and premise partition (d) of the reduced rule base.

to some knowledge about the process to be controlled. This knowledge can often be obtained from experienced process operators. By defining an appropriate structure, the design of a Multiple-Input-Multiple-Output (MIMO) FLC can be decomposed into the design of several Single-Input-Single-Output (SISO) decoupled FLCs (decentralized structure) and / or the design of several Multiple-Input-Single-Output (MISO) or smaller dimensional MIMO FLCs (hierarchical structure). At any stage of the hierarchy in a structured system, one can seek to reduce the number of input variables through *mathematical fusion*, or combination of the sensory input variables. An added advantage of such structured systems, is that their maintenance, e.g., tuning and extension, becomes much easier.

5.1 Mathematical fusion

Mathematical fusion, also called *sensory fusion*, is based on combining sensor signals (input variables) before presenting them as inputs to a fuzzy logic controller (FLC). The variables are often fused linearly, and various combinations of input variables can be possible. Some cases are shown in Fig. 8 to illustrate the method. The parameters of the fusion functions can be derived by physical considerations, designers knowledge or experience.

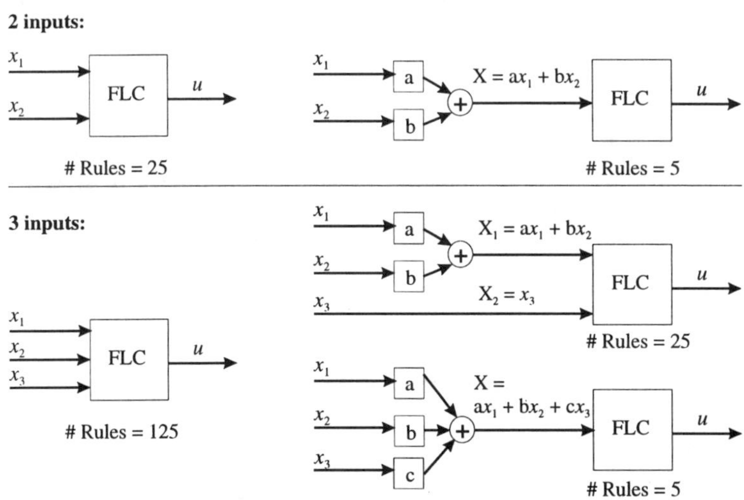

Figure 8: Mathematical fusion for 2 and 3 inputs. It is assumed that the domains of the inputs to the FLC are always partitioned by $m = 5$ linguistic labels each.

Not all variables can be trivially combined. In practice, when FLC problems are considered, usually only two variables are fused: generally these are the error and the change of the error.

5.2 *Decentralized structure*

In MIMO processes coupling effects can occur between the inputs and the outputs: one input can control more than one output, and one output can be controlled by more than one input. If the coupling effects are small, the design of the MIMO controller comes down to the design of a set of SISO controllers, each of them being designed for one pair of input–output variables. The MIMO control problem is then decomposed into a set of simpler SISO control problems.

Many approaches based on the approximation or decomposition of multi-dimensional fuzzy relations in two-dimensional ones have been studied [20,21,22]. Besides providing a much simpler representation, an approximation of the multidimensional relations by two-dimensional ones, also reduces the computational load. However, such approximation may lead to unsatisfactory results as some peculiarities of the process may be neglected. For this reason conditions have been investigated under which both presentations give equal results [22,23]. However, these studies concern only systems using the max–min composition operator. Moreover, the conditions are in general very strict, hampering their practical usefulness.

5.3 *Hierarchical structure*

The hierarchical fuzzy controller [24] is a structure where the number of rules will increase linearly (not exponentially) with the number n of system inputs. The total number of rules reaches its minimum value if every level contains only two input variables as illustrated in Fig. 9.

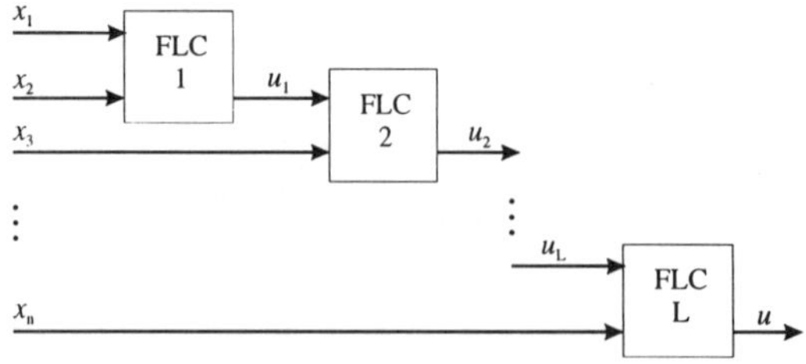

Figure 9: The hierarchical fuzzy controller.

Deciding where the variables are put into the hierarchy is an important and sometimes difficult process. Often the hierarchy can be based on knowledge of the system or on some kind of sensitivity analysis. In practice, the variables are classified according to their importance. In that way, the process followed in the hierarchical structure is very similar to that used by an operator: a first decision is made based on the most important variables and then the operator modifies it according to the other parameters.

A big advantage of the hierarchical FLC structure is that it is easy to add or subtract a variable without having to alter other rules in the rule base.

5.4 Multi-level rules

The structured hierarchical approach was introduced as a crisp structure with fuzzy rules [25,26] but has been generalized to fuzzy structures [27,28]. It is based on the following observation: *"a very complex multivariable system might have a behavior that depends locally on only a subset of the system's variables"*. This subset varies according to the domain of the input space under consideration. Thus, the input space is divided into many domains and for each region "a local rule base" can be written. The local rule bases, called *sub rule bases*, consider only a subset of the system's variables. Thus, the structured hierarchical approach reduces the necessary number of rules.

In the higher levels, called *meta levels*, some meta rules decide which sub rule bases to activate. All levels, excepted the bottom one are meta levels. The total input space is divided into domains and the domains form a partition of X,

$$X = \cup_{i=1}^{p} D_i \,, \tag{28}$$

where D_i are the domains. Every domain can be partitioned itself as, e.g.,

$$D_i = \cup_{j=1}^{p_i} D_{ij} \,. \tag{29}$$

On the bottom level the rules have the usual form. However, the premise part of the rules here, instead of considering all the system's variables $X = \{x_1, x_2, \cdots, x_n\}$ considers only a subset of X. Figure 10 illustrates the hierarchical structure, and show the following rules:

IF x_1 is A_{11} AND x_2 is A_{22} THEN consider domain D_1
IF x_1 is A_{12} AND x_2 is A_{22} THEN consider domain D_2
  Sub rule base of domain D_2 :
  R_2^1: IF x_5 is A_{51} THEN Y is ....
  R_2^2: IF x_5 is A_{52} THEN Y is ....
IF x_1 is A_{11} AND x_2 is A_{21} THEN consider domain D_3

IF x_1 is A_{12} AND x_2 is A_{21} THEN consider domain D_4
Sub rule base of domain D_4:
IF x_3 is A_{31} AND x_4 is A_{42} THEN consider domain D_{41}
Sub rule base of domain D_{41} :
R_{41}^1: IF x_6 is A_{61} AND x_7 is A_{72} THEN Y is ....
...
R_{41}^4: IF x_6 is A_{62} AND x_7 is A_{71} THEN Y is ....
...
IF x_3 is A_{32} AND x_4 is A_{41} THEN consider domain D_{44}.

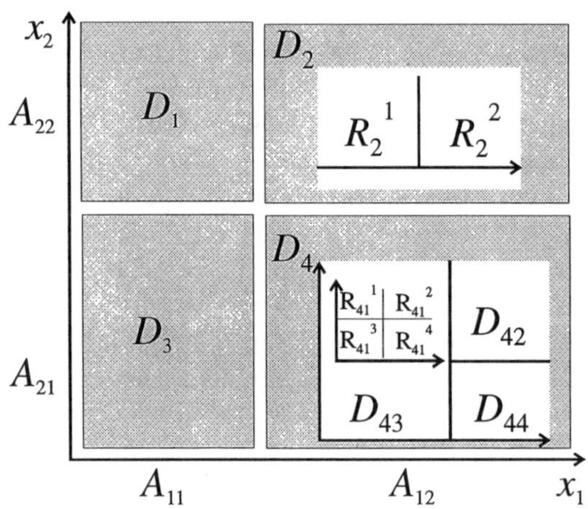

Figure 10: Illustration of a structural hierarchy with four domains on the highest meta level. Two rules are shown in domain D_2 and four rules are shown in sub domain D_{41}.

In the simpler form of the structured hierarchical approach, domains form a Boolean partition of the input space [26], such that one given situation corresponds to only one domain. Consequently, only one sub rule base is invoked. In the general approach [28], the input space partition is fuzzy. This means that more rule bases can be applicable at the same time. This fuzzy input space partition leads to a great flexibility for the choice of invoked sub rule bases but in the meantime raises two essential problems. The first is how to combine sub rule bases referring to (partially) different input variables. One way [28] is to find the least set containing all the variables involved in the different activated sub rule bases in order to define a subspace where all the rules can be represented by their cylindrical extension. The union of all extended rules will form a temporary rule base to be used for reasoning for the input at hand.

The other problem is how to treat contradictions in the resulting temporary rule base? How to evaluate nonconvex conclusions? A defuzzification method like the center of gravity method may lead to an unacceptable conclusion, depending on the application, and consequently, the maximum method has been suggested [28]. Another way to treat the problem is to consider separately each of the activated sub rule bases. Conclusions are calculated for each invoked sub rule base and are then ordered according to preference. The best solution is then selected.

6 Concluding remarks

In this chapter a number of complexity reduction methods for fuzzy modeling have been addressed. When facing multivariable complex systems it is important to reduce the problem's dimension if possible, and then try to structure the system. This splits the design of a multivariable fuzzy system into the design of several lower dimensional rule bases. These rule bases are easier to design both manually and in a semi-automated manner. When considering data-driven construction, one should take care that the rule base is properly initialized before starting adaptation, and avoid grid-like initialization that might produce irrelevant rules. Both the rule selection techniques and the fuzzy clustering method reviewed in this chapter can be used to obtain a good initial rule base of low complexity. After the construction/adaptation of a rule base, it can be sought further simplified by the similarity driven simplification method.

Acknowledgments

This work was supported by the Research Council of Norway.

References

1. M. Setnes, R. Babuška, U. Kaymak, and H. R. van Nauta Lemke. Similarity measures in fuzzy rule base simplification. *IEEE Transactions on Systems, Man and Cybernetics – Part B: Cybernetics*, 28(3):376–386, 1998.
2. B. G. Song, R. J. Marks II, S. Oh, P. Arabshahi, T. P. Caudell, and J. J. Choi. Adaptive membership function fusion and annihilation in fuzzy if-then rules. In *Proceedings FUZZ-IEEE/IFES'93*, pages 961–967, 1993.
3. C. T. Lin. *Neueal Fuzzy Control Systems with Structure and Parameter Learning.* World Scientific, Singapore, 1994.
4. C. T. Chao, Y. J. Chen, and T. T. Teng. Simplification of fuzzy-neural systems using similarity analysis. *IEEE Transactions on Systems, Man*

and Cybernetics – Part B: Cybernetics, 26:344–354, 1996.

5. M. Setnes, A. Koene, R. Babuška, and P. Bruijn. Data-driven initialization and structure learning in fuzzy neural networks. In *Proceedings FUZZ-IEEE'98*, pages 1147–1152, Anchorage, Alaska, May 1998.
6. R. Zwick, E. Carlstein, and D. V. Budescu. Measures of similarity among fuzzy concepts: A comparative analysis. *International Journal of Approximate Reasoning*, 1:221–242, 1987.
7. V. Cross. *An Analysis of Fuzzy Set Aggregators and Compatibility Measures.* Ph.D. thesis, Wright State University, Ohio, 1993.
8. M. Setnes. Fuzzy rule-base simplification using similarity measures. M.Sc. Thesis, Delft University of Technology, Dep. of El. Eng., Control Laboratory, Delft, the Netherlands, July 1995. (A.95.023).
9. T. Takagi and M. Sugeno. Fuzzy identification of systems and its applications to modelling and control. *IEEE Transactions on Systems, Man, and Cybernetics*, 15:116–132, 1985.
10. L. X. Wang and J. M. Mendel. Fuzzy basis functions, universal approximation, and orthogonal least-squares learning. *IEEE Transactions on Neural Networks*, 3(5):807–813, 1992.
11. J. Hohensohn and J. M. Mendel. Two-pass orthogonal least-squares algorithm to train and reduce fuzzy logic systems. In *Proceedings FUZZ-IEEE*, pages 696–700, Orlando, USA, 1994.
12. J. C. Bezdek. *Pattern Recognition With Fuzzy Objective Functions.* Plenum Press, New York, 1981.
13. D. E. Gustafson and W. C. Kessel. Fuzzy clustering with a fuzzy covariance matrix. In *Proceedings IEEE CDC*, pages 761–766, San Diego, USA, 1979.
14. R. Babuška. *Fuzzy Modeling for Control.* Kluwer Academic Publishers, Boston, 1998.
15. X. L. Xie and G. A. Beni. Validity measure for fuzzy clustering. *IEEE Trans. on Pattern Anal. and Machine Intell.*, 3(8):841–846, 1991.
16. U. Kaymak and R. Babuška. Compatible cluster merging for fuzzy modelling. In *Proceedings FUZZ-IEEE/IFES'95*, pages 897–904, Yokohama, Japan, March 1995.
17. M. Setnes and U. Kaymak. Extended fuzzy c-means with volume prototypes and cluster merging. In *Proceedings EUFIT'98*, pages 1360–1364, Aachen, Germany, September 1998.
18. M. Setnes, R. Babuška, and H. B. Verbruggen. Complexity reduction in fuzzy modeling. *Mathematics And Computers In Simulation*, 46(5–6):507–516, 1998.

19. G. J. Klir and B. Youan. *Fuzzy Sets and Fuzzy Logic: Theory and Applications.* Prentice Hall, New Jersey, 1995.
20. N. Boverie, D. Narishkin, J. Lequellec, and A. Titli. Fuzzy control of high order systems using a parallel structure of second order blocks. In *Preprints from IFAC World Congress*, pages 573–576, Sidney, Australia, 1993.
21. L. Jia and X. Zhang. Identification of multivariable fuzzy systems through fuzzy cell mapping. In *Preprints from IFAC World Congress*, pages 389–393, Sidney, Australia, 1993.
22. A.E. Gegov and P.M. Frank. Reduction of multidimensional relations in fuzzy control systems. *Systems and Control Letters*, 25:307–313, 1995.
23. A. Gegov. *Distributed Fuzzy Control of Multivariable Systems.* Kluwer, Dordrecht, The Netherlands, 1996.
24. G.V.S. Raju and Jun Zhou. Fuzzy logic process controller. In *IEEE International Conference on Systems Engineering*, pages 145–147, Pittsburgh, USA, August 1990.
25. M. Sugeno, Murofushi T, J. Nisho, and H. Miwa. Helicopter control based on fuzzy logic. In *Second Fuzzy Symposium on fuzzy systems and their applications to human and natural systems*, Tokyo, 1991.
26. M. Sugeno, M.F. Griffin, and A. Bastian. Fuzzy hierarchical control of an unmanned helicopter. In *Fifth IFSA World Congress*, pages 179–182, Seoul, 1993.
27. L. T. Kóczy and K. Hirota. Interpolation in structured fuzzy rule bases. In *Proceedings FUZZ-IEEE'98*, pages 402–405, San Francisco, 1992.
28. L. T. Kóczy and K. Hirota. Modular rule bases in fuzzy control. In *EUFIT'93-First European Congress on Fuzzy and Intelligent Technologies*, pages 606–610, Aachen, Germany, September 1993.

FUZZY DECISIONS FOR CONTROL SYSTEMS

UZAY KAYMAK
Shell International Exploration and Production
Research and Technical Services
P.O. Box 60, 2280 AB Rijswijk, the Netherlands
Email: U.Kaymak@siep.shell.com

The types of problems fuzzy decision making addresses are very similar in nature to the types of problems that are encountered in many control problems, such as expert control, supervisory control and task-oriented control. As a result, fuzzy control can benefit significantly from the techniques and methods developed for fuzzy decision making. This chapter considers the relationship between fuzzy control and fuzzy decision making. In particular, the discrete choice problem formulated according to Yager's fuzzy decision model is discussed, and its application to various control problems is studied. The applications considered include supervisory control, the selection of discrete control actions, the recognition of operational environment, fuzzy model-based predictive control, fuzzy aggregated membership control, satisficing control and the design of objective functions from expert knowledge.

1 Introduction

Since Mamdani's[1] application of fuzzy set theory to a control problem, fuzzy control has gained gradual acceptance as a valid control method. Today, fuzzy control is more and more recognized as one of the established tools for the control engineer to design control systems. Fuzzy control owes this success partly to the initial applications (e.g. Holmblad and Østergaard[2]), where the fuzzy controller has been used to implement the control strategy of experienced operators for controlling processes for which a mathematical model did not exist or could not be obtained easily. This made the automatic control of a number of processes possible, which could not be analyzed previously by using the techniques from conventional control engineering. Afterwards, fuzzy controllers have also been used as nonlinear PID controllers, where the inputs to the fuzzy system are the difference between the desired output and the actual output, its derivative and its time integral.

In addition to their desirable properties for implementing nonlinear controllers, fuzzy systems also possess characteristics to deal with uncertainty, vagueness and fuzziness that one often encounters in supervisory and task-oriented control. These control problems are characterized by subjective goals for the control system, flexible constraints and a mix of continuous and discrete control actions. As the automated control systems become more complex, the importance of such high level control systems is increasing. The control problem for supervisory control, task-oriented control and scheduling shows many similarities to the decision making problems. Fuzzy decision making can be used to enhance the existing fuzzy control techniques for dealing with these types of control. This chapter discusses several ways in which fuzzy decision making methods can be applied to control problems. The application of the methods considered extend the applicability of fuzzy control systems to a larger class of control problems. Attention is paid to the basic principles of the application of fuzzy decision making in control systems, without going into specific details. The reader is referred to the publications mentioned in the text for a more detailed consideration of the methods discussed.

The outline of the chapter is as follows. Sec. 2 describes the formulation of the fuzzy decision making problem following the Bellman and Zadeh [3] approach. In particular, the basic elements of the discrete choice problem are considered. Sec. 3 describes explicitly the relations between the control problems and the decision problems. It also formulates an optimal fuzzy control problem as fuzzy decision making. The main part of the chapter is found in Sec. 4, where the application of fuzzy decision making methods in control problems is explained based on a number of examples and generic cases. Finally, the conclusions are presented in Sec. 5.

2 Fuzzy decision making

This section explains the formulation of a fuzzy decision making problem and discusses the basic elements of a fuzzy decision problem. The decision making is formulated following the Bellman and Zadeh [3] approach. For more information on fuzzy decision making and its various elements the reader is referred to one of the many publications in this field. [4,5,6,7,8,9,10]

2.1 General formulation of decision making

In decision logic, a *decision* is formulated as a quintuple $(A, \Theta, \Xi, \kappa, D)$ [6]. The symbols herein are defined as follows.

- A is the set of alternatives or possible actions. The decision maker is expected to make a selection from this set by using the available information.
- Θ is the set of the states[a] of the environment in which the decision making is taking place. These states are usually not known, although

[a] The word "states" is used here in a different meaning than the "state" in control engineering. The states of the environment describe various conditions of the environment in which

a probability, possibility or plausibility distribution may sometimes be available. The states of the environment can not be controlled by the decision maker, but they must be dealt with in the decision making process.

- Ξ is the set of consequences. The consequences result from the choice of a particular alternative. An alternative can lead to several consequences all of which may need to be taken into account in multicriteria decision making. In that case Ξ is multidimensional.

- κ is a mapping $A \times \Theta \longrightarrow \Xi$ which specifies a resulting consequence for each element of the set of environment states Θ and each element of the set of alternatives A. Ξ is the set of consequences of the decision alternatives. It is assumed that Ξ is known completely and that a consequence can be calculated for each alternative–state pair. The space $A \times \Theta$ defines the solutions for the decision problem and it is sometimes also called *the solution space.*

- D is the decision function which is defined as $D : \Xi \longrightarrow \mathbb{R}$. It reflects the preference structure of the decision maker. The decision function D incorporates the goals of the decision maker. It induces a preference ordering on the set of consequences Ξ such that

$$\xi_i \succ \xi_j \text{ if and only if } D(\xi_i) > D(\xi_j), \tag{1}$$

  where $\xi_i, \xi_j \in \Xi$ and $\succ$ is the preference relation, i.e. consequence ξ_i is preferred to consequence ξ_j.

Note that since κ maps each alternative to a consequence, the decision function implicitly induces a preference ordering on A such that for alternatives $a_i, a_j \in A$

$$a_i \succ a_j \text{ if and only if } D(\xi_i) > D(\xi_j).$$

2.2 Bellman and Zadeh's formulation

Two important elements in decision making are the decision goals and the decision constraints. Fuzzy decision making essentially replaces the crisp goals and the constraints with their fuzzy equivalents. Fuzzy goals and constraints are defined as follows [3].

---

decisions are taken. However, this does not imply that the states give all the information regarding the environment.

Definition 1 *Let A be a given set of possible alternatives which contains a solution to a decision making problem under consideration. A fuzzy goal G is a fuzzy set on A characterized by its membership function*

$$\mu_G : A \longrightarrow [0,1],$$

which represents the degree to which the alternatives satisfy the specified decision goal.

Definition 2 *Let A be a given set of possible alternatives which contains a solution to a decision making problem under consideration. A fuzzy constraint C is a fuzzy set on A characterized by its membership function*

$$\mu_C : A \longrightarrow [0,1],$$

which constrains the solution to a fuzzy region within the set of possible solutions.

A fuzzy constraint is a generalization of a crisp constraint. Some alternatives in the solution set satisfy the constraints completely, while others violate them to some degree.

Realizing that a decision should satisfy the decision goals as well as the decision constraints, Bellman and Zadeh [3] suggested the following model for the fuzzy decision.

Definition 3 *Let A be a given set of possible alternatives which contains a solution to a decision making problem under consideration. Let G be the set of fuzzy goals for the decision, represented by the membership function* $\mu_G(a),\ a \in A$, *and let C be the set of fuzzy constraints represented by the membership function* $\mu_C(a),\ a \in A$. *Then the* fuzzy decision *F results from the intersection of the fuzzy decision goals and the fuzzy constraints, i.e.*

$$F = G \cap C. \tag{2}$$

The fuzzy decision is characterized by its membership function

$$\mu_F(a) = \mu_G(a) \wedge \mu_C(a) \ a \in A, \tag{3}$$

where $\wedge$ *denotes the minimum operation.*

Definition 4 *The optimal decision* a^* *in fuzzy decision making is the decision with the largest membership value, also called* the maximizing decision, *which is defined by*

$$a^* = \arg\max_{a \in A} \mu_G(a) \wedge \mu_C(a). \tag{4}$$

It is now generally accepted that the intersection operator can be replaced by a suitable fuzzy aggregation and the minimum operator can be replaced by another aggregation operator. In conventional decision models, the decision goals and the decision constraints are two separate entities that are also treated differently by the decision logic. In Bellman and Zadeh's model, the distinction between the goals and the constraints disappears. Both the goals and the constraints are represented by membership functions defined on the set of possible alternatives. The decision function (the conjunction in the model) makes an appropriate combination of the goals and the constraints. Because of the symmetry between the goals and the constraints, this model is sometimes called *a symmetric model.*

2.3 Discrete choice problem

A specific type of decision problem that is encountered often is concerned with the selection of a course of action out of a finite and countable set of possible actions, taking into account multiple criteria that are of importance for the decision. This type of decision problems is denoted as *discrete choice problems.* Yager[11] has proposed a decision model for the fuzzy discrete choice problem, which can be considered as a special case of Bellman and Zadeh's model. The principle elements of his approach to the discrete choice problem are the following.

1. The set of alternatives. The aim of the decision process is the selection of an alternative out of a set A of possible alternatives. This set has as its elements the possible outcomes $a_j, j = \{1, \ldots, n\}$ of the decision process that is under consideration, i.e.

$$A = \{a_1, \ldots, a_n\}.$$

2. The set of goals and/or constraints (decision criteria) upon which the decisions are based. In terms of the general formulation of a decision problem, the criteria represent different aspects of the set of consequences Ξ. Let the decision criteria form the set

$$C = \{c_1, \ldots, c_m\}.$$

   The selection of the criteria outlines the structure of the decision problem.

3. The judgments (ratings or membership values) per alternative per criterion which indicate the satisfaction of the criteria by the alternative. In fuzzy decision making, the judgment information is usually provided

in the form of membership values μ_{ij}, which can be obtained by using membership functions that represent the fuzzy criteria. The judgment information can be summarized in an *evaluation matrix* which looks like

$$\begin{array}{c} \\ c_1 \\ \vdots \\ c_m \end{array} \begin{array}{c} \begin{array}{ccc} a_1 & \cdots & a_n \end{array} \\ \begin{pmatrix} \mu_{11} & \cdots & \mu_{1n} \\ \vdots & \ddots & \vdots \\ \mu_{m1} & \cdots & \mu_{mn} \end{pmatrix} \end{array}. \tag{5}$$

When a membership function cannot be specified explicitly (e.g. when dealing with data that is subjective or difficult to measure), the membership values μ_{ij} can also be specified directly by the decision maker and filled into the evaluation matrix. Note that the judgments in this model are crisp numbers that are obtained from the membership of the alternatives to the fuzzy criteria.

4. The weight factors which indicate the importance that the decision maker attaches to various goals and/or constraints. Usually the decision criteria are not all equally important. In this case, the criteria should have different influence on the result of the decision. More important criteria should influence the outcome of the decision more than the less important criteria. Weight factors are used for modeling the different sensitivity of the decision problem to the different criteria.

5. A performance function which orders the set of alternatives according to the satisfaction of the goals and/or constraints. This is also called the *aggregation function.* The aggregation function

$$D^w(\mu_{1j}, \ldots, \mu_{mj}),\ j \in \{1, \ldots, n\} \tag{6}$$

is a function of the m membership values, m weight factors and possibly other (external) parameters. In fuzzy decision making, the aggregation function D is often a suitable aggregation operator with which fuzzy sets can be combined. The aggregation function in fuzzy decision making is also called *the decision function.* The decision function should reflect the aims of the decision maker. Hence, it can be interpreted as a suitable translation of the goals of the decision into a mathematical representation. Many possible decision functions have been suggested in literature. The aggregation function may consist of a single aggregation operator from the fuzzy set theory, or it can be a combination of these operators.

6. An ordering mechanism for the ranking of alternatives. The decision function assigns a value to each alternative which indicates the overall

suitability of an alternative. When the aggregated value is crisp, the ordering of the combined judgments corresponds to the ranking of the alternatives such that

$$D(a_k) > D(a_l) \iff a_k \succ a_l,$$

where $\succ$ represents the preference relation. The decision function D is then said to impose a preference ordering on the set of alternatives.

An extension of Yager's model with fuzzy evaluations is obtained when one uses type II fuzzy sets for specifying the fuzzy criteria. The evaluations in the evaluation matrix are then fuzzy sets (fuzzy numbers) themselves and the aggregation function operates on these fuzzy numbers. In this case, the decision function D is a mapping

$$D : \tilde{\mathcal{P}}([0,1]^m) \longrightarrow \tilde{\mathcal{P}}([0,1]), \tag{7}$$

where $\tilde{\mathcal{P}}([0,1])$ denotes the set of all fuzzy subsets of the unit interval. Hence, there is not a total ordering structure of the alternatives that uniquely determines the ranking. The partial ordering of the fuzzy evaluations must then be considered in order to determine the best alternative, as discussed in various publications.[12,13,14,15,16,17] This procedure is a decision making process itself and resembles the defuzzification procedure for the fuzzy controllers.

3 Control as a decision problem

Given the decision making approach from Sec. 2 and considering the explicit relationship between the control problems and the decision making problems, the fuzzy control problem can be formulated as a multistage decision problem in a fuzzy environment. As shown in Sec. 4, this formulation is quite important for some of the applications of fuzzy decision making for control purposes.

A one-to-one relation between control problems and decision problems can be established as follows.

1. In control problems, a number of control actions $u \in U$ are available to the controller. The controller must make a selection for the suitable control action out of the available ones for achieving the specified control objectives. The control actions thus correspond to the set of alternatives A.

2. There are a set of conditions which cannot be influenced by the control system. These are typically the disturbances that act on the system from the environment. They correspond to the set Θ in decision making.

3. The control actions and the disturbances act on the controlled system and lead to several consequences such as new states, error signals and so on. The system mapping that maps its manipulated variables and the disturbances to the outputs corresponds to the mapping κ in decision making.

4. The consequences of the control actions are evaluated within a time window by using various criteria which are typically norms of various signals or quantities derived from the consequences, such as the squared sum of various error signals. The consequences thus correspond to the set Ξ of consequences in decision making, and they are evaluated using several criteria.

5. The evaluations for the consequences of the control actions are combined by an objective function. The optimal control action is determined by minimization (or maximization) of the objective function. The objective function corresponds to the decision function D in decision making.

Consider, now, a single-input single-output deterministic system described by the time-discrete state transition equation

$$x(k+1) = f(x(k), u(k)) \tag{8}$$

where $x(k) \in X$ is an n-dimensional state vector, $u(k) \in U$ is the control action applied to the input, and $f : X \times U \longrightarrow X$ is a function describing the system. If the system is nonlinear, f is a nonlinear function, and it can be a time-dependent function in case of time-varying systems. Starting from a given state $x(0)$, the temporal evolution of the system under the control actions results in a trajectory $(x(0), u(0), x(1), u(1), \ldots, x(k), u(k), \ldots)$. At each time step k, the control applied to the system can be subjected to a fuzzy constraint $\mu_C^k(u(k))$ and a fuzzy goal $\mu_G^{k+1}(x(k+1))$ can be specified on the state attained as a result of the control applied. Let N be the termination time for the fuzzy control. The optimal decision problem for the controller is then the determination of the maximizing decision such that the fuzzy decision

$$\begin{aligned}\mu_D(u(0), \ldots, u(N-1)|x(0)) = \mu_C^0(u(0)) \circ \mu_G^1(x(1)) \circ \\ \circ \cdots \circ \mu_C^{N-1}(u(N-1)) \circ \mu_G^N(x(N))\end{aligned} \tag{9}$$

is maximized. The sign $\circ$ in Eq. (9) denotes a combination of fuzzy sets by a fuzzy set aggregation operator. Often t-norms are used for the aggregation in control systems. The multistage decision in this type of optimal control problem is then the determination of a sequence of control actions $u(0), \ldots, u(N-1)$ to maximize Eq. (9).

A similar formulation can also be given for systems with infinite termination time, fuzzy termination time and implicit termination time specified by a termination condition on the states x. Further, fuzzy systems and stochastic systems can also be considered in addition to the deterministic systems. For a detailed consideration of these type of systems, the reader is referred to Kacprzyk [18]. Note that the search space for the decision problem is given by the product space U^N. As the number of time steps considered increases, the size of the search space increases exponentially, and the optimization problem can take very long times to complete. Consideration of optimization methods in this approach is therefore important. Dynamic programming can be used [3,18] for determining the maximizing decision Eq. (9). Other discrete search methods such as branch-and-bound algorithm can also be used [18,19,20] provided that the control space U is quantized. One obtains an approximate solution to the optimization problem in this case. Applications of evolutionary computation methods such as genetic algorithms [21] have also been reported in the literature [22] to solve this type of optimization problem.

4 Applications of fuzzy decision making in control

The previous sections have considered the basic elements of a fuzzy decision making problem and the formulation of a control problem as a multistage fuzzy decision. This section reviews some ways in which the formulation of the fuzzy decision making is applied in control systems to address various control problems.

4.1 Selecting discrete control actions

In many control problems, control actions are selected from a continuous domain. In order to make use of digital computers, the continuous domain is discretized, as discussed in Sec. 3. However, there is a wide range of control problems in which the control actions are by nature discrete. Control theory often does not address this type of problems. The application of decision making methods is a natural candidate for dealing with systems where a control action needs to be selected from discrete alternatives. The suitability of various control alternatives can be evaluated based on a number of criteria, and the alternative (i.e. the control action) that best fulfills the control goals and the constraints can be applied to the system. Since many of the goals and constraints that one encounters in these problems will only be fulfilled partially by the control actions, fuzzy decision making can be applied.

An example of a common control problem that involves a selection from a discrete number of alternatives is the selection of a gear in automatic trans-

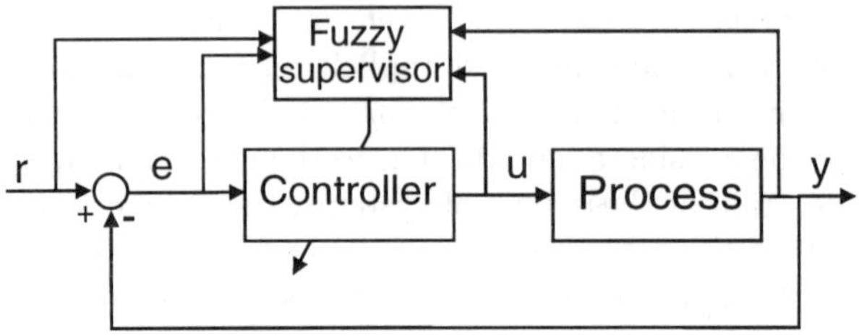

Figure 1: Fuzzy supervisory control.

mission systems. The automatic transmission systems of cars use a variety of signals to determine the moment for shifting a gear. At each decision moment, the decision logic can determine whether the gear should be shifted up, shifted down or should not be changed (no action). The shifting decision can be based on the speed of the vehicle, the number of revolutions per minute (r.p.m.) of the engine, etc. The decision criteria can then be, for example, a target range for r.p.m., or a range of desired vehicle speed for the current gear. The decision algorithm evaluates the available alternatives based on the criteria and selects the most appropriate control action after combining the information about the satisfaction of the decision criteria.

4.2 Supervisory control

A supervisory controller is a higher level controller which augments an existing controller so that the control objectives can be met, which would not be possible without the supervisory controller. Supervisory systems are characterized by the additional flexibility they bring to the control system. A supervisory controller can adjust the parameters of a low-level controller according to the process information, user preferences or operational environment. Figure 1 shows the general scheme for a fuzzy supervisory controller.

A form of a fuzzy supervisory system is described by Sousa *et al.* [23] The authors have implemented a supervisory controller that adapts the decision alternatives for a low level fuzzy controller. The fuzzy controller discretizes the control space U and selects the optimal control alternative from the available discrete control actions. In order to keep the computational load of the optimization acceptable, the number of discrete control actions can not be large, which may lead to limit cycles for slowly varying references or when approaching a steady-state situation. The fuzzy supervisor adapts the set of available control actions based on information about the state of the controlled system. A decision making approach is used for the adaptation. The decisions are based on two fuzzy criteria *small error* and *small output change*. The minimum operator is used as the decision function, i.e.

$$\mu_F(k) = \min(\mu_e(e(k)), \mu_y(y(k))), \tag{10}$$

where $e(k)$ is the error and $y(k)$ is the output at time step k, respectively. The decision alternatives for the direct controller are adapted based on a transformation of the aggregated result μ_D as

$$\gamma_k = 1 - \mu_F. \tag{11}$$

The alternatives are scaled (adapted) by the factor $\gamma(k)$. Sousa *et al.*[23] report improvements in the controller performance and a reduction of the limit cycles.

4.3 Identification of operational environment

Humans can successfully (learn to) control a large number of processes, be them linear, nonlinear, time varying or only partially known. Automated controllers have great difficulty in dealing with most of control problems that humans can control satisfactorily. The parking of a car is such a problem that humans find relatively easy to tackle, while automated controllers fail to accomplish the task satisfactorily. An important aspect of human control is that humans use a variety of control strategies and schemes, switching from one to the other smoothly as circumstances require. The operational conditions are recognized quickly and an appropriate combination of control schemes is then used. The modification of the control actions or controller parameters depending on the operational environment (as in gain scheduling) or the selection of the control strategy based on the operational environment improve the performance of automated controllers significantly. Fuzzy decision making can be used to determine the operational environment from a set of possible alternatives.

An example of a control problem where the recognition of the environment is made by a fuzzy decision making system is the Nissan automatic transmission system described by Chiu.[24] The rules for switching gears in an automatic transmission system can be made to follow humans' strategy if the controller can recognize what kind of operational environment the car is in at any given time. For example, the accelerator input tends to vary slowly on a highway, while it shows large variations on a winding road. The rules for switching gears in these two distinct conditions are then also different. By using membership functions that represent the decision criteria for accelerator input, variance in accelerator input and the vehicle speed, the fuzzy decision is able to determine the degree to which the available evidence supports each of the possible environments (decision alternatives).

4.4 Rule-based objective functions

The selection of the decision function is an important aspect of the fuzzy decision making problem. The decision function reflects the subjective preferences of the decision maker and an appropriate combination of the decision criteria for the specific decision problem. Sometimes, the control problem may require a more complicated aggregation of criteria than is possible by using the aggregation operators from the fuzzy set theory. As discussed before, the aggregation of the decision criteria criteria corresponds to the objective function in an optimal control problem. However, it is not possible, in many cases, to specify the objective function explicitly in a mathematical form, while heuristic rules may be available for assessing the system response. The heuristic rules thus represent a complicated aggregation behavior expressed in terms of linguistic rules. These rules are then used to rank the possible control actions.

An example of the rule-based assessment of control actions and the use of decision logic for selecting the optimal control actions from the rule-based objective functions is an automatic train control system described by Yasunobu and Miyamoto.[25] In this system, the expected consequence of a control action and its desirability in terms of a number of criteria such as the smoothness of the ride and the accuracy of the stopping position are coded as a set of linguistic rules. The fuzzy system can then determine which control action should be applied, based on the rule-based objective function.

4.5 Fuzzy aggregated membership control

A decision function which combines m criteria is a mapping $\mathbb{R}^m \longrightarrow \mathbb{R}$. The overall mapping is defined by a combination of one dimensional membership functions, which show the degree of satisfaction of a single criterion. In this sense, a fuzzy decision system can be used as a direct fuzzy controller, since fuzzy controllers are nonlinear systems which implement a nonlinear control law (mapping) between the controller inputs and the controller outputs. Kaymak *al.*[26] have introduced a type of nonlinear controller called *fuzzy aggregated membership (FAME)* controller, which is based on the aggregation of fuzzy sets using decision functions . The fuzzy sets are defined on the domain of discourse of the individual input variables of the controller, and they indicate the degree of satisfaction of a particular control goal by the input variables. The nonlinear control surface is formed by a suitable aggregation of the fuzzy sets. The nonlinearity of the control surface is determined explicitly by the selection of the membership functions and the aggregation operator. Only one fuzzy set per variable is defined, which reduces the number of controller parameters compared to a conventional fuzzy controller.

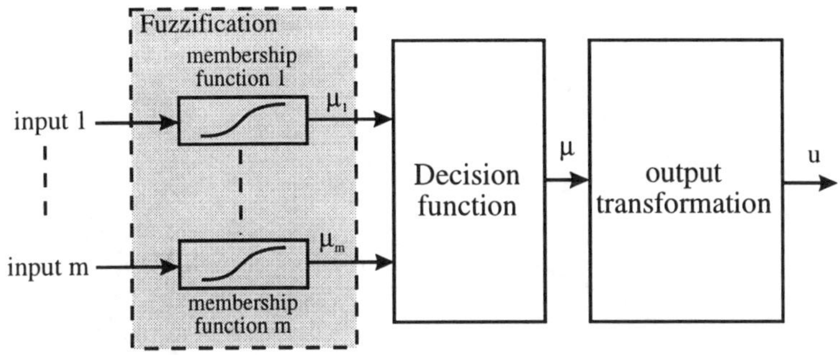

Figure 2: Structure of a fuzzy aggregated membership controller.

The proposed controller is obtained in the following way. A particular structure for the controller inputs and the outputs are selected depending on the specific control problem. For example, if the characteristics of the controlled process requires the use of a nonlinear PD controller, the error signal e and its derivative Δe should be selected as the inputs of the controller. Then, one membership function for each variable is defined. The membership function is defined in such a way that the resulting control surface corresponds to the characteristics of the controlled process (e.g. gain factors). This membership function describes a particular property for the variable, such as "large error signal." The measured values for the inputs (e.g. error and error derivative) are converted into a membership value by using the membership functions. In this manner, different variables are brought into a decision space within which various quantities can be combined using a decision function. The decision function, which should reflect the goals of the controller, combines the membership values from different inputs into a single decision value. Finally, the output of the controller is determined by transforming the aggregated membership value into a control value. Figure 2 shows the structure of the proposed FAME controller. The steps that a FAME controller takes for calculating the output can be summarized as follows.

1. Determine the membership values μ_i for each input variable using the measured values of the variables and the membership functions defined for each variable. Here, the membership function is a mapping $\mathbb{R} \longrightarrow [0, 1]$, and it need not have a height of one or a bounded support.

2. Determine the aggregated membership value μ using the decision function and possibly the weight factors

$$\mu = D^w(\mu_1, \ldots, \mu_m). \tag{12}$$

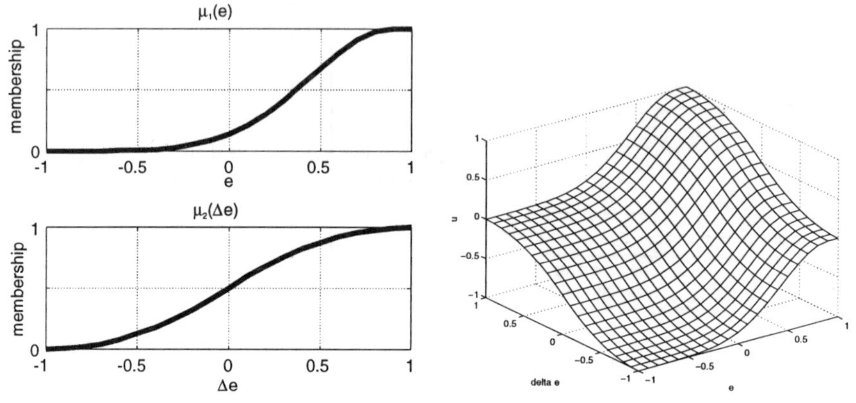

(a) Membership functions *(b) Control surface*

Figure 3: A nonlinear PD-controller implemented using arithmetic mean as the fuzzy decision function. (a) Input membership functions. (b) Control surface of the controller.

3. Calculate the controller output u using an appropriate transformation of μ

$$u = g(\mu). \tag{13}$$

This transformation is typically a scaling of the closed interval $[0, 1]$ to the interval of operating values by using a linear mapping.

Figure 3 depicts a nonlinear PD controller that is obtained by using the proposed scheme. The controller has as inputs the error signal e and its derivative Δe. The membership functions that are defined for the two inputs and the resulting control surface of the controller are shown in the figure. The aggregation operator that is used is the arithmetic mean. Note that a linear PD controller would have been obtained if the membership functions for the inputs had been trapezoidal.

4.6 Fuzzy model-based predictive control

Model-based predictive control (MBPC) is a general control methodology that has been widely applied in industry, [27] and the theory of model-based predictive control has been studied extensively in the literature. [28,29] MBPC is characterized by the use of a process model to predict the process output at

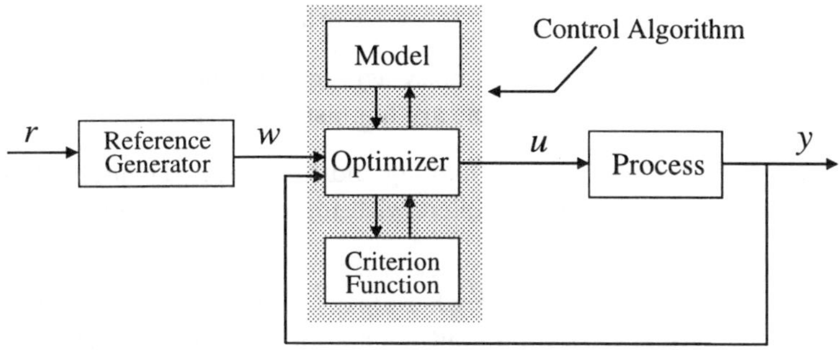

Figure 4: Schematic illustration of model-based predictive control scheme.

future discrete time instants over a specified prediction horizon, given the process inputs and the desired reference output. The sequence of future control actions is computed by minimizing an objective function while satisfying a set of operation constraints. The optimization is repeated at every sample instant with the new available process data, and each time only the first control action in the calculated sequence is applied to the process (*receding horizon principle*). The optimization problem considers only a finite prediction window into the future determined by a *prediction horizon* H_p. In order to reduce the search space, a *control horizon* $H_c \leq H_p$ is usually defined, such that the control signals remain constant outside the control horizon, i.e.

$$u(k+i) = u(k+H_c-1), \quad i = H_c, \ldots, H_p - 1.$$

Figure 4 illustrates the concept. In conventional MBPC, linear combinations of squared differences between the desired and the predicted sequences, and the control effort are generally used as the objective function, because they lead to tractable, convex optimization problems for linear systems. However, since the objective function reflects the control goals, it is advantageous to have additional freedom for specifying more complicated control goals. This additional freedom can be achieved by choosing a different representation of the objective function, e.g. as a combination of fuzzy goals and constraints, in a fuzzy decision making approach.

Fuzzy multi-criteria decision making is an approach that translates objectives and constraints to predictive control in a transparent way. Fuzzy predictive control combines the model-based predictive control approach with methods from fuzzy multicriteria decision making.[10,20] The reasons for combining model-based predictive control with fuzzy decision making are the following.

- Fuzzy goals and constraints can be used in optimization, so that the goals of the designer are represented in a natural manner.

- Knowledge of the behavior of fuzzy aggregation operators can be used to design objective functions that are linguistically transparent to the user.

- Methods from fuzzy decision making and conventional control theory can be combined to obtain controllers with improved response.

Note that this definition of fuzzy predictive control differs from the terminology used in some of the literature,[30,31] where fuzzy predictive control denotes the use of fuzzy models as crisp input–output mappings in a conventional model-based predictive control setting.

The formulation of the fuzzy predictive control scheme is very similar to the formulation of the fuzzy control scheme in Sec. 3, but the optimization takes place within the prediction horizon. Considering q decision goals $G_1, \ldots, G_q$ and r decision constraints $C_1, \ldots, C_r$, the aggregated fuzzy decision is based on the membership function μ_π for the control policy π, given by

$$\mu_\pi = D\left(\mu_{G_1}, \ldots, \mu_{G_{\tilde{q}}}, \mu_{C_1}, \ldots, \mu_{C_{\tilde{r}}}\right), \tag{14}$$

where D is the decision function that combines the criteria, and $\tilde{q}$, $\tilde{r}$ denote that the goals and the criteria are repeated H_p times in order to consider each time step within the prediction horizon.

The membership functions for the decision criteria quantify how much the system satisfies the criteria given a particular control sequence, and they bring various quantities into a unified domain (e.g. the unit interval). The use of the membership functions introduces additional flexibility for expressing the control goals and it leads to increased transparency as it becomes possible to explicitly specify what type of system response is desired. For instance, it becomes possible to penalize errors that are larger than a specified threshold more severely.

After the membership functions are defined, they can be combined by using various decision functions such as the parametric aggregation operators from the fuzzy set theory. Usually, the additional parameter of the decision function influences the optimization results in a way that cannot be expressed by the weight factors, providing additional flexibility. The increased number of parameters provides the control engineer with increased flexibility for specifying the objective function. In addition to the prediction and control horizons, and the weight factors, the definition of the membership functions for the fuzzy criteria can also be used to tune the objective function. The parameters of the aggregation operators can be tuned by optimization, by heuristic rules or by trial-and-error.

Kaymak *et al.*[32] have compared the performance of fuzzy predictive control with that of conventional model-based predictive control. Significant performance improvements can be obtained, especially when the system inputs are constrained. For example, a simulated non-minimum phase, open-loop unstable linear system described by the transfer function

$$G(s) = \frac{s-1}{s^3+s^2+s+2} \tag{15}$$

is considered. A fuzzy model-based predictive controller is designed by using Yager's t-norm for the aggregation, and its performance is compared to that of a conventional model-based predictive controller that uses a quadratic objective function. The control space is discretized and the optimal control sequence is determined by an enumerative search. The control horizon H_c is limited to two time steps in order to keep the computational load low. To further reduce the computational load, a two-step optimization approach is used where first a rough solution is found by using a coarse discretization of the control space, followed by the calculation of a finer solution around the rough solution. At each step, the total range is divided into 11 steps. The control horizon H_c is chosen equal to two and and the prediction horizon H_p equal to six with a constraint on the rate of change of the input of $|\Delta u| \leq 0.5$. It is required that the controller can bring the system to any level in the interval $[-3, 3]$. Figure 5 shows the response of the system for several step references. It is clear that the predictive controller with fuzzy criteria can improve the speed of the response considerably, while avoiding the overshoot.

4.7 Satisficing control

One generalization of the fuzzy model-based predictive control is obtained through satisficing control. The satisficing control theory is inspired by the theory of satisficing decisions, where the goal is not finding the best (optimal) decision, but finding an acceptable decision without disproportionally large effort. The satisficing control is characterized by two elements that are evaluated for all alternatives.

1. The gains to be made by trying to improve on the control goal.

2. The cost of undertaking this activity.

If the cost of improving the control response exceeds the gains that can be made, better control actions are not searched further.

Goodrich *et al.*[33] discuss the satisficing control theory based on conventional quadratic objective functions. Their approach is related to optimal

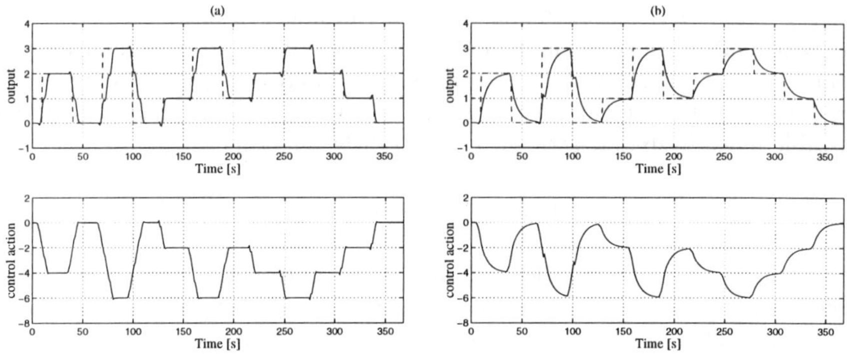

(a) Fuzzy criteria *(b) Conventional criteria*

Figure 5: Step responses for the linear system.

control in case of an infinite control horizon, but it can also be used within the receding horizon scheme of predictive control. When using conventional objective functions, Goodrich *et al.* [33] have shown that closed forms of controllers can be obtained, and hence this type of control can be implemented without large demands on the computational capacity of the controllers. In the fuzzy model-based predictive control scheme, the objective function can better be adjusted to express more complicated goals by using fuzzy sets and fuzzy aggregation operators. However, closed forms of controllers may not be obtained, and thus optimization using a receding horizon principle must be adopted as discussed in Sec. 4.6. Two fuzzy decisions can be specified, with different sets of fuzzy criteria for the decision making. One fuzzy decision quantifies the gains to be made from improving the control goal, while the other one quantifies the cost of the improvement. The two fuzzy decisions can then be combined by using another fuzzy set aggregation operator, such as a residual implication operator [34] $\mathcal{I}(v, z)$ given by

$$\mathcal{I}(v, z) = \max(0, v - z). \tag{16}$$

In Eq. (16) $v \in V$ represents the fuzzy decision for the gains, and $z \in Z$ represents the fuzzy decision for the costs.

4.8 Expert design of objective functions

One of the properties of fuzzy predictive control is that it provides a means to translate transparently the linguistic goals and preferences of the control engineer into an objective function that can be used for optimization. Often, human experience and knowledge is available for controlling a process. Especially for systems which are difficult to control, the heuristic used by humans may provide good control response. A disadvantage of the conventional model-based predictive control paradigm, however, is that the control strategy of a human cannot be translated into a simple objective function that the predictive controller can use. Fuzzy decision making provides a means for this translation by the use of many types of decision functions from fuzzy decision making.

Humans use a variety of methods to achieve the control goals, whereby the goals set for the controller as well as the controller parameters can be made time and state dependent. Therefore, the translation of a human's control strategy into a decision function involves dealing with time-varying goals and the adaptation of the parameter values. Kaymak and Sousa [35] have applied the fuzzy decision theory to design an objective function for fuzzy predictive control that leads to a mimicking of the control strategy of an experienced gantry crane driver. In this control problem, a container must be transported by the gantry crane, while restrictions are imposed on the swing of the load. The control strategy of a crane driver consists of two consecutive parts.

1. Minimize the load swing on achieving a maximal trolley speed.

2. Minimize the load swing on bringing the trolley to rest.

Minimizing the load swing can be described by "small swing angle" and "small swing angle speed". These goals do not change during the transport. There are also two different goals concerning the trolley speed, each of which becomes dominant at different stages of the transport. At the beginning of the transport, achieving a large trolley speed is the controller's goal, which gradually changes to obtaining a small trolley speed at the end of the transport. The authors have combined these criteria by using Yager's t-norm. [36] A weight factor approach is used. In this approach, different criteria are weighted differently, and the weight factors are adapted dynamically during the transport according to the distance of the trolley from the desired position. Initially, the weight factors for reducing swing angle are small, and they are increased in size gradually as the trolley is displaced. A fourth membership function regarding the position error of the trolley is added on approaching the desired position, and its weight is increased as the trolley approaches the desired position. These measures decrease the steady-state error and reduce the oscillations during the transient phase.

5 Conclusions

The application of fuzzy decision making methods in control systems has been considered. The problems addressed by fuzzy decision making and many control problems that involve interaction with humans, supervisory control, selection from discrete control alternatives share many similarities. Hence, fuzzy decision making can also be applied to these problems. The use of fuzzy decision making methods for control problems extend the application of control systems to a larger class of control problems. A number of possibilities for the application of fuzzy decision making in control systems has been discussed with references to the publications in literature. It is shown that fuzzy decision making methods can enhance the existing control techniques in applications such as supervisory control, selection of a control action from discrete alternatives, fuzzy model-based control, use of fuzzy aggregation functions to design nonlinear controllers, heuristic specification of objective functions and the design of objective functions from expert knowledge. Such applications illustrate the power of fuzzy decision making to synthesize controllers in areas that remained largely outside the scope of conventional control theory.

References

1. E. H. Mamdani. Applications of fuzzy algorithms for control of simple dynamic plant. In *Proceedings IEE*, number 121, pages 1585–1588, 1974.
2. L. P. Holmblad and J. J. Østergaard. Control of a cement kiln by fuzzy logic. In M. M. Gupta and E. Sanchez, editors, *Fuzzy Information and Decision Processes*, pages 389–399. North Holland, New York, 1982.
3. R. E. Bellman and Lotfi A. Zadeh. Decision-making in a fuzzy environment. *Management Science*, 17(4):141–164, December 1970.
4. Ching-Lai Hwang and Abu Syed Md. Masud. *Multiple Objective Decision Making: methods and applications*, volume 164 of *Lecture Notes in Economics and Mathematical Systems*. Springer-Verlag, Berlin, 1979.
5. Ronald R. Yager and Dimitar P. Filev. *Essentials of Fuzzy Modelling and Control*. John Wiley, New York, 1994.
6. Michel Grabisch, Hung T. Nguyen, and Elbert A. Walker. *Fundamentals of Uncertainty Calculi With Applications to Fuzzy Inference*, volume 30 of *B: Mathematical and Statistical Methods*. Kluwer Academic Publihers, Dordrecht, 1995.

7. G. J. Klir and B. Yuan. *Fuzzy Sets and Fuzzy Logic: theory and applications.* Prentice Hall, Upper Saddle River, 1995.
8. H.-J. Zimmermann. *Fuzzy Set Theory and its Application.* Kluwer, Boston, third edition, 1996.
9. Freerk A. Lootsma. *Fuzzy Logic for Planning and Decision Making.* Kluwer, Dordrecht, 1997.
10. Uzay Kaymak. *Fuzzy Decision Making with Control Applications.* Ph.D. thesis, Delft University of Technology, P.O. Box 5031, 2600 GA, Delft, the Netherlands, November 1998.
11. Ronald R. Yager. Fuzzy decision making including unequal objectives. *Fuzzy Sets and Systems*, 1:87–95, 1978.
12. Sjoerd M. Baas and Huibert Kwakernaak. Rating and ranking of multiple-aspect alternatives using fuzzy sets. *Automatica*, 13:47–58, 1977.
13. Didier Dubois and Henri Prade. Ranking fuzzy numbers in the setting of possibility theory. *Information Sciences*, 30:183–224, 1983.
14. G. Bortolan and R. Degani. A review of some methods for ranking fuzzy subsets. *Fuzzy Sets and Systems*, 15:1–19, 1985.
15. J. J. Buckley. Ranking alternatives using fuzzy numbers. *Fuzzy Sets and Systems*, 15:21–31, 1985.
16. Tian-Shy Liou and Mao-Jiun J. Wang. Ranking fuzzy numbers with integral value. *Fuzzy Sets and Systems*, 50:247–255, 1992.
17. Oscar P. Dias Jr. Ranking alternatives using fuzzy numbers: a computational approach. *Fuzzy Sets and Systems*, 56:247–252, 1993.
18. Janusz Kacprzyk. *Multistage Fuzzy Control: a model-based approach to fuzzy control and decision making.* Wiley, Chichester, 1997.
19. J. M. Sousa, U. Kaymak, P. M. Bruijn, and H. B. Verbruggen. Branch-and-bound optimization in predictive control with fuzzy decision functions. In *Proceedings of IFAC International Workshop on Artificial Intelligence in Real-Time Control*, pages 481–486, Kuala Lumpur, Malaysia, September 1997.
20. João Miguel da Costa Sousa. *A Fuzzy Approach to Model-Based Control.* Ph.D. thesis, Delft University of Technology, Delft, April 1998.
21. David E. Goldberg. *Genetic Algorithms in Search, Optimization, and Machine Learning.* Addison–Wesley, New York, 1989.
22. C. Onnen, R. Babuška, U. Kaymak, J. M. Sousa, H. B. Verbruggen, and R. Isermann. Genetic algorithms for optimization in predictive control. *Control Engineering Practice*, 5(10):1363–1372, 1997.
23. J. M. Sousa, M. Setnes, and U. Kaymak. Adaptive decision alternatives in fuzzy predictive control. In *Proceedings of the IEEE World Congress*

on Computational Intelligence, pages 698–703, Anchorage, Alaska, May 1998.

24. Stephen L. Chiu. Using fuzzy logic in control applications: beyond fuzzy PID control. *IEEE Control Systems Magazine*, 18(5):100–105, October 1998.
25. S. Yasunobu and S. Miyamoto. Automatic train operation by predictive fuzzy control. In M. Sugeno, editor, *Industrial Application of Fuzzy Control*, pages 1–18. North Holland, Amsterdam, 1985.
26. U. Kaymak, H. R. van Nauta Lemke, and P. M. Bruijn. Using decision functions for nonlinear controller design. In *Proceedings of the Fifth International Conference on Fuzzy Systems*, pages 1788–1793, New Orleans, USA, September 1996.
27. J. Richalet. Industrial applications of model based predictive control. *Automatica*, 29:1251–1274, 1993.
28. E.F. Camacho and C. Bordons. *Model predictive control in the process industry*. Springer Verlag, London, UK, 1995.
29. Ronald Soeterboek. *Predictive control: a unified approach*. Prentice Hall, New York, USA, 1992.
30. Robert Babuška. *Fuzzy Modeling and Identification*. Ph.D. thesis, Delft University of Technology, P.O. Box 5031, 2600 GA, Delft, the Netherlands, January 1997.
31. H. te Braake, R. Babuška, and E. van Can. Fuzzy and neural models in predictive control. *Journal A*, 35(3):44–51, October 1994.
32. U. Kaymak, J. M. Sousa, and H. B. Verbruggen. A comparative study of fuzzy and conventional criteria in model-based predictive control. In *Proceedings of the Sixth IEEE International Conference on Fuzzy Systems*, pages 907–914, Barcelona, Spain, July 1997.
33. M. A. Goodrich, W. C. Stirling, and R. L. Frost. A theory of satisficing decisions and control. *IEEE Transactions on Systems, Man and Cybernetics, Part A*, 28(6):763–779, November 1998.
34. Bernard de Baets and Radko Mesiar. Residual implicators of continuous t-norms. In *Proceedings of Fourth European Congress on Intelligent Techniques and Soft Computing*, volume 1, pages 27–31, Aachen, Germany, September 1996.
35. U. Kaymak and J. M. Sousa. Model based fuzzy predictive control applied to a simulated gantry crane. In *Proceedings of 2nd Asian Control Conference*, pages III–455 — III–458, Seoul, Korea, July 1997.
36. Ronald R. Yager. On a general class of fuzzy connectives. *Fuzzy Sets and Systems*, 4:235–242, 1980.

FUZZY TECHNIQUES IN FAULT DETECTION, ISOLATION AND DIAGNOSIS

P. M. FRANK, T. MARCU
Gerhard Mercator University of Duisburg,
Department of Measurement and Control,
Bismarckstr. 81 (BB), D-47048 Duisburg, Germany
E-mail: p.m.frank@uni-duisburg.de

This contribution reviews the state of the art of the application of fuzzy logic to model-based fault diagnosis of technical processes. The fuzzy approach is motivated by being able to operate with complex systems and imprecise information in a comprehensive and transparent manner. Emphasis is placed upon the use of different fuzzy models for symptom generation and fuzzy logic decision-making for symptom evaluation. For the latter task a concept using the assistance of human expertise is also described. The enumeration of some successful applications illustrates the efficiency of the approach.

1 Introduction

Fault tolerance of automatic control systems is gaining more and more importance. This is due to the increasing complexity of modern control systems and the growing demands for quality, cost efficiency, availability, reliability and safety. The tolerance to faults can be achieved by either passive or active strategies. The passive approach makes use of robust control strategies to ensure that the closed-loop system becomes insensitive with respect to faults. In contrast, the active approach provides fault accomodation, i.e. the reconfiguration of the control system when a fault has occured. Whilst robust control can tolerate small faults to a certain degree, the reconfiguration concept is absolutely inevitable when serious faults occur that lead to a failure of the whole system. In order to achieve fault accomodation, a number of tasks have to be performed. A most important and difficult one is the early diagnosis of the faults. Besides this, fault diagnosis is needed as a part of the supervision of control systems that incorporate artificial and computational intelligence.

The most efficient fault-diagnostic strategy is the so-called model-based approach. It makes use of either analytical or knowledge-based models or combination of both along with analytical or heuristic reasoning. In the case of complex systems, one is faced with the problem that no, or no sufficiently accurate mathematical models are available. The use of knowledge-based techniques, eihter in the framework of diagnosis expert systems or in combination with the human expert, is then the only feasible way. In this case, fuzzy logic is of particular importance, because it allows to operate with unprecise information, linguistic variables and rules similarly to the natural way of thinking of the human operator. All this motivates an increasing consideration of fuzzy logic in fault diagnosis and supervisory systems, known as the domain of "fuzzy supervision" in contrast to "fuzzy control".

The present paper reviews the stage of this development along with the enumeration of some successful applications. One starts with the basic concepts and tasks of fault diagnosis. The concept of the fuzzy logic approach is then outlined together with a number of feasible solutions to the fault diagnosis problem. A first group refers to symptom generation based on fuzzy output observers where the plant model is implemented by either fuzzy qualitative simulation, or fuzzy relational modelling, or fuzzy functional modelling. Next, symptom evaluation using fuzzy logic is described. Two different evaluation concepts are outlined, namely automatic decision-making without or with the assistance of the human expert. Different strategies to construct and check the inference mechanism are discussed as well. Final remarks conclude the paper.

2 Problem Statement

Faults in an automated system can occur in both the plant and control units, hardware and software. The first category, faults in the plant, is considered in the sequel. Considering a plant of an automatic control system, the faults can affect its different parts as illustrated in Figure 1. Three categories of faults can be distinguished, namely actuator faults, component faults, i.e. faults in the framework of the process, and sensor faults. The faults can commonly be described as input signals.

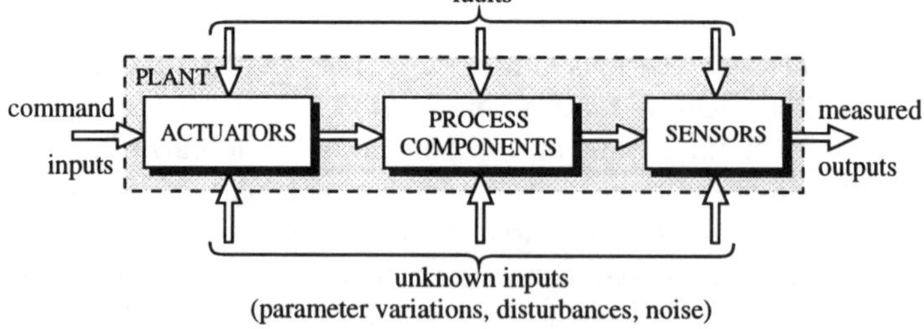

Figure 1: Definition of faults and unknown inputs in the plant of the process.

In addition, there is always modelling uncertainty due to unmodelled disturbances, noise and control model mismatches. This may not be critical for the system operation. It may, however, obscure the detection of faults by releasing false alarms. All these different types of uncertainty are taken into account as unknown inputs.

2.1 The Basic Concept of Fault Diagnosis

The basic goals of fault diagnosis are to detect and isolate occuring faults and to provide information about their size and source. This has to be done on-line, in the face of the existing unknown inputs and with as few false alarms as possible. As a result, the overall concept of fault diagnosis consists of the three sub-tasks: fault detection, fault isolation, and fault analysis.

The fault diagnosis process involves the following main steps:

1. Symptom generation, i.e. the generation of signals or symptoms which reflect the faults. The isolation of different faults requires properly directed signal vectors or properly structured sets of symptoms.
2. Symptom evaluation (fault classification), i.e. logical decision-making on the time of occurence and the location of a fault.
3. Fault analysis, i.e. determination of the type of fault and its size and cause.

The first two steps constitute the concept of fault detection and isolation (FDI). Among the most important properties of an FDI system is that of robustness. This implies the maximization of the detectability and isolability of faults under the constraint of minimization of false alarm rate.

The backbone of modern FDI systems is the so-called model-based residual generation. The actual behaviour of the system is compared with that of a nominal fault-free model driven by the same inputs. The result of comparison leads to a class of signals, called residuals. Basically, the dynamic behaviour of a system can be described either by a quantitative (analytical) model or by a knowledge-based model. In the latter case, the system response is classified either by heuristic symptoms or by a qualitative description making use of the knowledge about the system in terms of rules and facts.

A perfect analytical model, if available, represents the deepest and most concise knowledge of the process. However, precise analytical models are in practice hardly or almost never available. In this case, knowledge-based models are the only realistic alternative, allowing one to exploit as much knowledge about the process as available. The structural diagram of residual generation and evaluation is given in Figure 2. The third step, fault analysis, requires in general either a human operator or an artificial knowledge-based expert, referred to as "diagnosis expert system".

A selection of the comprehensive literature on fault diagnosis techniques refers to a number of books[1,2] and survey papers[3–6], and the literature cited

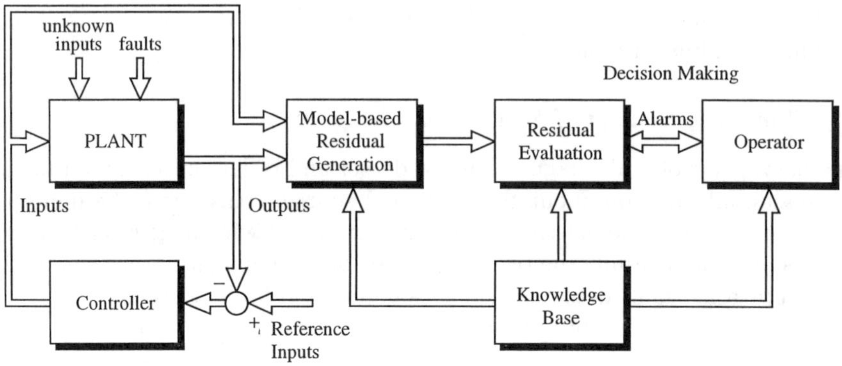

Figure 2: The concept of model-based fault detection and isolation.

therein. The following presentation is dedicated to the model-based approach using qualitative models based on fuzzy logic.

2.2 The Fuzzy Logic Approach

Knowledge-based methods can be applied in all three phases of fault diagnosis, namely residual generation, residual evaluation, and fault analysis. However, in this case the phases are not always as clearly separable as with the analytical appproach. The basic strategies for diagnosis using soft modelling were described in [7], followed by a number of papers from the artificial intelligence community. Early contributions providing a bridge between diagnosis using hard and soft models were made in [8,9].

Knowledge-based diagnosis techniques fall into two main categories:

1. symptom-based,
2. qualitative model-based.

The former group makes use of heuristic symptoms, knowledge about process history, or statistical knowledge. Their evaluation is organized in the framework of diagnosis expert systems. If the symptoms are considered in connection with the inputs of the system, one speaks of symptom-model-based approach. Typical for the latter group is that the knowledge is derived in terms of facts and rules from the description of structure and behaviour (first principles) of the process. Compared to analytical models, this information, however, may be incomplete and uncertain.

A fuzzy logic system (FLS) has the main property of being able to simultaneously handle numerical data and linguistic knowledge [10,11]. On the one hand, fuzzy systems are universal approximators that perform deterministically a nonlinear mapping of an input data vector into scalar outputs. On the other hand, fuzzy systems are knowledge-based systems consisting of linguistic If-Then rules that can be designed using the knowledge of experts. This duality allows for qualitative knowledge to be combined with quantitative data in a complementary way. Fuzzy set theory and fuzzy logic establish the specifics of such a nonlinear mapping.

The application of this unique framework to the design of FDI systems mainly consists of fuzzy system identification for residual generation and fuzzy reasoning for residual evaluation. Such fuzzy systems provide a rather transparent representation of the system under study even if it is nonlinear, based on the linguistic interpretation in the form of rules. In this way, process data can be translated into the model and analysed in a manner similar to the way of human thinking. Moreover, the rules extracted from data can be validated by experts and combined with their prior knowledge. A more or less complete system model which describes the real process can thus be obtained.

The generic structure of a fuzzy logic system as widely used in engineering applications [10–12] is depicted in Figure 3. It contains three components: fuzzifier, fuzzy inference system (FIS), and defuzzifier. The fuzzy inference system represents the core of the FLS. It is built of three conceptual components: rule base and data base, which constitute the knowledge base, and inference engine. Since the FIS operates with fuzzy sets, it is interfaced with the numerical environment by means of fuzzifier and defuzzifier, respectively. Such a framework is based on the well-established theory of fuzzy reasoning [13].

The rules describe the functioning of the system and may be provided by experts or can be extracted from numerical data. The uncertainties of the variables and parameters are expressed in terms of membership functions. In either case, engineering rules reflecting the constraint of causality are expressed as a collection of If-Then statements. Each of them consists of a premise part (If ...) and a conclusion part (Then ...). In the premise part, multiple linguistic fuzzy sets of one or multiple variables can be related with each other by logical operations.

The inference engine maps fuzzy sets into fuzzy sets. It handles the way in which the rules are combined. The fuzzifier maps crisp numbers into fuzzy sets. This is needed in order to activate rules which are in terms of linguistic variables. The latter ones have fuzzy sets associated with them. In many applications, crisp numbers must be obtained as the output of an FLS. The defuzzifier maps output sets into crisp numbers. There is a large number of

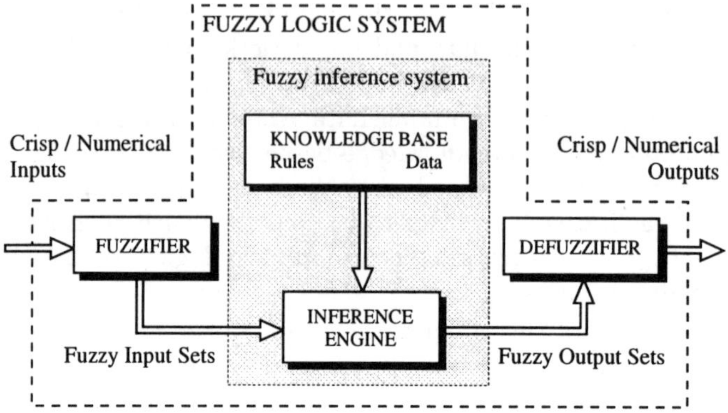

Figure 3: The generic fuzzy-logic system.

possibilities to build an FLS. The application of some of them to the design of FDI systems is described in the next sections of the paper.

The design of an FLS involves mainly methods based of fuzzy-logic and approximate reasoning[11]. However, the current trend is to optimize the design by taking into account the ideas of computational intelligence or soft-computing, namely artificial neural networks and evolutionary algorithms of genetic type. While fuzzy logic offers a methodology for dealing with inaccuracy and approximate reasoning, neural networks (NN's) provide tools for (universal) approximation, learning an adaptation. Finally, genetic algorithms (GA's) represent the framework for systemized stochastic search and optimization.

A broad overview reflecting this tendency is given in [14], along with an increasing number of contribution papers. Fourteen areas of research are revealed there, among them neuro-fuzzy systems [12], genetic fuzzy systems [15], and genetic neuro-fuzzy systems [16]. The primary objective of using such integrated approaches is to develop fuzzy systems that are able to learn [17]. Such a hybrid technique has to possess a good learning capacity, a better learning time than that of pure GA's, and less sensitivity to local extremes than NN's. In addition, it has to generate a fuzzy knowledge base which has a linguistic representation and a low degree of computational complexity [16], as shown in Figure 4.The application of these methodologies in the field of fault diagnosis is, however, in an incipient stage, excepting neuro-fuzzy systems [18].

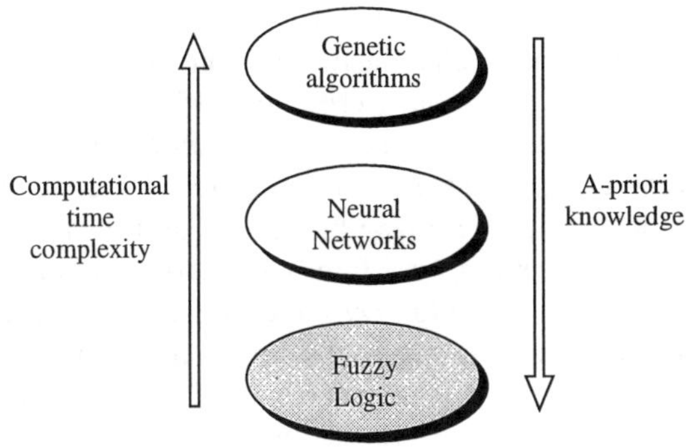

Figure 4: Computational intelligence/soft computing techniques.

3 Fuzzy Symptom Generation

3.1 The Fuzzification Concept

Symptoms can be generated based on the estimation of the process outputs using observers. The approximating signals are then compared with existing measurements, leading to so-called residuals. In the ideal case, these must be different from zero when a fault occurs and zero otherwise. If the system behaviour is described in stochastic terms, a Kalman filter instead of a Luenberger observer is used for the estimation of the process outputs in the nominal (error-free) mode of operation. Different schemes of estimators have lead to successful robust diagnosis [3]. However, most approaches are only applicable to linear systems. Robust nonlinear observers have only been developed for particular classes of nonlinear dynamic systems [5,6].

As opposed to the analytical approach, the knowledge-based or qualitative models make use of heuristic qualitative knowledge instead of differential equations. Linguistic variables specify the input and output signals using linguistic terms. This enhances the robustness of the model versus unknown or time-dependent parameters of the system. The relationship between the input or output variables may be described in different ways by:

1. fuzzy qualitative models, using a rule base;
2. fuzzy relational models, using a set of parameters that are determined during an identification stage based on a learning data set;
3. fuzzy functional models, using several local submodels to describe the system behaviour in the environments of different operating areas.

These approaches are described in the sequel. They lead to different variants of what may be called fuzzy (output) observer.

It should however be noticed, that for FDI purposes, one needs a representative model of the plant. This means that one needs only that part of the model which reflects the faults of interest and, with respect to robustness, is not or only weakly affected by disturbances and modelling uncertainty. In this way, the models for FDI can be simpler than those for control[19].

The transfer behaviour of a fuzzy functional unit is strictly deterministic. It can therefore be represented by a family of static nonlinear characteristics, relating the inputs to the outputs. Dynamic behaviour can be introduced by means of external dynamic units, i.e. by including temporal shiftings of the variables used in the rules. On the other hand, both the rules and the membership functions are assigned by the available knowledge of the functioning and the uncertainty of the process. Typical for the membership functions is that they provide a quantization of the physical variables by a family of fuzzy sets. Hence, the fuzzy approach to system modelling can be interpreted as a special kind of nonlinear interpolation between the fixed points of full membership of the different fuzzy sets. This may also be termed as knowledge-based interpolation.

After synthesizing a fuzzy observer, under certain circumstances the FLS can be expressed mathematically, e.g. as a linear combination of fuzzy basis functions[10]. The results from the application of sensitivity theory to fuzzy systems could then be used by the designer of a fuzzy FDI system[20]. Firstly, this permits to assess the relative robustness to system parameters of fuzzy models. Secondly, it is possible to determine which parameters of the fuzzy system itself affect the response of the system to particular faults and disturbances. This information could be used in two ways:

1. as a guideline to the designer as to decide which parameters should be tuned when attempting to improve the system manually;
2. as the basis for a mathematical optimization of the system parameters, given certain constraints or specifications.

3.2 Fuzzy Qualitative Observer

The method of fuzzy-qualitative model-building can be considered as an extension of conventional mathematical modelling. It is based on two categories of knowledge: the structural description of the system and the knowledge about the behaviour of each individual part. The latter includes physical laws abstracted from former experience in different contexts. A constraint model is built based on two categories of elements: qualitative variables and constraints. The qualitative variables are functions of time and represent the physical parameters of the system. The constraints define the way in which those parameters are related with each other[21].

The qualitative variables are defined in a quantity space. This is generated through the quantification of the set of real numbers, i.e. an alternating sequence of points and open intervals. At any moment of time, a qualitative variable is described by a set of qualitative values: qualitative magnitude and its derivative, i.e. direction and rate of change. A fuzzy quantity space is obtained[22] by replacing the limits between adjacent values with fuzzy transitions. The variables of the dynamic systems and their derivatives are represented by fuzzy sets.Both strength and sign information are represented by fuzzy relations holding for two or more variables. The constraints are composed of arithmetic, functional and derivative primitives.

The description of the behaviour of subsystems are integrated within the model in the form of If-Then rules or in the form of qualitative differential equations (QDE). The rules describe linguistically the relation between the qualitative variables. The QDE's can be interpreted as a generalization of conventional differential equations. For example, they can be defined by the term $DERIV\{y\} = f(x)$, where $DERIV\{\cdot\}$ denotes the qualitative time derivative, and f represents a set of monotonically increasing or decreasing functions.

To infer the behaviour of the complete system from its structure and the behaviours of its parts, qualitative simulation is applied to the QDE and its given initial conditions. A qualitative state is here defined as a tuple of qualitative values associated with the process variables. A qualitative behaviour is then defined by a sequence of states, i.e. a state chain. The behaviour can thus be described by a graph consisting of the possible future states of the system. The simulated possible behaviour is in general not unique, but could take any path through the graph starting at the initial state, as shown in Figure 5.

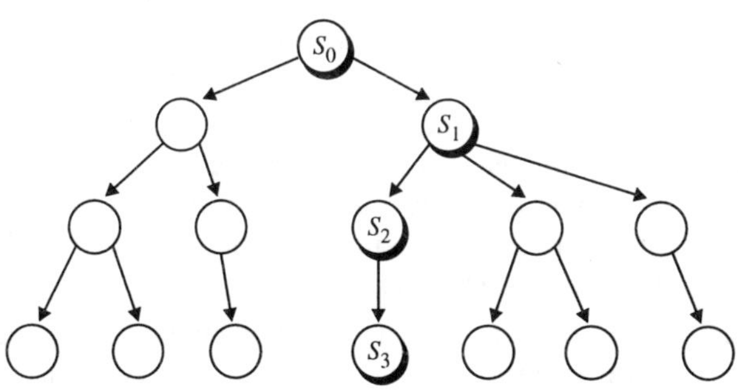

Figure 5: Graphical description of behaviours.

The qualitative simulation proceeds by determining all possible transitions in qualitative values permitted to each argument (parameter). One checks then the consistency between each of the large combinations of qualitative transitions and the known information about the system, excluding the inconsistent (impossible) ones. The check-exclusion process is known as filtering. Several kinds of techniques are: constraint filtering, temporal filtering, global filtering, and observation filtering [23]. Complete state descriptions are then generated from the filtered tuples and these new states are assigned as children states of the current state. If more than one qualitative change is possible, the current state has multiple successors, and hence the simulation produces a branching tree of states.

A fuzzy qualitative observer, as shown in Figure 6, used for residual generation, makes use of qualitative simulation on the basis of conventional filtering techniques to perform observation filtering [23].

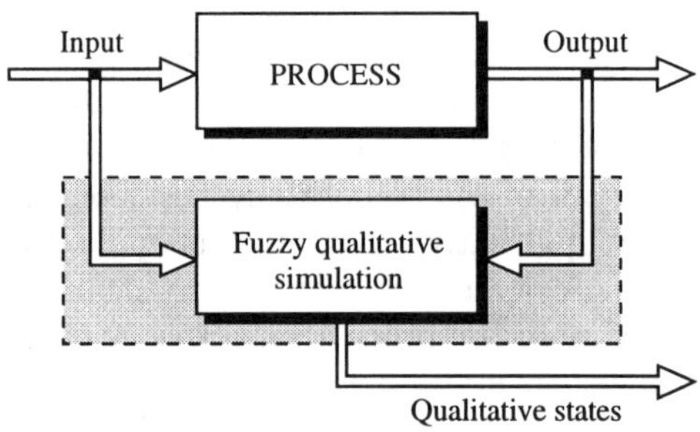

Figure 6: Fuzzy qualitative observer.

The main concept of the latter is that the simulated qualitative behaviour of a variable must cover its counterpart of the measurement obtained from the system itself. Otherwise, the simulated behavioural path is inconsistent and can be eliminated. A fault causes a deviation of the system output in such a way that its counterpart of the estimated output is no more consistent. In other words, a fault will produce an empty set of qualitative estimated states, which is impossible in a fault free-case.

The approach has been applied to component fault diagnosis of a laboratory setup three-tank system in [19]. Different observer-based schemes are presented in [23]. Those qualitative observers are driven by the process inputs and different sets of process outputs. In this way, qualitative dedicated and generalized observer schemes can be designed. They have been applied to sensor fault diagnosis of a spring-block system.

3.3 Fuzzy Relational Observer

The fuzzy relational observer [24] is based on the principles of fuzzy relational modelling. They make use of a relational matrix that represents all rules describing the dynamic behaviour of a system. For this purpose, the system is decoupled into several multiple-input single-output subsystems. The complete structure of a fuzzy output observer is shown in Figure 7.

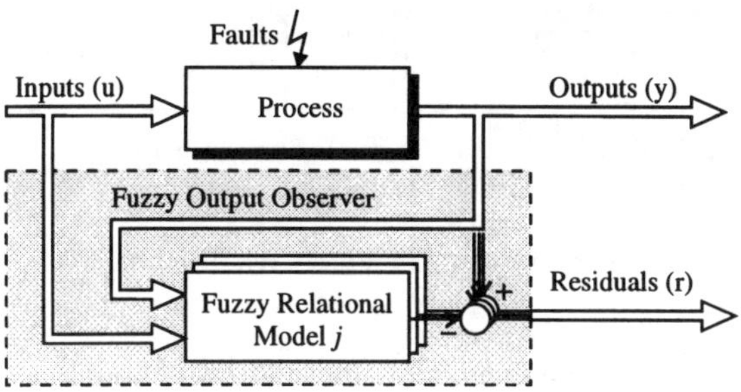

Figure 7: Application of the fuzzy relational observer to residual generation.

Each fuzzy relational model is designed for the estimation of one process output signal. The structure of these submodels is represented in Figure 8, for the case of a single-input single-output system. The submodels consist of four main parts: a dynamic block that memorizes the previous samples of relevant input-output data, a combination block in the form of the fuzzy Cartesian product, the fuzzy relation itself, and the interfaces between measurement and fuzzy spaces, namely fuzzification and defuzzification.

The functional correlation between process inputs and outputs is represented in the discrete-time input-output form by a fuzzy relational equation:

$$Y(k) = R_0 \circ X(k) \tag{1}$$

$$X(k) = U(k) \times \ldots \times U(k-\kappa_u) \times Y(k-1) \times \ldots \times Y(k-\kappa_y) \tag{2}$$

where U and Y are linguistic variables of the system input u and output y signals, respectively. The operator $\times$ is a t-norm (minimum, product) and the relational operator $\circ$ is a t-norm/t-conorm (minimum/maximum, product/bounded sum). The time delays κ_u and κ_y are either determined by an identification procedure, or they are known a-priori. To obtain the relational matrix of the nominal (fault-free) model R_0, a fuzzy relational equation has to be solved. However, in general, the set of solutions is empty [24,25]. The alternative approach is to find a solution R_a whose output is an estimate $\hat{Y}$. The

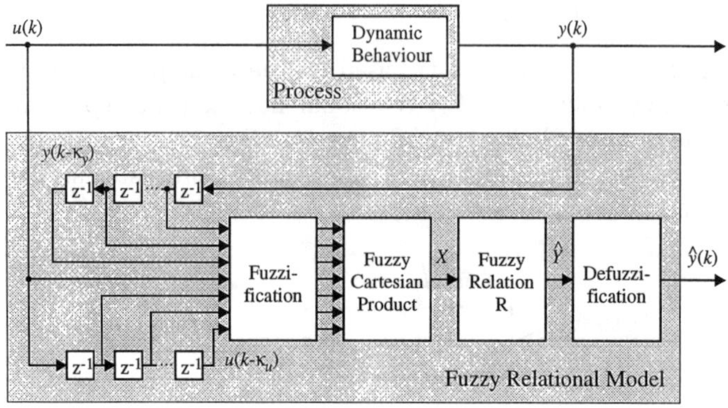

Figure 8: Structure of a fuzzy relational model.

latter must be closed to the output of the relational model (1), in the sense of a given criterion. This could be the minimization of the squared error:

$$J(R) = \frac{1}{2} \sum_{k=1}^{N} [y(k) - \hat{y}(k)]^2 \tag{3}$$

where N is the number of elements in the learning set, $y(k)$ is a given value and $\hat{y}(k)$ is the predicted value obtained after defuzzification.

The approach has been applied to the sensor fault diagnosis of a laboratory three tank system in [24]. An application study of a front-wheel driven vehicle has been investigated in [25]. The first results, obtained by the application of fuzzy observers to the detection of oversteering, show the ability of fuzzy relational models to the modelling of complex dynamic systems. A comparison regarding the application of analytical and fuzzy relational observers to sensor fault diagnosis of a winding-machine is presented in [26]. The results show a faster detection and a better isolation of faults by the fuzzification of the observer for residual generation.

3.4 Fuzzy Functional Observer

The fuzzy functional model approximates a nonlinear dynamic process by piece-wise linear models [27,28]. Each local model is valid in a sub-region of the input space. The partitions where the local models are valid are not crisp, but fuzzy. Therefore the validity of one local model for a certain point in the input space is expressed by a weighting function continuously defined over the whole input space and scaled from zero (not valid) to one (fully valid). The model output is computed as a weighted sum of all local models (hyperplans). This leads to a nonlinear interpolation between the local model outputs. The model can be interpreted as a Takagi-Sugeno fuzzy model[29] with product operator as t-norm. The rule antecedents describe fuzzy regions in the input space. The rule consequents are crisp functions of the model inputs.

The nonlinear dynamic discrete function f with m inputs u_i, $i = 1, \ldots, m$, and one output y:

$$\begin{aligned} y(k) = \; & f(\underline{x}(k)) \quad , \qquad \underline{x}(k) = [x_1(k), \ldots, x_n(k)] \\ \underline{x}(k) = [\; & u_1(k-1) \; , \ldots, u_1(k - \kappa_{u,1}), \ldots, \\ & u_m(k-1) \; , \ldots, u_m(k - \kappa_{u,m}), \\ & y(k-1) \quad , \ldots, y(k - \kappa_y)] \end{aligned} \tag{4}$$

is approximated by piece-wise linear models:

$$\begin{aligned} & \text{Rule } R_i : \\ & \text{IF } x_1(k) \text{ is } A_{i,1} \text{ AND } x_2(k) \text{ is } A_{i,2} \text{ AND } \ldots \text{ AND } x_n(k) \text{ is } A_{i,n} \\ & \text{THEN } y_i(k) = w_{0,i} + w_{1,i}x_1(k) + w_{2,i}x_2(k) + \ldots + w_{n,i}x_n(k), \end{aligned} \tag{5}$$

$i \in \{1, \ldots, M\}$, where $w_{0,i}$, $w_{1,i}$, ..., $w_{n,i}$ are the parameters of the i-th linear regression model, x_1, ..., x_n are its inputs, and $A_{i,1}$, ..., $A_{i,n}$ are the fuzzy sets defined on the universe of discourse of the inputs. The rule consequents represent linear difference equations. Choosing the product operator as t-norm, the output of the fuzzy system with M rules is aggregated as:

$$y(k) = \sum_{i=1}^{M} y_i(k)\Phi_i(\underline{x}, \underline{c}_i, \underline{\sigma}_i) \tag{6}$$

where Φ_i is the normalized Gaussian weighting function for the i-th model with the center $\underline{c}_i$ and standard deviation $\underline{\sigma}_i$.

In the sequel, Φ_i instead of $\Phi_i(\underline{x}, \underline{c}_i, \underline{\sigma}_i)$ is used for the sake of simplicity. Equation (6) can also be written in the form:

$$\begin{aligned} y(k) &= \left(\sum_{i=1}^{M} w_{0,i}\Phi_i\right) + \left(\sum_{i=1}^{M} w_{1,i}\Phi_i\right) x_1(k) + \ldots + \left(\sum_{i=1}^{M} w_{n,i}\Phi_i\right) x_n(k) \\ &= w_0 + w_1x_1(k) + \ldots + w_nx_n(k) \end{aligned} \tag{7}$$

where each parameter w_i is a nonlinear interpolation between the local model parameters $w_{i,j}$. This can be interpreted as a dynamic linearization[30]. Figure 9 illustrates this approach.

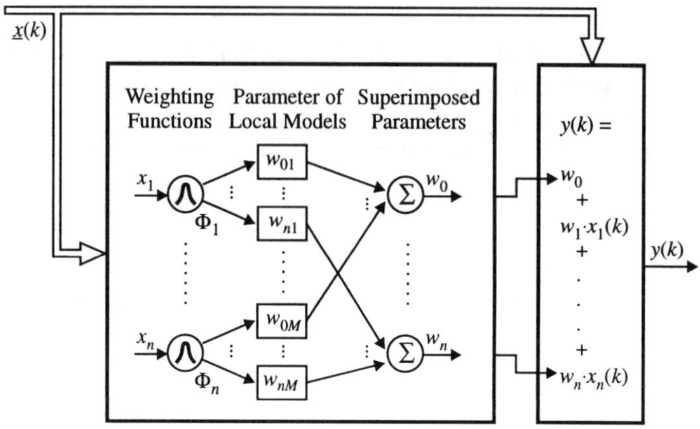

Figure 9: Fuzzy functional model.

The algorithm of local linear model tree (LOLIMOT), proposed in [27], is applied for fuzzy model identification. The fuzzy functional observer is then designed based on a multi-model approach [30] as shown in Figure 10. A separate model for each subprocess is identified and used for the generation of symptoms. The output residuals are used to generate a set of structured symptoms in a way that each residual depends on a different set of inputs and components of the process. The pattern of deviated and undeviated residuals therefore indicates the location of the faults.

The feasibility of LOLIMOT algorithm for system identification has been demonstrated in [27]. Real world data from a cooling blast pilot plant have been used. That plant shows a complex nonlinear dynamic behaviour. The fuzzy functional observer has been used to generate symptoms for fault diagnosis of an electro-pneumatic valve in a closed control loop [28]. The approach has also been applied to an integrated scheme comprising fault detection, fault diagnosis and reconfigurable control of a heat exchanger [30]. An adaptive predictive control scheme based on fuzzy process models is used there for the nonlinear control of an output temperature. By means of four fuzzy models of different sub-processes, sensor faults can be detected and diagnosed. This information can be used for the reconfiguration of the controller. Hence, the process control can be resumed with partly degraded performance.

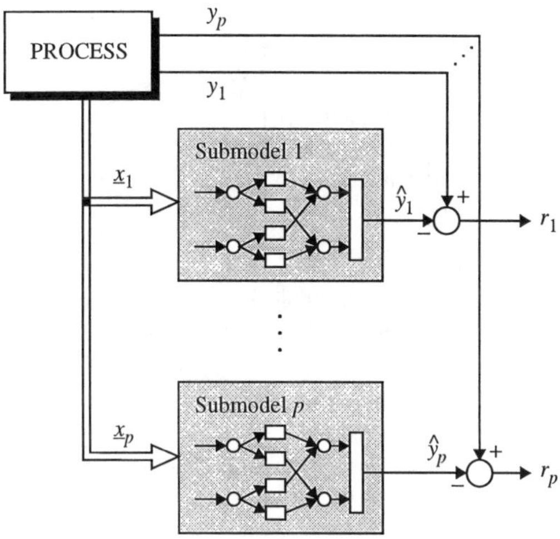

Figure 10: Fuzzy functional observer based on multi-model approach.

4 Fuzzy Symptom Evaluation

The symptom evaluation is a logic process of decision-making that transforms quantitative knowledge into qualitative statements, e.g. "yes" or "no". The goal is to decide whether and where a fault has occured under avoidance of wrong decisions, i.e. false alarms. This can also be seen as a classification problem as illustrated by Figure 11. The task is to match each pattern of the symptom vector with one of the preassigned classes of faults or the fault-free case. A variety of well established decision-making algorithms exists: threshold logic, pattern recognition, fuzzy decision-making, or neural networks [6,31,32].

The main difference between pattern recognition and diagnosis lies in the fact that the former deals with a finite set of classes, while the latter must adapt its decision rule to the emergence of a new class [33]. Therefore, three main tasks must be achieved, namely the recognition of known classes or states, the detection and learning of emerging states, and the on-line detection of evolutions. These problems cannot be solved in the context of classical pattern recognition. The classes are there described by a characteristic function whose value is 1 for any pattern that belong to a class, and 0 otherwise. Besides, a fuzzy class is described by a continuous function taking values in the interval $[0, 1]$.

The fuzzy-logic framework allows for a unified symptom representation and evaluation, either analytic or heuristic symptoms are used [18]. Symptoms

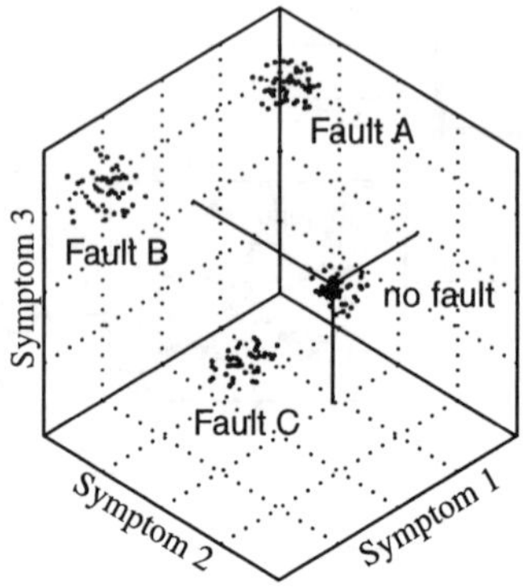

Figure 11: Illustration of the fault classification problem.

represented by residuals are considered in the sequel. The fundamental difficulty with this kind of symptom evaluation is that residuals are normally uncertain, corrupted by noise, disturbances and modelling uncertainty. For the FDI procedure, the residuals must meet the ideal conditions of being zero in the fault-free case and different from zero in the case of a fault.

In practice, due to the modelling uncertainty and the measurement noise, it is necessary to assign thresholds larger than zero in order to avoid false alarms. The latter implies a reduction of fault detection sensitivity. With the choice of the threshold, a compromise between fault decision sensitivity and false-alarm rate can be achieved. The design of a fuzzy-logic system for residual evaluation can be divided into the three tasks of fuzzification, inference and presentation of fault alarms. They are described in the sequel.

4.1 Residual Fuzzification

The fuzzification describes the assignement of a suitable number of fuzzy sets to each residual component r_i, $i = 1, \ldots, p$. The task is of major importance since it influences the isolation properties of the whole scheme of residual evaluation. Supposing that r_i is composed of s fuzzy sets $r_{i,k}$, $k = 1, \ldots, s$, then the following mapping is performed:

$$r_i \to r_{i,1} \circ r_{i,2} \circ \ldots \circ r_{i,s}, \quad r_i \to [0, 1] \tag{8}$$

where the operator $\circ$ means the fuzzy composition operator. The evolutions of the membership functions can be assigned on the basis of heuristic process knowledge, statistical distribution functions, subjective knowledge or by learning with the aid of neural networks. The idea of the latter is to transform the membership functions into a neural net, then to teach it, and finally to transform it back into the fuzzy domain.

The principle of residual fuzzification is presented in Figure 12. The fuzzification of residual can also be seen as a procedure for the generation of a robust threshold. In the mentioned illustration, two fuzzy sets are used to characterize the absence and the presence of a fault. The free parameter a_0 has to be assigned proportional to the noise amplitude and the effects of modelling uncertainties. The parameter δ can be chosen as the variance of the process noise, disturbances and influence of time varying modelling errors.

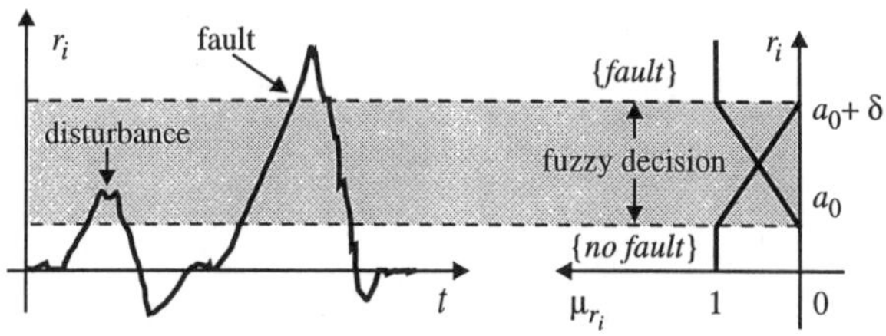

Figure 12: Principle of residual fuzzification and fuzzy (robust) threshold.

The membership functions μ_{r_i} shown in Figure 12 can be considered as the simplest form of fuzzification of the threshold or of the residual. In general, that specification can be extended to using more fuzzy sets. A related approach is that of adaptation of membership functions of the residuals and related rules[34]. This mechanism is necessary when the operational conditions of the process are changing. If one performs the defuzzification operation, an adaptive threshold is obtained. This represents, however, another approach to fuzzy evaluation of residuals.

The underlying idea is illustrated in Figures 13 and 14. The threshold is adapted depending on the changes in the values of process input u and output y, in terms of rules among fuzzy sets that are specified by proper membership functions. The resulting relation for the fuzzy adaptive threshold is given by:

$$J(u, y) = J_0 + \Delta J(u, y) \tag{9}$$

The first term $J_0 := J_0(u_0, y_0)$ denotes the constant (nominal) threshold of the operational point (u_0, y_0), where only the influence of the stationary disturbances (e.g. measurement noise), but no faults, are taken into account. The increment $\Delta J(u, y)$ denotes the effects of modelling errors, due to the deviation of the process from its operational point, in If-Then rules between the residuals which are characterized by membership functions.

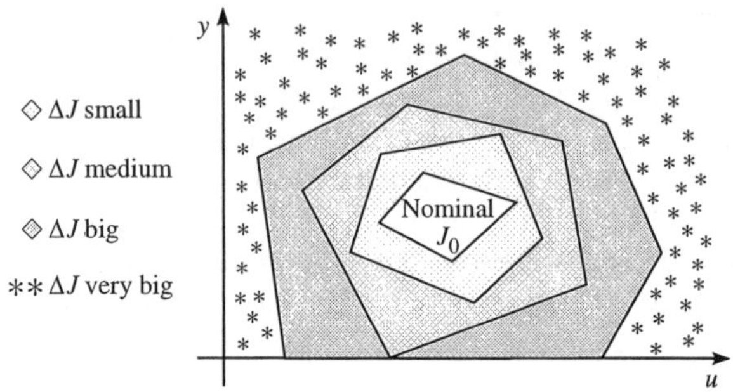

Figure 13: Fuzzy threshold adaptation.

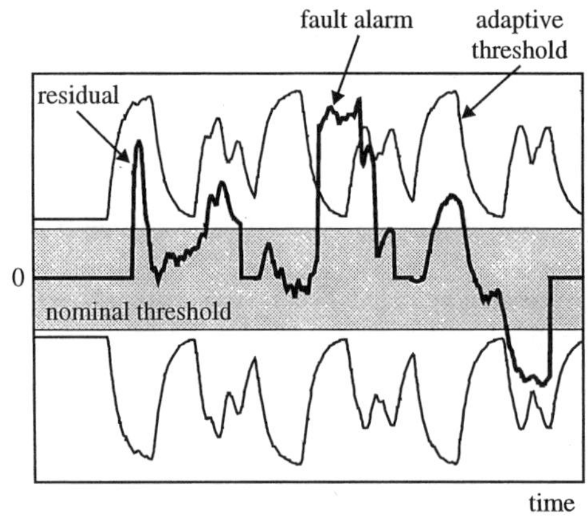

Figure 14: Detection of faults in the presence of uncertainties.

4.2 Inference Mechanism

In general, the task of fault decision is to conclude from a set R of residuals, $r_i \in R$, onto $f_j \in F$ of the set F of possible faults. In order to solve this problem, one has to make fuzzy conditional statements, from a composition of all combinations of fuzzy sets, $r_{i,k}$, of all residuals, Equation (8). The rules to be evaluated are of the type:

$$\begin{aligned} &\text{IF } (effect = r_{i,1}) \text{ AND IF } (effect = r_{i,2}) \ldots, \\ &\text{THEN } (cause = f_j) \end{aligned} \tag{10}$$

where f_j represents the j-th fault in the system. These rules form the knowledge base of the resulting diagnosis expert system.

Evaluating these rules, it becomes possible under certain conditions to find out for each combination of residuals $[r]$ a fault $[f]$ that is responsable for this combination of residuals. In other words, the fuzzy conditional statement is a mapping of the residuals onto the faults with the aid of the rules in the knowledge base. This can be illustrated by directed graphs or fault-trees [35], as shown in Figure 15. The way through the fault tree goes against the direction of the causality, namely from below to the top. In addition, contrarily to the traditional methodology, where the connections between the faults (causes) to the residuals (effects) constitute a crisp mapping, these connections are now fuzzy.

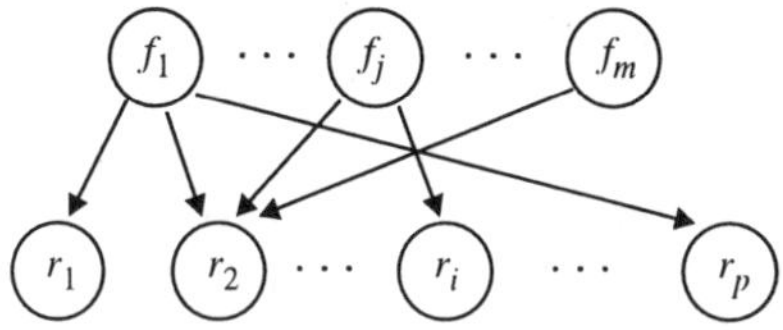

Figure 15: Representation of fault decisions using a fault tree; f_j faults, r_i residuals.

The knowledge base designed for a complex process must also include the connections between the different types of sub-systems [36]. Each one of them is modelled either analytically or heuristically. The connections between these submodels can be described by qualitative or heuristic knowledge. This is often easier, because some qualitative or heuristic description of the plant or the interconnection between submodels is normally known. This knowledge can be expressed by fuzzy rules in order to describe the faulty behaviour of the system in a fuzzy manner.

In order to make sure that there is always a unique solution of the fuzzy inference mechanism, one has to check the consistency of the rules. This can be done simultaneously with the inference. An approach to solve the problem of consistency checking has been suggested in [37]. It is based on the use of so-called fuzzy switching functions. They are mappings of $[0,1] \to [0,1]$ described by fuzzy formulas based on a fuzzy algebra.

The main steps of the algorithm are the following:

1. Define the number of faults which are of interest.
2. For each residual component, two fuzzy sets have to be used as an initial definition. They are *no fault* or *normal*, and *fault* or *not normal.*
3. Based on the defined fuzzy sets and the considered faults, the resulting number of rules have to be generated.
4. The rules are transformed into fuzzy switching functions.

5. Prove whether it is possible or not to distinguish between all faults.
6. If the distinction is possible, all faults can be detected and isolated. In this case, the procedure is terminated.
7. If a perfect distinction is not possible, redefine one or more fuzzy sets differently. E.g. instead of the fuzzy set *not normal*, two fuzzy sets *slightly deviating* and *strongly deviating* may be discriminated.
8. The procedure is repeated until a unique distinction is possible.

This approach has been applied to a part of a power plant, namely a high-pressure pre-heater (HPPH) [36]. That system is responsible for cooling the steam and plays an important role for the efficiency of the whole power plant. An analytical model exists only for the HPPH, but not for the connections between the different HPPH's. The inference mechanism processes set of residuals generated by analytical observers and parallel models. The presented method has also been applied to an aerobic waste-water plant [38]. The chemistry inside such process is highly nonlinear and very complex. Finally, a nonlinear model of 15-th order has been used to describe the process adequatelly. The quantitative residual vector obtained from an observer scheme as well as the qualitative description of the faulty behaviours are the inputs of the fuzzy inference system, in order to detect and isolate the faults.

Another approach to the evaluation of the fault symptom patterns is performed using a fuzzy classification tree [28]. That method of self-learning classification tree (SELECT) combines the learning ability of neural networks with the transparency of a fuzzy system. The resulting tree can be augmented with a-priori knowledge. The procedure consists of the following steps:

1. Creation of membership functions, e.g. by unsupervised clustering [39].
2. Selection of a fuzzy rule for the easiest separable fault.
3. All data belonging to the fault situation which can be diagnosed with the rule from the step 2 is removed from the training set. Thereby, a new data set is created containing one class less than before. The algorithm returns to the step 2 and iteratively selects new rules for the other fault classes. A sequential classification tree is created, where different fault possibilities are tested one by one.
4. Adding a-priori known rules to the top of the tree.
5. Optimization of parameters yielding to optimal classification accuracy.

Figure 16 shows an example of the tree structure that evolves from the SELECT procedure.

The approach has been applied to fault diagnosis of an electro-pneumatic valve, namely a flow control circuit which is a part of an industrial scale thermal plant [28]. The fuzzy functional observer that has been described previously has

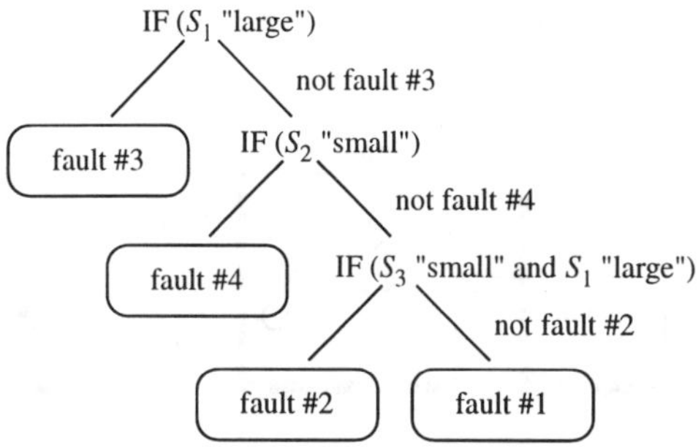

Figure 16: Example of classification tree (SELECT); S_i symptoms, i=1,2,3.

been used to generate the symptoms. Six faults are detected in the closed loop operation of the flow circuit. Since there is an overlap of some of the faulty classes, two diagnostic trees have been designed. A more complex example of application of the SELECT method refers to sensors and component fault diagnosis and reconfiguration of a heat exchanger [33].

4.3 Presentation of Fault Alarms

The final step in FDI is the proper representation of the fault situation by the operator who has to make the final yes/no decision. This can be done by means of an automatic defuzzification procedure [40]. However, one can refrain from that computerised method [36,38]. Typical for the fault detection problem is that the output consists of a number of fault indication signals, one for each fault and the normal operation, respectively. They are still in a fuzzified format and each can be called fuzzified fault indication signal (FFIS). An FFIS represents actually the degree of compatibility for the possible faults, in case of a fault. No fuzzy sets for the output of the inference have to be assigned. Thereby, the final decision about a fault alarm is left to the operator.

More specifically, instead of using the standard format of the statement

$$\ldots \text{ THEN } fault = big \tag{11}$$

where *big* is defined as one of a number of fuzzy sets that characterizes the output, the following format of the statement is used:

$$\ldots \text{ THEN } fault = fault_j \tag{12}$$

where $fault_j$ is the only existing fuzzy set of fault f_j. This applies in a similar way to all faults under consideration. Here, the fuzzy set $fault_j$ has a degree of membership which is identical to the aggregated output of the evaluated residuals (see Figure 17).

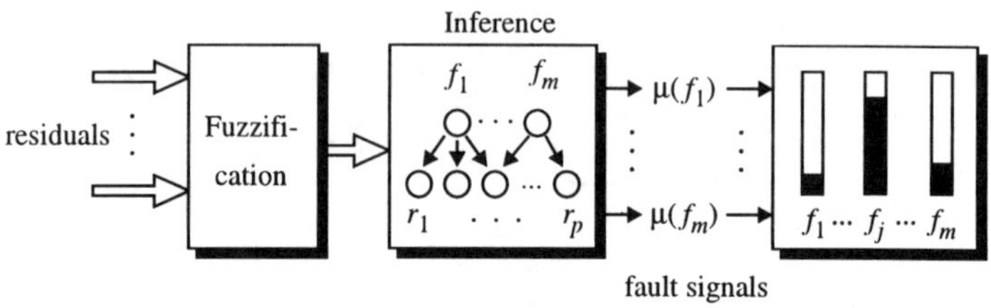

Figure 17: Residual evaluation and fault alarm presentation.

The membership signal corresponding to a certain class ω_j, $\mu_j := \mu(f_j)$, is therefore an one-dimensional time function composed of stable phases interrupted by abrupt changes. Oscillations of the signal around its mean value reflect internal movement of the system within the class [33]. A decreasing value indicates that the system is leaving the state j, while an increase reflects a movement toward this class. An evolution from a state ω_j to a state ω_i introduces a decrease of μ_j, possibly associated with an increase in μ_i, if ω_i is a known class. An evolution to an unknown state ω_x will introduce a decrease of all μ_k, $k = 1, \ldots, m$.

The presentation of fault alarms could therefore be done in form of bar charts as shown in Figure 17, even with coloured bars ranging from green to red or in gray tones. The height of the bars would then reflect the moving average of the degree of compatibility. In case a fault occurs, which was not considered in the design procedure, the bar chart would also give information about the possible cause of that fault.

5 Conclusions

The present paper outlines different ways of application of fuzzy logic systems to the main tasks of fault diagnosis, namely symptom generation and symptom evaluation. The former is based on various approaches to fuzzy system modelling in the framework of an observer configuration. The latter uses fuzzy reasoning either autonomously or assisted by the decision-making of the human operator. The development of the theoretical methodology in this area is accompanied by an increasing number of successful applications in laboratory and industrial plants.

One should admit, however, that the acquisition of knowledge and its systematic evaluation and adaptation may often be a rather difficult task. In this respect, the integration with the other methods of soft computing, neural networks and genetic algorithms, seems to be appealing for an optimal design of fuzzy logic systems. It is expected that these results would produce significant results in the fault diagnosis domain as well.

Based on the results obtained by the development of robust analytical model-based fault detection concepts and the progress of qualitative model-based techniques in recent years, there is a broad methodological background for robust fault-diagnosis and for practical applications aiming at fault tolerant control systems. The current trend is to integrate both analytical and knowledge-based approaches. In this respect, fuzzy logic and the other techniques of computational intelligence provide adequate support and tools.

Acknowledgment

The authors gratefully acknowledge the support of this work by the Alexander von Humboldt Foundation, Germany, by awarding the second author with a research fellowship.

References

1. R.J. Patton, P.M. Frank and R.N. Clark (Eds.), *Fault Diagnosis in Dynamic Systems*, Prentice-Hall, Englewood Cliffs, NJ (1989).
2. J. Gertler, *Fault Detection and Diagnosis in Engineering Systems*, Marcel Dekker, New York (1998).
3. P.M. Frank, Fault diagnosis in dynamic systems using analytical and knowledge-based redundancy, *Automatica* **26**, 459-474 (1990).
4. R. Isermann, Fault diagnosis of machines via parameter estimation and knowledge processing, *Automatica* **29**, 815-836 (1993).
5. R.J. Patton, Robust model-based fault diagnosis: the state of the art, Prep. *IFAC Symp. on Fault Detection, Supervision and Safety for Technical Processes - SAFEPROCESS*, Espoo (1994).
6. P.M. Frank, Quantitative and qualitative model-based fault diagnosis, *European Journal of Control* **2**, 6-28 (1996).
7. R. Milne, Strategies for diagnosis, *IEEE Trans. on Systems, Man, and Cybernetics* **17**, 333-339 (1987).
8. M. Ulieru and R. Isermann, Design of a fuzzy logic based diagnostic model for technical processes, Proc. *Int. ASME Conf. on Fuzzy Signals and Systems*, Geneva (1992).
9. P.M. Frank and N. Kiupel, Fuzzy supervision and application to lean production, *Int. J. Systems Sci.* **24**, 1935-1944 (1993).

10. J.M. Mendel, Fuzzy logic systems for engineering, *Proc. of the IEEE* **83**, 345-376 (1995).
11. R. Babuska and H.B. Verbruggen, An overview of fuzzy modeling for control, *Control Engineering Practice* **11**, 1593-1606 (1996).
12. J.-S.R. Jang and C.-T. Sun, Neuro-fuzzy modeling and control, *Proc. of the IEEE* **83**, 378-405 (1995).
13. H.J. Zimmermann, *Fuzzy Set Theory and Its Applications*, Kluwer Academic Publishers, Boston, MA (1991).
14. W. Pedrycz (Ed.), *Fuzzy Evolutionary Computation*, Kluwer Academic Publishers, Boston, MA (1997).
15. Y.H. Joo *et al*, Fuzzy systems modeling by fuzzy partition and GA hybrid schemes, *Fuzzy Sets and Systems* **86**, 279-288 (1997).
16. M. Russo, FuGeNeSys - A fuzzy genetic neural system for fuzzy modeling, *IEEE Trans. on Fuzzy Systems* **6**, 373-388 (1998).
17. O. Nelles, FUREGA - Fuzzy rule extraction by a genetic algorithm, Prep. *European Congress on Intelligent Techniques and Soft Computing - EUFIT*, Aachen (1996).
18. M. Ayoubi and R. Isermann, Neuro-fuzzy systems for diagnosis, *Fuzzy Sets and Systems* **89**, 289-307 (1997).
19. P.M. Frank, E. Alcorta-Garcia and B. Köppen-Seliger, Modelling for fault detection and isolation, Prep. *ESF Third Joint COSY Workshop*, Budapest (1997).
20. P. Klotzek, T. Dalton and P.M. Frank, Application of sensitivity theory to fuzzy logic based FDI, Prep. *EC-INCO Copernicus IQ2FD Workshop*, Kazimierz Dolny (1998).
21. B. Quipers, Qualitative simulation, *Artificial Intelligence* **29**, 289-338 (1986).
22. Q. Shen and R. Leitch, Fuzzy qualitative simulation, *IEEE Trans. on Systems, Man, and Cybernetics* **23**, 1038-1064 (1993).
23. Z. Zhuang and P.M. Frank, Qualitative observer and its application to fault detection and isolation systems, *Proc. Instr. Mechn. Engnrs.* **211**, I, 253-262 (1997).
24. P. Amann and P.M. Frank, On fuzzy model-building in observers for fault diagnosis, Proc. *IMACS World Congress*, Berlin (1997).
25. P. Amann *et al*, Fuzzy observer for fault detection in complex systems applied to detection of critical driving situations, Prep. *ESF Third Joint COSY Workshop*, Budapest (1997).
26. R. Querelle *et al*, Fault diagnosis on a winding-machine, Prep. *IFAC Symp. SAFEPROCESS*, Kingston upon Hull (1997).
27. O. Nelles and M. Fischer, Local linear model trees (LOLIMOT) for non-

linear system identification of a cooling blast, Prep. *EUFIT*, Aachen (1996).

28. D. Füssel, P. Balle and R. Isermann, Closed loop fault diagnosis based on a nonlinear process model and automatic fuzzy rule generation, Prep. *IFAC Symp. SAFEPROCESS*, Kingston upon Hull (1997).
29. T. Takagi and M. Sugeno, Fuzzy identification of systems and its application to modeling and control, *IEEE Trans. on Systems, Man, and Cybernetics* **15**, 116-132 (1985).
30. P. Balle *et al*, Integrated control, diagnosis and reconfiguration of a heat exchanger, *IEEE Control Systems* **18**, 52-63 (1998).
31. T. Marcu, Pattern recognition techniques using fuzzily labeled data for process fault diagnosis, *J. Appl. Math. and Comp. Sci.* **6**, 815-840 (1996).
32. T. Marcu and L. Mirea, Robust detection and isolation of process faults using neural networks, *IEEE Control Systems* **17**, 72-79 (1997).
33. M.A. Peltier and B. Dubuisson, A fuzzy diagnosis process to detect evolutions of a car driver's behaviour, Prep. *IFAC Symp. SAFEPROCESS*, Espoo (1994).
34. H. Schneider and P.M. Frank, Observer-based supervision and fault detection in robots using nonlinear and fuzzy logic residual evaluation, *IEEE Trans. on Control Systems* **4**, 274-282 (1996).
35. J. Chang, S. DiCesare and G. Goldbogen, Failure propagation trees for diagnosis in manufacturing systems, *IEEE Trans. on Systems, Man, and Cybernetics* **21**, 767-776 (1991).
36. N. Kiupel *et al*, Fuzzy residual evaluation concept, Prep. *IEEE Int. Conf. on Systems, Man, and Cybernetics*, Vancouver (1995).
37. N. Kiupel and P.M. Frank, An algorithm for a filter design for fuzzy supervision, Prep. *World Automation Congress*, Montpellier (1996).
38. N. Kiupel and P.M. Frank, Fuzzy supervision for an anaerobic wastewater plant, Prep. *IEEE Conf. on Computational Engineering in System Applications*, Lille (1996).
39. J.C. Bezdek and S.K. Pal, *Fuzzy Models for Pattern Recognition*, IEEE Press, New York (1992).
40. M. Ayoubi, Fuzzy systems design based on a hybrid neural structure and application to the fault diagnosis of technical processes, *Control Engineering Practice* **4**, 35-42 (1996).

PART III: APPLICATIONS

HIGH LEVEL PROCESS CONTROL IN THE CEMENT INDUSTRY BY FUZZY LOGIC

JENS-JØRGEN ØSTERGAARD
FLS Automation A/S
Høffdingsvej 77, DK-2500 Valby, Denmark
E-mail: jjoe@flsautomation.com

This paper outlines the basic principles of automatic high level process control in the cement industry by fuzzy logic as it is implemented in the ECS/FuzzyExpert system from FLS Automation A/S (FLSA) in Denmark. The background for using fuzzy control in the cement industry and other process industries is described, and some of the general practical problems of installing high level process control equipment are discussed. A design methodology is described for development and implementation of high level control strategies, and finally, some typical results are presented from industrial application of fuzzy control in the cement industry.

1 Introduction

Process control in general and rotary kiln control in particular was the first application of the theory of fuzzy sets, which was introduced in 1965 by Prof. L.A. Zadeh[1] . The first industrial application was developed by FLS Automation A/S (FLSA)[2], and put into operation at a rotary kiln in 1979. FLSA is a member of the FLS Industries Group which is one of the largest industrial groups in Denmark, and a major supplier of equipment to the cement industry and other related industries.

Fuzzy control is based on the theory of fuzzy sets, and in the mid 1970s it was presented to the process industry as a method for process optimization based on experience from manual operation, and it was pointed out that this was in contrast to other optimization techniques which were based on mathematical modeling of the process. The new method was attractive because the industry had experienced many unsuccessful projects on application of model based optimization techniques. Many industrial processes are so complex that sufficiently accurate mathematical models cannot be derived and therefore, more heuristic methods like fuzzy control are alternatives for implementation of automatic control.

2 The cement kiln process

Figure 1 shows a simplified example of a rotary cement kiln process . The kiln process consists of a preheater tower, a rotating steel tube and a cooler. The preheater is fed with raw materials which is a mixture of limestone, clay and other materials depending on the type of cement. The mix of materials have been grinded to a fine powder and prepared to have a certain chemical composition. The process dimensions are very large. The height of the preheater tower may by over 100m, kiln diameter more than 6m and normally, the length of the kiln is more than 70m.

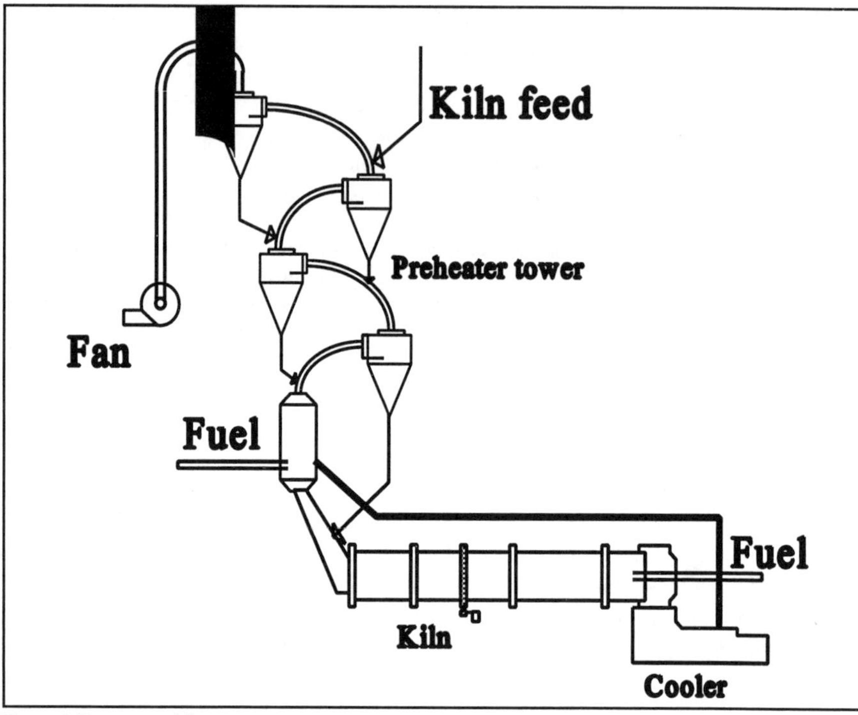

Figure 1 The cement kiln process

The amount of raw materials fed to the kiln depends, of course, on the process dimensions, but a medium size cement kiln is fed with about 300 t/h of raw materials. Normally, the kiln has fuel burners at the lower stages of the preheater and at the kiln end opposite to the preheater. The most common types of fuels are coal, lignite and coke or natural gas, but also a wide variety of waste products like shredded rubber tyres and waste fuel oil are used as fuels in cement kilns. The kiln process is very energy consuming, as a medium size kiln consumes about 25 t/h of coal. Air for combustion of the fuel is pulled through the kiln system by large variable speed fans.

The rotating kiln tube has an inclination of a few degrees and it is rotating at a speed is about 3 rpm, which results in a transportation time from preheater inlet to cooler outlet of about 1.0-1.5 hours. The process thus involves both long time constants and large time delays, which make automatic control more difficult.

The process is equipped with conventional PID controllers for maintaining kiln feed, kiln speed, fuel, and air flows at desired set point values, and the high level control tasks consist in coordination of these set point to fulfil the overall objectives about production, consumption of fuel and raw materials, and about the quality requirements to the end product, the so-called cement clinker.

3 The start

One of the very first papers on application of the fuzzy set theory to process control was presented by S. Assilian and E.H. Mamdani[3] at the The IFAC/IFIP International Conference on Digital Computer Applications to Process Control, held in Zürich, Switzerland, in 1974. As a postscript to this paper on learning controllers, fuzzy logic was proposed as an alternative approach to human-like controllers. The last remark of the concluding section says:

> "Greater dividends can be expected however if information is processed in a higher- level framework of conceptional representation. The potentiality of this approach has been demonstrated by successful implementation of the fuzzy logic controller and further research in this area is strongly recommended."

FLS had already started to work on automatic kiln control a few years earlier than 1974, and the approach taken was based on experience with manual control of the kiln as the control strategy was composed of simple decision tables on how to adjust the controller set points in different situations.

Mamdani's early experiments with rule based control of a steam engine[3] in combination with the rules of thumbs for manual control of a cement kiln[4] resulted in the first high level process control strategy based on fuzzy logic. The obvious link between the rule based approach of fuzzy logic and high level control is illustrated by Figure 2, where the control instructions for the cement kiln operators use the everyday terms LOW, OK, SLIGHTLY etc., which also are some of the basic primary terms that are represented by membership functions in the theory of fuzzy sets.

4 High level control

The first industrial application of fuzzy set theory to process control is named high level control, which may be defined as: "**Coordination of control loop set points, which normally is done by a human operator**".

A high level process control strategy works on the same level as the human operator. It takes over a part of, or all of the operator's job of controlling the process. High level control thus introduces an extra layer between the operator and the conventional control system in the process control pyramid as shown in Figure 3.

Case	*Condition*	*Action to be taken*	*Reason*
10	BZ OK OX low	a. Increase I.D. fan speed	To raise back-end temperature and increase oxygen percentage for action 'b'

	BE low	b. Increase fuel rate	To maintain burning zone temperature
11	BZ OK	a. Decrease fuel rate slightly	To raise percentage of oxygen
	OX low		
	BE OK		
12	BZ OK	a. Reduce fuel rate	To increase percentage of oxygen for action 'b'
	OX low	b. Reduce I.D. fan speed	To lower back-end temperature and maintain burning zon temperature
	BE high		
13	BZ OK	a. Increase I.D. fan speed	To raise back-end temperature
	OX OK	b. Increase fuel rate	To maintain burning zone temperature
	BE low		
14	BZ OK	NONE. However, do not get overconfident,	
	OX OK	and keep all conditions under close observation.	
	BE OK		
15	BZ OK	When oxygen is in upper part of range	
	OX OK	a. Reduce I.D. fan speed	To reduce back-end temperature
	BE high	When oxygen is in lower part of range	
		b. Reduce fuel rate	To raise oxygen percentage for action 'c'
		c. Reduce I.D. fan speed	To lower back-end temperature and maintain burning zone temperature
16	BZ OK	a. Increase I.D. fan speed	To raise back-end temperature
	OX high	b. Increase fuel rate	To maintain burning zone temperature and reduce BE low percentage of oxygen
17	BZ OK	a. Reduce I.D. fan speed slightly	To lower percentage of oxygen
	OX high		
	BE OK		

Figure 2. Extract from Peray's textbook for kiln operators.

The basic idea of high level process control based on fuzzy logic is to formulate sets of rules for automatic operation based on practical experience and knowledge about manual control of the process.

The terms of the rules, i.e. LOW, SMALL OK etc. are represented by membership functions, and the rules are evaluated by inference as described in the theory of fuzzy sets. Examples of fuzzy control rules are:

IF Temp is High and Pressure is Ok THEN Medium Flow
IF Temp is Ok and Pressure is Ok THEN Small Flow
IF .....

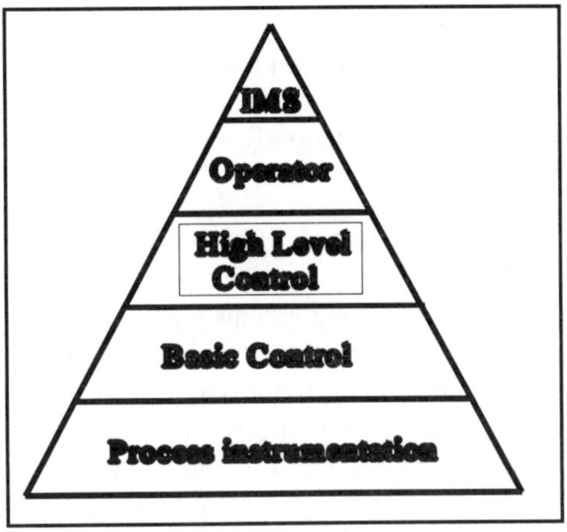

Figure 3. Process Control Pyramid

The terms High, Ok, Small and Medium are represented by membership functions, and the result of the rules are calculated by inference, e.g. the Max. -Min. logical product. In fuzzy logic, the condition of a rule is fulfilled to a certain degree, and each rule will influence the result of the set of rules in accordance with its grade of fulfilment. The result, in other words is a compromise between one or more conclusions where each conclusion has a certain weight.

This heuristic approach for design of control strategies for automatic process control is useful, and probably the only practical solution, when the process is only partly known, difficult to describe by a mathematical model, if few measurements are available, or if the process is highly nonlinear.

High level control systems may be integrated with the rest of the control system in different configurations. High level kiln control has a configuration as shown in Figure 4, where the operator selects between manual adjustments of set points and automatic set point coordination based on fuzzy control rules.

The existing control loops are still active, and the high level control strategy makes adjustments of the controller set points in the same way as the operator used to do,

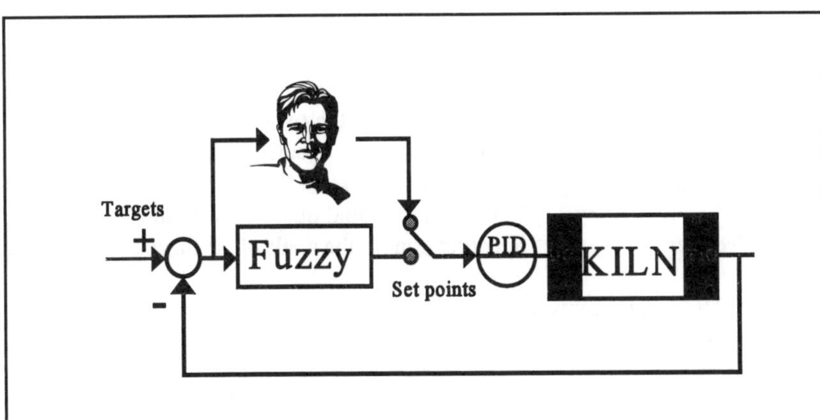

Figure 4 Fuzzy kiln control configuration

before the automatic control strategy was installed. It is up to the operator to decide whether manual or automatic control is likely to produce the best possible operation of the kiln which, of course, may create conflict situations between the operator and the fuzzy controller, a conflict which has to be addressed very seriously when working with high level process control.

5 The FLS design procedure

For an operator, control of the process consists in achieving various goals, which are more or less precisely defined: maximum output, minimum consumption of raw materials and energy, high product quality, safe process operation, etc.

Different processes have different control objectives but, in general, good process control may be defined through a list of control objectives which should be fulfilled as much as possible. The concept of control objectives is a key element in a high level control strategy. For a cement kiln, typical control objectives are:

- stable operation,
- good cement clinker quality,
- High production
- complete combustion,
- low fuel consumption.

As control objectives are frequently in conflict, high level co-ordination means approaching optimal conditions in succession of importance. Priorities, in other words, have to be assigned to the various control objectives, specifying which objectives are considered the most important to fulfill. Control objectives with different priorities are thus the basic elements of the design of a high level control strategy.

The elements of the FLS design procedure for a process control strategy are the following:

Utility programs:	Calculations concerning the actual process condition.
Control groups:	Arrangements of the overall control strategy into groups of control objectives.
Priority management:	Determines the extent to which the control actions to fulfil the individual objectives should be executed.
Control objectives:	Specifications of the goals of the fuzzy control strategy.

Examples of Utility programs are the so-called state indices, which combine various

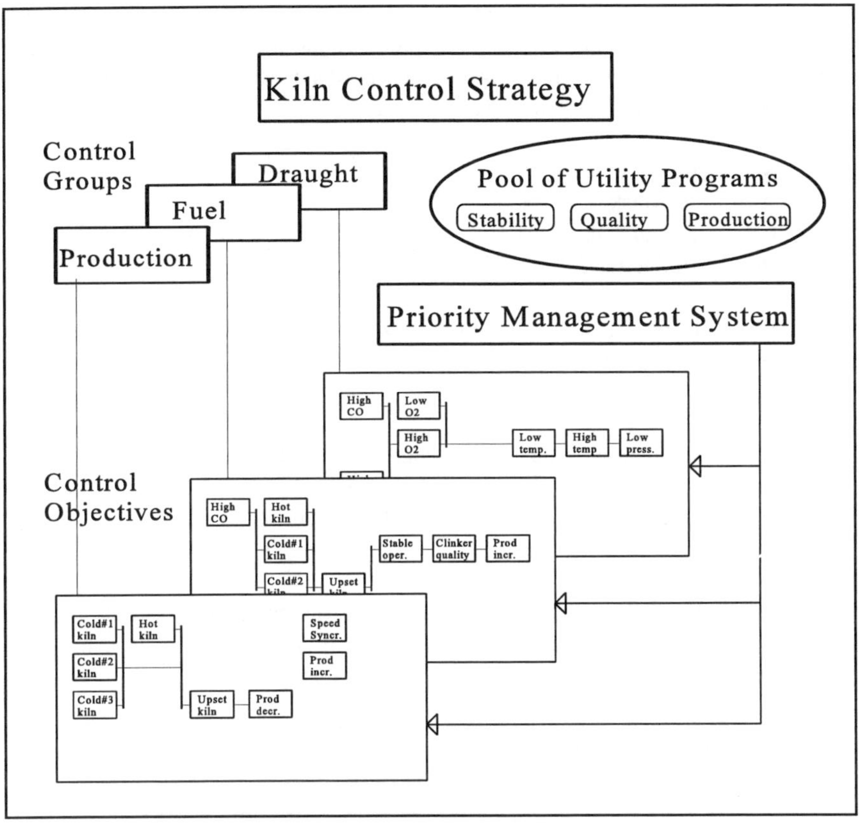

Figure 5 The FLS structure of a high level kiln control strategy

measurements into a single key figure about the actual state of the process. The degree of process stability, the product quality, the production level, etc., are all typical examples of state indices for a kiln control strategy. The state indices are important to the structure of the FLS design scheme for a high level control strategy, as they form the basis for dividing the overall strategy into control groups which can be treated independently. The state indices are used to co-ordinate control actions from the various control groups.

Figure 5 shows a diagram of the FLS control strategy structure, and design of a control strategy consists in filling in the structure with specific state index calculations, control groups and control objectives, whereas the priority management system is a fixed module in the system. For a new process, it is perhaps not so clear which are the most appropriate indices and control groups. This, however, is not a major problem as the structure facilitates a stepwise implementation of control objectives and index calculations concurrent with more and more process control knowledge being available.

Every control objective is implemented in accordance with the so-called objective module, which consists of the following four tasks:

- deviation task
- rule task
- output task
- timing task

Figure 6 shows the objective module diagram, which is used for implementation of all types of control objectives.

Deviation task

The deviation calculation evaluates the degree to which the objective is fulfilled. Normally, the calculation results in a "fuzzy value" between -1 and +1, which expresses how far the actual process situation is from fulfilment of the objective.

Rule task

The rule block holds the set of control rules for fulfilment of the specific objective. Normally, the rule block is formulated as a set of fuzzy control rules, but other techniques may also be used, like PID, neural nets, mathematical models, etc.

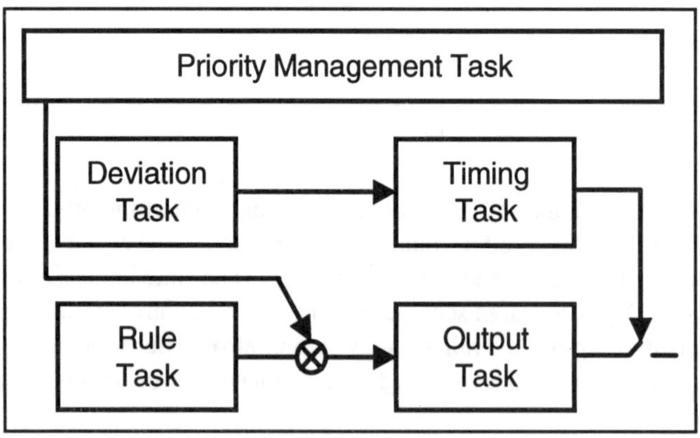

Figure 6. The Control Objective Module

Priority management

The rule block for each

control objective results in control actions which are multiplied by a weight factor between 0 and 1. The smaller the weight factor, the more the control actions are suppressed.

The weight factor for a given objective is a function of the deviation values for the higher priority objectives, which ensures that control actions to fulfill a given objective are reduced if higher priority objectives are not fulfilled.

Output program
The output program evaluates process constraints and selects among alternative control actions based upon the actual index values. The logic for selecting alternative adjustments may be fuzzy or non-fuzzy depending on whether a gradual or an either/or switch is the most appropriate. In most cases, the fuzzy logic approach gives the best control performance, simply because no process operates with sudden changes between alternative control actions.

Timing calculation
The timing calculation determines when control actions are to be executed. It is just as important as the rule block for determining the proper function of the control strategy. Also the timing calculation is normally fuzzy in the sense that the time interval between control actions changes gradually as a function of the deviation value. The larger the deviation, the more frequent the control actions.

6. Results

Fuzzy control in general yields the following advantages:

- Tuning of rule base, rather than mathematical equations
- Stepwise development
- Gradual transition between upset control and optimization
- Suitable for solving conflicting objectives
- End user involvement

The results which have been obtained with FuzzyExpert systems are typically as shown in the table below.

ECS/FuzzyExpert Results		
Production Increase	**Fuel savings**	**Quality improvement**
3 - 5 %	**2 - 4 %**	**30%**

References

1. L.A. Zadeh, Outline of a new approach to the analysis of complex systems and decision processes, *IEEE Trans. Systems Man Cybernet.* **SMC-3** (1) (1973) 28-44.
2. Østergaard, Jens-Jørgen, "Fuzzy Control of Cement Kilns - A Retrospective Summary", Proceedings of EUFIT'93, pp 552 - 558.
3. Assilian, S., and E.H. Mamdani. "Learning Control Algorithms in Real Dynamic Systems." Proceedings, IFAC/IFIP International Conference on Digital Computer Applications to Process Control. Springer: Zurich, Berlin: 1974.
4. Peray, K.E., and J.J. Wandell. The Rotary Cement Kiln. New York: Chemical, 1972

FUZZY LOGIC FOR PAPER AND PULP MANAGEMENT

HANS HELLENDOORN
*Siemens Nederland N.V., P.O. Box 16068, NL-2500 BB, The Hague
The Netherlands, e-mail: Hans.Hellendoorn@siemens.nl
Control Engineering Laboratory, Delft University of Technology
P.O. Box 5031, NL-2600 GA Delft, The Netherlands*

In paper and pulp industries very large and complex machines are used. On these machines there are hundreds of sensors that measure all kinds of process variables. The gigabytes of data that are delivered by these sensors are usually thrown away after some time although they might contain valuable information about the process and contain saving potential for the company. The process of extracting (fuzzy) information out of raw data is called data mining. We present a methodology for data mining in large industrial data sets. We show the most important steps in the process of extracting information out of the data: (1) Preprocessing the data, (2) Reducing the data, (3) Modeling the data, (4) Rule extraction, and (5) Drift analysis. We also show some examples of industrial applications where fuzzy logic based data analysis plays a role.

1 Introduction

Fuzzy logic [4] and sensor analysis build a good combination as experience in paper industry, chemical industry, car industry, traffic control, etc. shows. There are several reasons for the increasing interest in this area. The main reasons are the increased use of sensors in any industrial area and the wish to describe and understand processes in an easy rule-based way. In particular in areas where many input parameters play a role, fuzzy logic can help to obtain more transparency in the description of the processes. Another argument for the use of fuzzy logic in sensor analysis is the imprecision that is allowed: perhaps a purely mathematical approach with many formulas might do better, but to handle such a theory often demands too many assumptions, that weaken this approach severely.

In this chapter we will briefly describe the general method of data analysis, and show with the help of some examples the use of it.

2 The project

A paper company in the Netherlands presented us data from their production process. We used sensor data from 240 sensors along the production line, including measurements of temperatures, pressures, levels, concentrations, flows, viscosities, etc. Furthermore, we got quality data in the form of sixteen measurement values for each tambour, the paper rolls at the end of the production line. We got data from 3,300 tambours; as it takes some time until a tambour is ready, one can imagine that we had a tremendous amount of data available.

The purpose of the project was to extract information out of the process and to give advises how to save chemicals and how to optimize the process. This included a model of the process, advises how to save chemicals, a description of the state of the machine, and a selection of parameters that have a dominant influence on the quality of the paper.

The hardest problem within such a project is to deliver *valuable* information. It is fairly easy to generate thousands of fuzzy rules that represent dependencies between variables, but most of this information is trivial and useless for the customer. Actually this happened in the very beginning of our project when a university from Belgium delivered a big pack of paper with thousands of rules to the customer. This almost killed the project. In a later stage Siemens was able to extract the really valuable information out of the data. It turned out that a small number of selected rules helped him to save a lot of money.

The customer was also interested to get information about the state of the machine. Such a machine can have a length of over one hundred meters, which makes it hard to tell something about its state. It is nevertheless of high importance to drive the machine in its optimal state. Our system is, for example, able to tell the user at which date he has stopped the machine to do maintenance, this clearly comes out of the Sammon-mapping later on this figure. Before the maintenance the system was on the right side of the dots cloud, afterwards it was on the left.

3 The Methodology

Although each project in which data analysis plays a role is different, there are some similarities in the development steps of these projects. Of course, there are differences too, time-dependent data can be treated in a completely different way than time-independent data. Usually we follow the steps presented below in data mining projects [1].

We will describe a five step approach to data analysis, viz.

1. Preprocessing the data, also called data reconciliation: The data comes from sensors that do not always function well. Hence the data should be 'cleaned'.

2. Reducing the data. Usually there are many sensors generating megabytes of data. The system should reduce the number of inputs and the amount of data.

3. Modeling the data: There are several ways to model the data. This model can be used forecast and control process behavior.

4. Rule extraction: The behavior of the process is described in terms of causal rules and presented to the operator in an understandable form.

5. Drift analysis: It is particularly interesting for the operator to know in which state the machine is situated.

It is important to observe that in the first three and partly in the fourth step *one has to put information into the process.* Information means data *and* know how from people who work on the plant or have experience with the system under concern. In the fourth and fifth step it is then possible to *extract information out of the process.*

4 The Preprocessing

Most data that are being used in industrial processes are noisy and erroneous. Usually, sensors are drifting, they may fall out, they often deliver strange values due to obscure reasons, they may be inconsistent and deliver values outside of the defined domain, etc. This means that usually the first step in data analysis is to remove those data that must be failures, which is a hard job in practice—one has to discuss with the operator or expert to check the ranges of all sensor inputs, including their expected means and standard deviations. Better than removing data items is to find reasonable substitutes by, e.g., filtering techniques so that one is not forced to remove valuable input/output vectors. Furthermore, filtering plays an important role in smoothing data. The comparison of expected ranges c.q. standard deviations with actual ones is an intensive job that can only partly be automated. The importance of *visualization* should be stressed here, it is important to produce pictures of two- or three-dimensional connections between variables.

4.1 The process data

The first data analysis step takes place in the sensors at the automation system on the plant. Here it is checked whether the sensors perform well, and a first filtering on the data takes place. The process data can, for example, be saved on a PC via the interface WIN-TM. At the same time the process data can be, for example, be registered in fixed interval times (e.g. 1 min.) on a PC in a dBase-file. In this way a data base with, e.g., the following process data comes into existence:

- Measurements (e.g., pressure, temperature, flow)

- Process information (information about local controllers, adjustments, opening widths, etc.)
- Machine adjustments (e.g., on/off)
- Quality data of the product

These quality data are divided into online and off-line data. Off-line data can be delivered by a laboratory that at fixed time intervals or process times checks the quality of the product from the plant. Besides the quality there can be off-line data about the product amount, the kind of process that was used, and economical information about the plant like product losses, special events, hazards, and product sums. It is important to map the non-numerical inputs on a numerical, possibly fuzzy scale, that takes into account the kind of off-line data.

4.2 Cleaning the data

In the further process of data analysis each datum can be decisive for the resulting final model of the process. Therefore it is important to carefully observe the measured data with regard to sensor failures and runaways. First of all it is necessary to define the interval ranges of each sensor. This means that from each sensor it has to be known its upper and lower value and, if possible, its expected mean value and standard deviation. Only with this information it is possible to distinguish regular from irregular inputs and, if the process data are available on a time scala, to substitute expected input data for faulty data using a (fuzzy) filtering algorithm or another intelligent device. The following actions should be carried out:

- Remove columns of constants from the data set, or substitute them by expected values using a nonlinear or linear filter.
- Remove lines out of the data set that contain at least one value described by NaN (Not a Number). Here again, if filtering is possible and useful, substitute the NaN by an assumed and realistic value. The same applies to the case where at least one measured value is situated outside of the predefined measurement interval.

To mark measurements as NaN can be performed automatically, but has usually to be performed manually using an editor. In some projects within Siemens we used special drift recognition programs to observe how the data behave with regard to the predefined expected mean values of the sensors.

4.3 *Normalizing the data*

As each sensor and other inputs are defined on domains that are typical to their physical behavior it is useful to map them onto normalized domains, e.g., $[-1,1]$ or $[0,1]$ (see Fig. 1 with the interval $[-4,4]$).

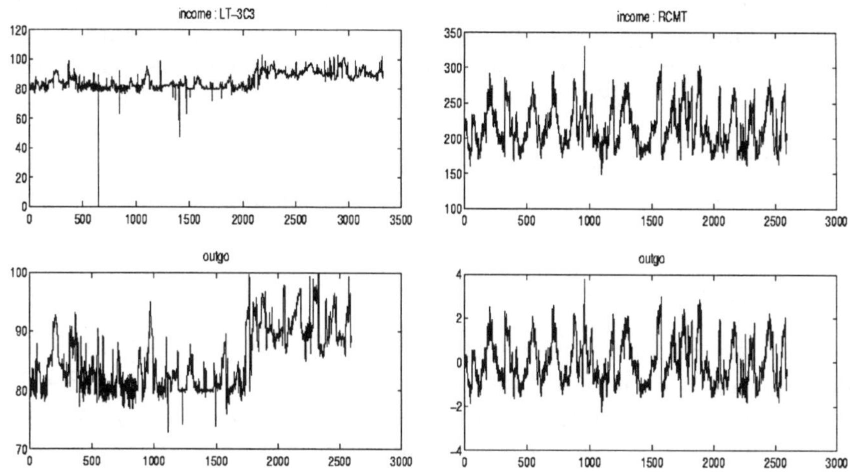

Figure 1: Left: the result of cleaning the data. Right: the result of normalizing the data.

This should prevent the fact that sensors defined on numerical larger domains than other ones get higher weights in the further data processing. The mathematical mapping of the standard deviation on the normalized domain should be regarded.

Babuška [2] has shown in his PhD-thesis that clustering with the fuzzy c-means algorithm in a multi-dimensional environment produces completely different result for normalized and non-normalized data. The Gustafson-Kessel algorithm, however, is almost independent from normalized and non-normalized conditions.

5 Data Reduction

Although there are cases where almost no data are available and one has to check carefully how each data item can be used perhaps twice to learn or check the desired model, usually it is the case that there are too many sensor inputs and enough data. An important question is which inputs are of relevance to the product quality of a plant or the diagnostic statement about a system. There are several methods that support this step. First, a correlation analysis gives information about the influence of inputs on the output, although one has to be careful in weighting correlation tests, they can only find a linear correspondence between variables. Secondly, a ranking test gives information about the importance of each parameter, in particular when we mix some random values amongst the input parameters. Thirdly, it is important to carry out a principal component test based on the correlation analysis. This usually greatly supports the reduction of the dimensions and gives a good insight in the importance of sensor inputs.

We will now in a couple of sections deal more closely with the above mentioned steps. A complete description can be found in [1].

5.1 Correlation analysis

There are several ways to deal with correlation coefficients. The first step should be to prepare a matrix with all mutual correlation values (cf. Fig. 2).

This can be done via the covariance matrix. In systems with several hundreds of inputs and outputs it often turns out that there are large clusters of sensors that are mutually correlated with values of 0.9 or even 0.99, e.g., in the case that temperatures are calculated at several places along the plant. A correlation matrix can be a helpful means to reduce the number of inputs/outputs that have to be considered for further study of causal relationships between variables.

One can also take one or several important output parameters and calculate the correlations of all inputs with this output. This is useful if one wants to find out which inputs have highest influence on one of the outputs, e.g., the quality of the product (Fig. 3).

These inputs will play an important role in later data analysis steps. Usually, only a small number of inputs have really significant correlation values with these outputs, i.e., one can considerably reduce the number of variables to be regarded more closely with this method.

The disadvantage of the correlation method is obvious for anyone who has been working with this method. If the data points of two inputs are not on a straight line, but, e.g., in a circle or on some very curvy line, the correlation value, a one-dimensional scalar, does not contain any valuable information. We present three different ways to uncover important relationships between input and output values, viz. the ranking test, PCA, and polynomial regression.

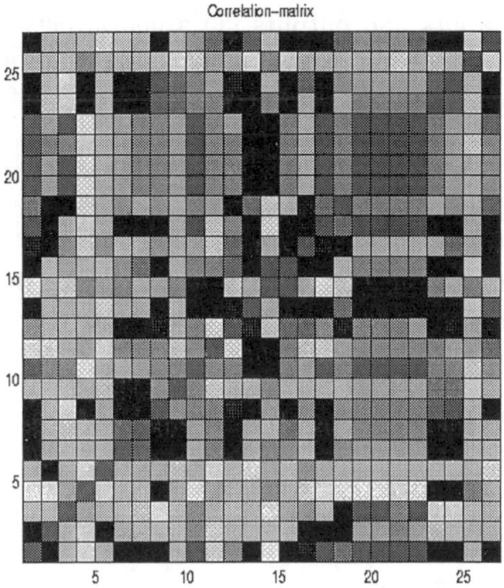

Figure 2: The correlation matrix.

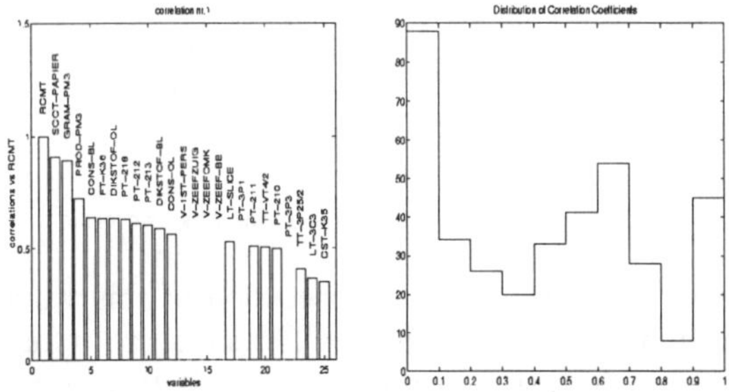

Figure 3: RCMT has correlation 1 with itself. Then follow other parameters.

5.2 Ranking

This methods presumes the existence of an RBF-model, a model of the relationships between inputs and outputs via a neural network with radial basis functions. If such a modeling exists one can substitute each variable consecutively by its mean value. Then one has to calculate for each variable the difference between the error of the RBF-model with the mean value taken as input and the model with the actual value of this input, i.e.,

$$\mathit{diff}(x_i) = E(\ldots, \bar{x}_i, \ldots) - E(\ldots, x_i, \ldots) \tag{1}$$

One has to sum up all $\mathit{diff}(x_i)$ to achieve one scalar difference value. One can now compare these error values for each input to obtain a ranking of the importance of each variable (see Fig. 4).

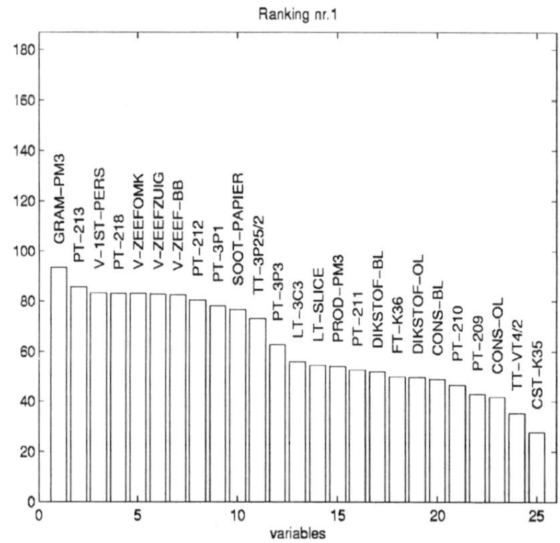

Figure 4: A ranking of several parameters.

One has to be careful in the interpretation of this value. It is certainly true that those inputs that have high differences are influential parameters in the process. On the other hand, however, if two or more variables are highly correlated (an event that often occurs), than removing one variable (substituting it by its mean value) will not result in any significant difference in the error values. Hence low values in the ranking test should be considered with care. Actually, this method cannot be used without the correlation test presented above.

5.3 Principal component analysis

Principal component analysis (PCA) is a very well known method that has been described in many places in literature. The user presents the number n of orthogonal axes that have to be obtained. Usually, n is 2 or 3, although this small number means the loss of a lot of valuable information. The resulting space is usually hard to interpret because the axes tend to be complex combinations of the original axes, depending on the data structure. The process takes the following steps: (1) Normalize the data, (2) Save the data in a matrix X, (3) Calculate the square matrix $A = X^T X$, and (4) Calculate the eigenvalues and eigenvectors of A. In a normalized data domain the size of the eigenvalue shows the importance of the attached eigenvector (cf. Fig. 5). Therefore, one should take the two or three most important eigenvectors as a

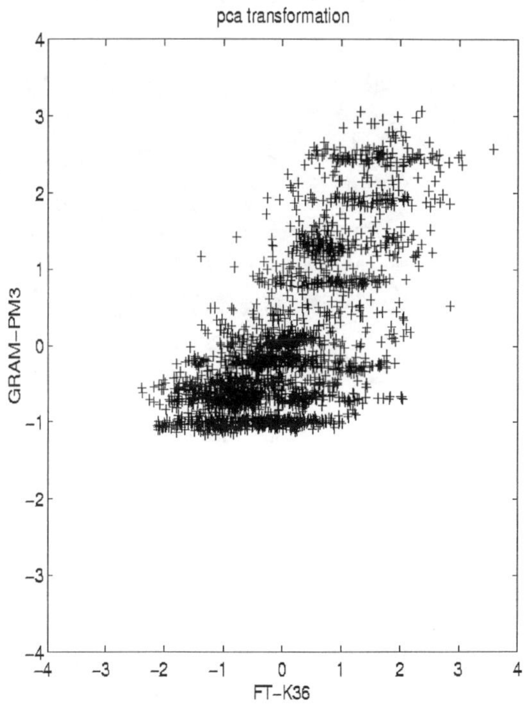

Figure 5: The location of two axes after principal component analysis.

new base of a reduced space frame. Furthermore, one should observe which inputs (considerably) contribute to these eigenvectors and to which degree.

These inputs turn out to be the most influential parameters in the process. Hence, in an indirect way, this is also a useful method to determine the most important or most influential parameters of the process.

5.4 Polynomial regression

Polynomial regression can be regarded as an intelligent correlation method. The correlation method attempts to put a straight line through the data points of two inputs or outputs. With polynomial regression it is possible to generate a polynomial curve that approximates the data points (Fig. 6). In practice, polynomials of degree 3 or 4 are the highest values one should take. More complex relationships are too hard to understand for the user.

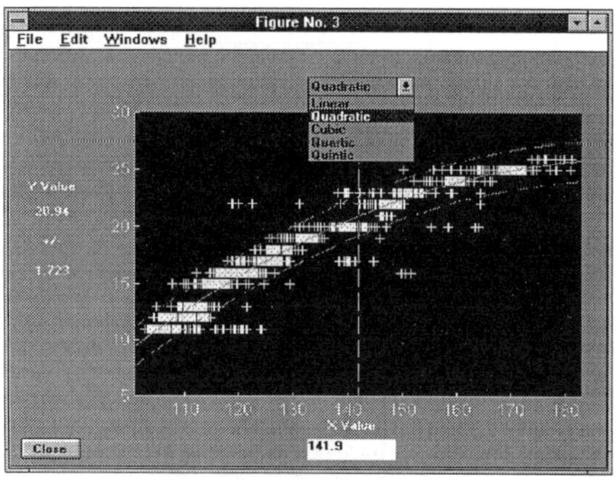

Figure 6: Polynomial regression with a quadratic polynomial.

6 Modeling

Based on the available data a model should be made that contains the dependencies between variables The easiest and surely not the worst way is to build a radial basis function (RBF) network. Usually, a simple RBF-network has a good performance in representing the input-output connection.

Another method that is frequently being used is fuzzy clustering [3]. There is fuzzy clustering of the input domain, of the output domain, of the relation between a number of inputs and outputs, etc. Furthermore, there is point clustering, line clustering, and also clustering around more complex structures like *c-elliptotypes* [11].

Thirdly, to gain insight into the model one can use Kohonen or similar mappings, e.g., the Sammon mapping. These mappings show the structure of the data in a two- or three-dimensional figure, and help to show the timely behavior of processes visually, which is important, as sensors as well as plants tend to drift during the time. Process operators are particularly interested to observe the state of their process during the time.

6.1 RBF-networks

It is not our purpose to give a detailed description of neural networks here. We will only deal with a special kind of neural networks, viz. networks with radial basis functions. Radial basis functions are usually multi-dimensional Gauß-functions. RBF-networks mainly play a role as cluster algorithms in a supervised learning environment. In a three layer neural network of this kind the network output y is equal to

$$y = NN(\mathbf{x}) = \frac{\sum_i w_i(\mathbf{x}) b_i(\mathbf{x})}{\sum_i b_i(\mathbf{x})} \tag{2}$$

w_i can be regarded as the output of a fuzzy rule; b_i as the input of a fuzzy rule, b_i can be any function but is usually chosen as the Gauß-function

$$b_i(x) = \kappa_j \exp\left(-\sum_{j=1}^{L} \frac{(x_j - c_{ij})^2}{2\sigma_{ij}^2} \right) \tag{3}$$

Figure 7 shows the results of training an RBF-network.

It can be easily shown that RBF-networks can be mapped on a special class of fuzzy systems and vice versa. Martinetz and Hollatz [9] present the following conditions:

1. The number of basis functions is equal to the number of fuzzy if-then rules.
2. The fuzzy rules are singleton or Takagi-Sugeno rules.
3. The membership functions in the fuzzy system are Gauß-functions.
4. The and-operation in the fuzzy system is performed by multiplication.
5. Both the RBF-network and the fuzzy system use the same defuzzification method.

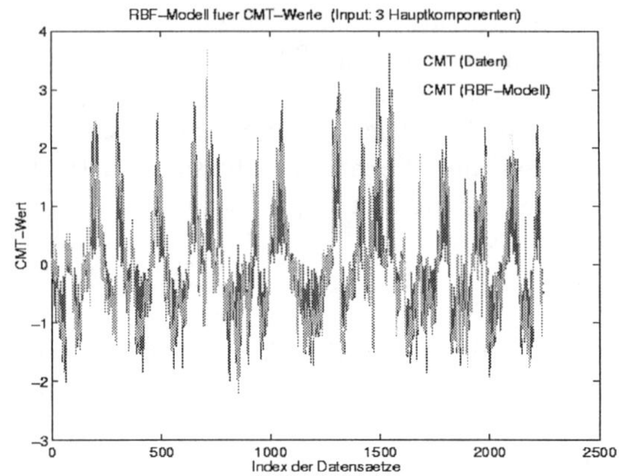

Figure 7: The result of training an RBF-network.

An RBF-network that satisfies these criteria can easily be mapped into a fuzzy system, which allows the interpretation of the contents of the neural network. Martinetz and Hollatz [9] used this system on a steel rolling mill. This resulted in rules like:

1. **If** b is *low*, T is *high*, and ϵ is *medium* f then decrease the rolling force by -28.9 MN.

2. **If** V is *medium*, b is *high*, and ϵ is *high* bf then increase the rolling force by 14.7 MN.

3. **If** d is *high* **then** decrease the rolling force by -6.8 MN.

4. **If** b is *high*, T is *low*, and ϵ is *high* **then** increase the rolling force by 12.6 MN.

Hence RBF-networks are one way to model the process and to obtain information out of process data.

6.2 Fuzzy clustering

Fuzzy clustering has been described in many papers and books [3], most recently in the dissertation of Babuška [2]. Fuzzy clustering is somewhat similar to RBF-networks, it takes the data and learns out of the data points a fixed number of fuzzy classes. Each point may belong to a certain degree to one or more classes. These classes may be centered around a point, but also, more interestingly, along a straight line or even along an elliptotype. When a fuzzy clustering exists, one can easily obtain fuzzy rules out of these clusters.

The cluster centers and the membership degree of data to each cluster center are calculated in an iterative process using

$$v_i = \frac{\sum_{k=1}^{n} u_{ik}^{m} x_k}{\sum_{k=1}^{n} u_{ik}^{m}} \tag{4}$$

where v_i are the cluster centers, x_k the data points, and

$$u_{ik} = \sum_{j=1}^{c} \left(\frac{d_{ik}}{d_{jk}}\right)^{-\frac{2}{m-1}} \tag{5}$$

is the membership degree to cluster i, and d is a distance function. This may be the distance to a point as well as the distance to a line (Fig. 8) or a more complicated mathematical function. In the elliptotype-case the distance function becomes fairly complicated, as one can imagine.

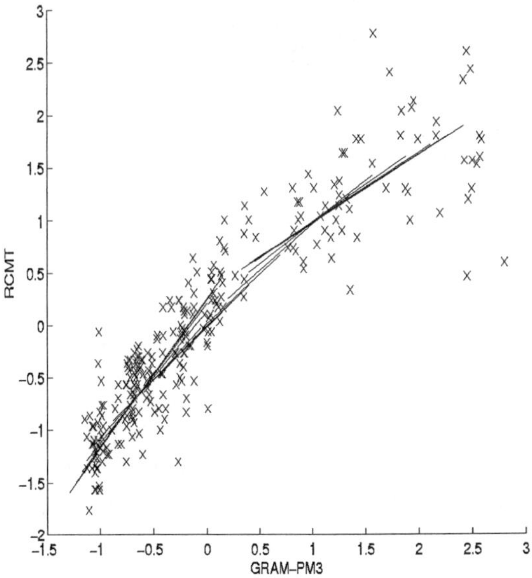

Figure 8: The resulting lines after fuzzy clustering.

The order of the data points given into the iteration procedure influences the location and size of the clusters.

6.3 *Kohonen mapping*

In [8] Kohonen proposed a method to use unsupervised neural network learning to map an n-dimensional space-frame to, e.g., a 3-dimensional (Fig. 9) or a 2-dimensional (Fig. 10) mapping, preserving as much as possible the structure in the data as well as the distances between neighboring data points. Therefore one needs an appropriate neighboring function. The visualization from 2-dimensional Kohonen mappings may give some insight into the structure of the data.

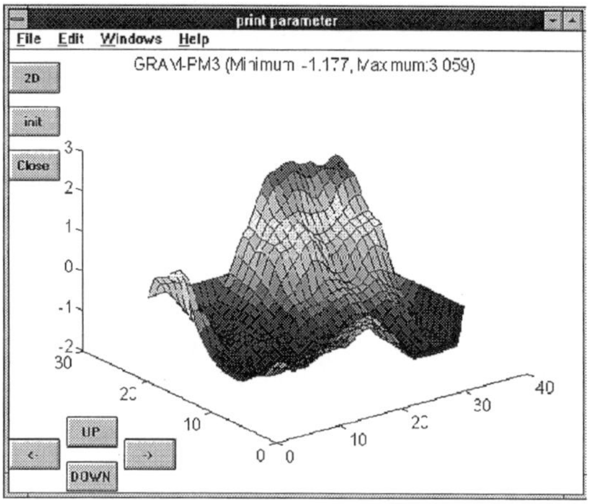

Figure 9: A three dimensional Kohonen mapping result.

It is particularly interesting to focus on one crucial output value. If one produces several maps with different clustered input variables and the same output one can compare the relationship between these variables and the output. In some cases it is possible with this method to save expensive chemicals and achieve good quality output. Of course, this presumes know how from experts from the plant. In a real life project we had some very interesting results in this area, that led to interesting savings for the plant owner.

6.4 *Sammon mapping*

A Sammon mapping [5] (cf. Fig. 11) has the same goal as a Kohonen mapping but uses a different way to achieve this goal. Kohonen uses a neural network in which all connections between variables are summarized, in a Sammon mapping gradient descent methods are used to map the n-dimensional space onto an

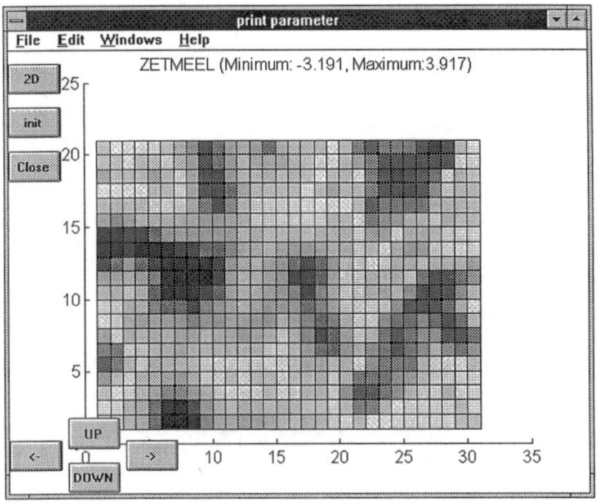

Figure 10: A two dimensional Kohonen mapping result.

$(n-1)$-dimensional and so on until the 2-dimensional space is reached. Goal is to preserve as much as possible the Euclidean distances between the data points. The choice of the energy function in this process is crucial.

There are the following steps:

1. There is a n-dimensional space and a corresponding 2- (or 3-)dimensional space.

2. There are distances d in the n-dimensional space frame and corresponding distances d^* in the 2-dimensional space.

3. Calculate
$$E(m) = \frac{1}{\sum_{i<j} d^*_{ij}} \sum_{i<j}^{N} \frac{(d^*_{ij} - d_{ij}(m))^2}{d^*_{ij}} \tag{6}$$
a measure for the differences in the distances in the two spaces after the m-th iteration cycle.

4. The adaptation rule in the m-the iteration cycle is given by a kind of linearization process, viz. $y_{pq}(m+1) = y_{pq}(m) - \alpha \cdot \Delta_{pq}(m)$. α is the adaptation rate that has to be determined empirically, Δ is given by
$$\Delta_{pq}(m) = \frac{\partial E(m)/\partial y_{pq}(m)}{\left|\partial^2 E(m)/\partial y_{pq}(m)^2\right|}. \tag{7}$$

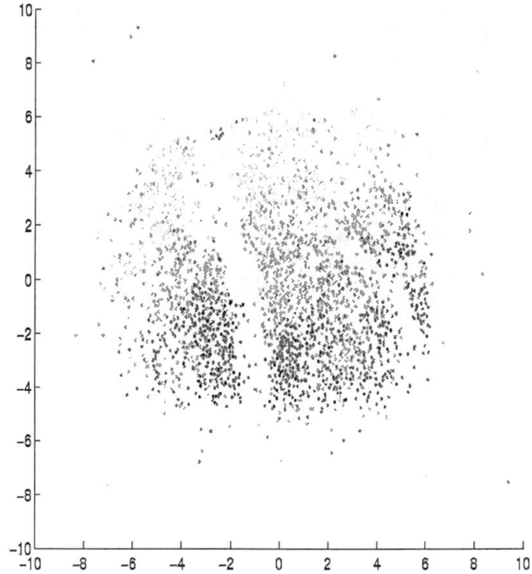

Figure 11: The result of a Sammon mapping.

7 Rule extraction

Another step is to extract information in the form of fuzzy rules out of the process or system. There are several ways to do this. One is via RBF-networks, it is not so hard to translate an RBF-network into a fuzzy system. A disadvantage of this approach is the complexity of the rules that arises. Each rule has its own premises with newly defined fuzzy sets, which makes it hard to understand the meaning of the rules. Another method is building rules by clustering techniques. The fuzzy clustering of input data, for example, corresponds to the input fuzzy sets. Using this method delivers rules that are usually of great interest to the customer. In some cases these rules immediately led to important savings, because relations that were not known until that time became apparent. Figures 12 and 13 show some screen dumps of our tool after RBF-training and fuzzy clustering.

7.1 Rule extraction from RBF-networks

We have seen before how the extraction of rule from RBF-networks proceeds. We have to point to a problem, however, that in cases of many inputs, a

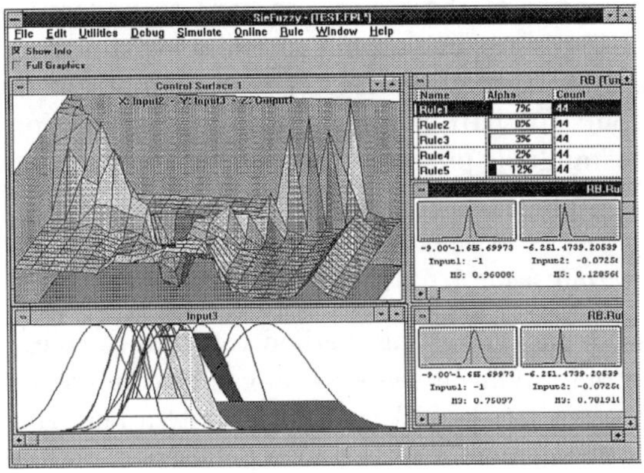

Figure 12: The resulting fuzzy system after all the described steps.

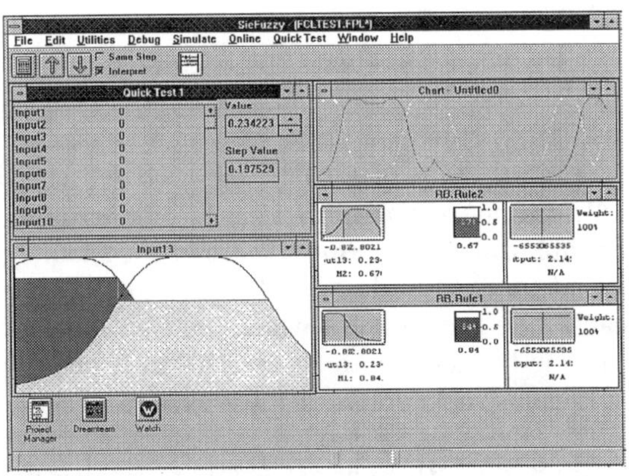

Figure 13: Some variables of the final fuzzy system.

PCA is needed to reduce the number of inputs. This PCA then produces, say, three main axes, that are usually hard to interpret linguistically. Rules produced on the basis of these axes are hard to understand for the people from the plant. Therefore, one should carefully explain them the rules. If one is able to determine a small number of really important inputs, it is better to use these inputs for RBF-modelling and rule production. This results in rules with understandable premises, the conclusions remain crisp, usually very precise values.

7.2 Rule extraction from fuzzy clustering

Fuzzy clustering is an interesting method to produce rules, because one can derive a number of local linear centers that can be combined via Takagi-Sugeno rules [4]. This means that the Takagi-Sugeno system performs an interpolation between the cluster centers. As this interpolation is performed inside of the system, the rules remain understandable. Produced rules in one of our projects had the following form:

$$\text{If Input}_1 \text{ is } A_1 \text{ then Output} = 1.41955 \times \text{Input}_1 + 0.170627.$$
$$\text{If Input}_1 \text{ is } A_2 \text{ then Output} = 0.536239 \times \text{Input}_1 + 0.52224.$$

The two clusters A_1 and A_2 represent two important domains of Input_1.

8 Drift Analysis

Drift analysis is easy when one knows the natural state of the plant. In this case, one has to describe this state and calculate the difference between the current state and the predefined state. In many cases, however, one does not know the standard state of a plant or a system, in this case one has to follow carefully the movements in the data. If there are jumps in several variables that are abnormal with respect to time, one should give a signal. This usually implies that one has to do a re-identification. This re-identification is dangerous, because one has to know which data to chose for this process. If one uses data that is too local with respect to time, one might be learning a system that has nothing in common with the system as a whole. If a new drift occurs, the system cannot be controlled appropriately. A solution is to save some standard data and use them as extra data to re-identify the system. This is dangerous too, because systems have natural drifts and might never recur to such a 'standard' state. Careful analysis is needed in this case.

1. W. Ader and A. Nörling. Extraktion von Fuzzy-Regeln zur Analyse technischer Prozesse, MSc-thesis, Munich, 1995.

2. R. Babuška. Fuzzy Modeling and Identification, PhD-thesis, Delft, 1996.
3. J.C. Bezdek. Pattern Recognition with Fuzzy Objective Function Algorithms, New York, Plenum Press, 1981.
4. D. Driankov, H. Hellendoorn, and M. Reinfrank. An Introduction to Fuzzy Control, Heidelberg, Springer Verlag, 1993.
5. R. Duda and P. Hart. Pattern Classification and Scene Analysis, New York, Wiley, 1973.
6. H. Hellendoorn and R. Palm. Fuzzy Systems at Siemens R&D, Fuzzy Sets and Systems, 63(1994)245–269.
7. H. Hellendoorn and R. Baudrexl. Fuzzy-Neural Traffic Control and Forecasting, Proc. of the FUZZ-IEEE/IFES'95 Conference, Yokohama, 1995, pp. 2187–2194.
8. T. Kohonen. Self-organized Formation of Topologically Correct Feature Maps, Biological Cybernetics, 43(1982)59–69.
9. T. Martinetz and J. Hollatz. Neuro-Fuzzy in der Prozeßautomatisierung, Neuro-Fuzzy – Grundlagen und Anwendungen in der industriellen Automatisierung, (Ed. K.W. Bonfig), Renningen-Malmsheim, expert verlag, 1995, pp. 135–144.
10. R. Palm, D. Driankov, and H. Hellendoorn. Model Based Fuzzy Control, Heidelberg, Springer Verlag, 1997.
11. T. Runkler. Iterative *c*-Elliptotype Clustering for Efficient Data Analysis, Workshop Fuzzy-Neuro-Systeme (Ed. R. Isermann), Darmstadt, 1995.

IMPLEMENTING FUZZY CONTROL IN THE MANUFACTURING OF WASHING POWDERS

MAGNE SETNES
Control Engineering Laboratory, Delft University of Technology
P.O.Box 5031, 2600GA Delft, The Netherlands

This chapter describes a PC-hosted fuzzy controller for the control of a large scale industrial process concerning the production of washing powders. The controller consists of several fuzzy rule bases that model the control actions of experienced process operators in response to different quality deviations of the product. A fuzzy supervisor is used to activate the various rule bases when corrective measures are needed. In this way, a controller is obtained capable of controlling the process within quality bounds narrower than those applied by human operators. Development, implementation and test results are reported.

1 Introduction

Henkel-Ecolab, situated in Nieuwegein (NL), produce washing powders for the industrial market. Industrial users put high demand on the quality of the washing powder. They typically apply volume based automated dosage of powder in their processes, and a stable relation between volume, weight and washing effect is important.

Figure 1: Pulverization tower

Most washing powders are refined from a half-fabricated compound powder, and the quality demands on the final product translates to this compound. The compound is produced from a paste in a spray drying process that takes place in a pulverization tower under the influence of high temperature[1]. Later, thermally instable ingredients like perfume and enzymes are added to the compound resulting in the final product.

A fuzzy controller for the production of compound powders has been developed and implemented using SieFuzzy, a CAE tool for fuzzy systems design[2]. A PC connected to the plants process control system hosts the controller which acts as a support system for the human operators.

The fuzzy controller consist of a supervisor and several rule bases concerning possible control actions in event of a quality deviation of the product. The supervisor determines the occurrence of such a quality deviation, taking into account the current product in production, the process dynamics, and various constraints. It activates the processing of the relevant rule bases to infer the action to undertake in a given situation.

In the following, the development and implementation of the fuzzy controller is described. Section 2 shortly describes the production process and presents the control goals. Section 3 treats the development of the fuzzy controller followed by implementation issues in Section 4. Section 5 presents results from testing and validation of the system, and Section 6 summarizes the chapter.

2 Process description and control goals

A distributed control system (DCS) is used in the production. Human operators supervise the process from the control room, and give the necessary setpoints for the low-level PID controllers. Figure 2 gives an overview of the process and the variables to be controlled by the fuzzy controller.

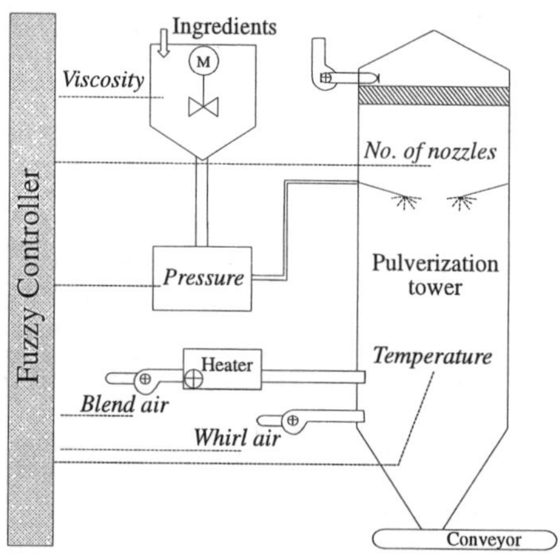

Figure 2: Overview of the process. Controlled variables are printed in *italics*.

2.1 Production phases

The operators recognize two separate phases in the production: batch preparation and pulverization. During *batch preparation* fluid and solid ingredients are mixed and heated to form a thick paste. The *viscosity* can be controlled, but the response time is long.

During *Pulverization* the paste is sprayed into the drying tower. The spray *pressure* and the *number of nozzles* are controlled. Falling down the tower, the paste is dried by the *blend air*, an up-going hot air flow. A second airflow, the *whirl air*, bounces the grains against the walls of the tower, resulting in a smaller grain size and increased weight. The amount of blend- and whirl air can be controlled, along with the temperature in the tower. Response to the control actions in this phase is fast, and they are regarded as "first choice" if corrective measures become necessary.

2.2 Process states

Three qualitative states can be distinguished in a production run. First there is the *start-up* state, in which the goal is to bring the process to a certain throughput with a stable quality as fast as possible. Then the process enters the production- or *steady-state* in which high quality powder is being produced. The operators tries to maintain a stable quality of the powder throughout this state. At the end of a production run comes the *shut-down*. The operators apply different control rules for all three states of the production. Correspondingly, the knowledge base of the fuzzy controller has three rule bases, one for each state, as shown in Fig. 3.

2.3 Control goals

Liter-weight (g/l) and *moistness* (%) are the two most important quality factors of the washing powder. The desired liter-weight describes a relation between the volume and the washing effect for the product. To ensure customer satisfaction, this relation should be constant, since mostly volume based dosage is used in applications. Also, variations in volume of packed bags causes variations in pallet heights and troubles transport and storage. During production, the operator aims for stable liter-weight and moistness following the recipe.

The maximum percentage of water that can be bounded in the powder is determined for each product by its recipe. The aim for the moist level during production is this maximum level, minus a safety margin. When the maximum level is exceeded, the product will clog, causing trouble in automatic dosing systems, or even during the production. If the moist level is too low, the powder will dust, leading to dosage problems and environmental pollution. By accurate control of the moist level, the safety margin can be minimized, without the risk of clogging. In this way, more water and less expensive ingredients are used, and the risk of pollution is decreased.

3 The fuzzy controller

The controller design consist of three main blocks: an input/fuzzification block, a fuzzy system, and an output/defuzzification block, illustrated in Fig. 3. The fuzzy system is further divided into a supervisory part and a knowledge base. The knowledge base consist of three separate rule bases in which operator knowledge corresponding to the three main production states is encoded in fuzzy if-then rules. Each of the four parts of the fuzzy system is a rule-based system of its own with internal hierarchies and inference processes.

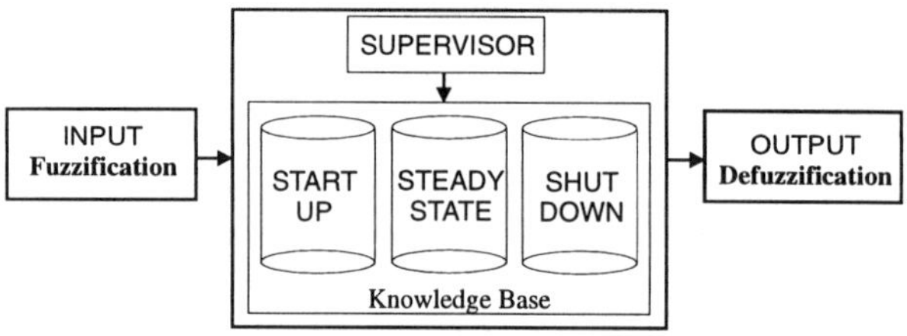

Figure 3: Structure of the controller design.

3.1 Hierarchies

The hierarchies enable fuzzy transitions between separate rule bases depending on the operating state (start-up, steady-state and shut-down) and between different sets of rules in each rule base depending on the product state (current product quality). The use of hierarchies makes it possible to decompose the reasoning and distribute it in different rule bases. This increase the transparency of the system and makes maintenance and extensions more easy.

The *"operating state – rule base"* form the main hierarchy. Depending on the operating state, determined by the supervisor, corresponding rule bases are activated. In each rule base, a secondary hierarchy is formed by *"product state – rule sets"*. Here the current product state is determined to activate suitable sets of rules. The hierarchy is illustrated in Fig. 4.

The reason for deciding upon the state of the product inside the rule bases, and not in the supervisor, is that the different operating states does not only have different control rules, but also different product state descriptions. For instance, in steady-state, the product state is described by its quality deviation partitioned into nine regions like *heavy, wet, dry,* etc. as shown in Fig. 4 and Fig. 6. During start-up or shut-down, however, the control goals are different. Thus, other premises are applicable for selecting control rules in these states.

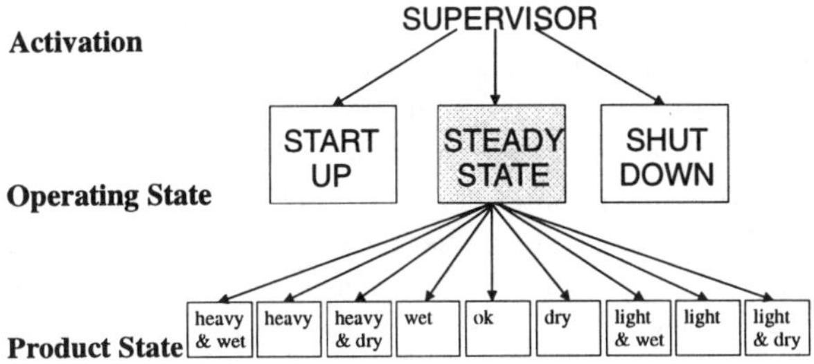

Figure 4: Decision hierarchy in the fuzzy controller.

3.2 Supervisor

The supervisor decides to which degree (membership) the production can be said to be in each of the three main operating states ***start-up, steady-state*** and ***shut-down***. Based on the current product quality, a decision is then made whether an advice is needed or not, and for which type of deviation a corrective control action should be inferred: deviation in weight and/or deviation in moist. Further, the supervisor estimates the degree to which a control action is constrained from being increased or decreased. This information is passed on to the rule bases for reasoning about corrective actions, while the process' memberships in the three main phases are passed on to the output part where it is used in weighted aggregation of the advices generated by each of the corresponding rule bases. The structure of the supervisor is illustrated in Fig. 5, and the main parts are described below.

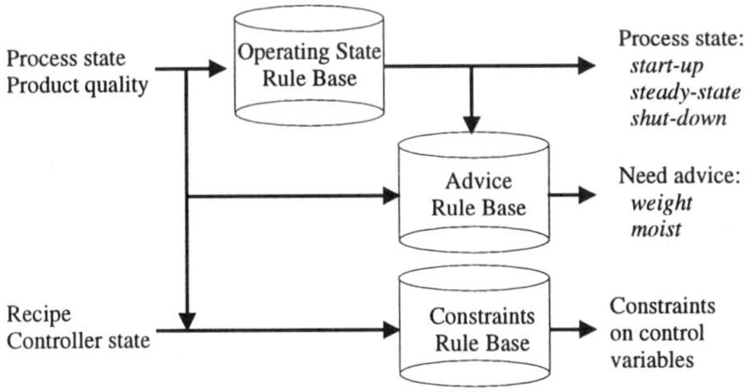

Figure 5: The supervisor decides on the generation of advices, activation of knowledge bases, and constraints on the controller actions.

The Operating State Rule Base concerns the operating state of the process. The outputs of this rule base are three fuzzy variables, *start-up, steady*, and *shut-down*, whose values correspond to the process' membership in the respective operating states. This is a linguistic rule base with qualitative rules of thumb about the operating states that have been acquired from the operators.

The Advice Rules Base has linguistic rules to determine when a control action is necessary. The inputs to this rule base are current product quality (error), the change in error, and qualitative information about the operating state of the process. The latter is passed from the *Operating State Rule Base* and is necessary since the strategy of the operators, when to apply a control action, differs with the state of the process. The dynamics of the process is implemented by means of fuzzy rules operating on the error states and the change in error. When the process enters an error region, the size of the error and the derivatives decide whether a control action should be taken or not, using rules like:

IF state is *steady* AND weight is *Low* AND ...
Δ-weight is *Not Pos.* THEN weight control,
IF state is *steady* AND weight is *Very Low* THEN weight control,

where the linguistic terms *Low, Not Positive*, etc., are defined by means of fuzzy sets. For instance, if the deviation is small, and the change of the deviation is in the direction of the desired value, it is not necessary to take any action. When needed, the supervisor activates the rule bases that in turn infers the control actions. To make sure that no new advice are given during the time a previous action needs to take effect, a look-up table specifies a "wait-time" for each control action. As with all the other rules, these "wait-times" have been obtained from experienced operators.

The Constraints Rule Base has as inputs the state of the process and the controller variables. The outputs are the degrees to which the control variables may be increased or decreased, that is, the degree to which a constraint is applicable. Also in this rule base, the rules are simple linguistic rules of thumb, elicited from the human operators. As an example, consider two rules constraining the variation of the pressure during steady-state production:

IF state is *steady* AND pressure is *Not High* AND ...
Batch-Prep.-Time is *Not Long* THEN pressure may *increase*,
IF state is *steady* AND pressure is *Not Low* ...
THEN pressure may *decrease*.

3.3 Steady-state rule base

The knowledge base contains one rule base for each state of the production (see Fig. 3). Each fuzzy rule base models the actions taken by the process operators in the corresponding state for different conditions of the compound powder. The degrees to which each model is applicable are determined in the supervisor. The inputs to the rule bases are the product quality by means of deviations in weight and moistness, and the current constraints on all control variables.

For the steady-state, both product quality input variables are partitioned by three fuzzy sets, representing *negative, about zero,* and *positive* deviations from desired values. By combination, this gives an input space with nine fuzzy error regions. The operators were interviewed about how they would control the process to compensate for these errors. It was found that for each error there are several actions possible. These possible actions were ordered in a priority list. Higher priority was given to alternatives that increase the production capacity and have short response times. Also, depending on the process-state (e.g. constraints like "valve cannot open more"), some actions (e.g. "open valve further") may not be taken. The next action on the priority list is then taken instead. Figure 6 shows the fuzzy partitioning of the rule base antecedent space, where $R_1, R_2, \ldots, R_8$ refers to the sets of control actions suitable to compensate for the quality deviations in each region.

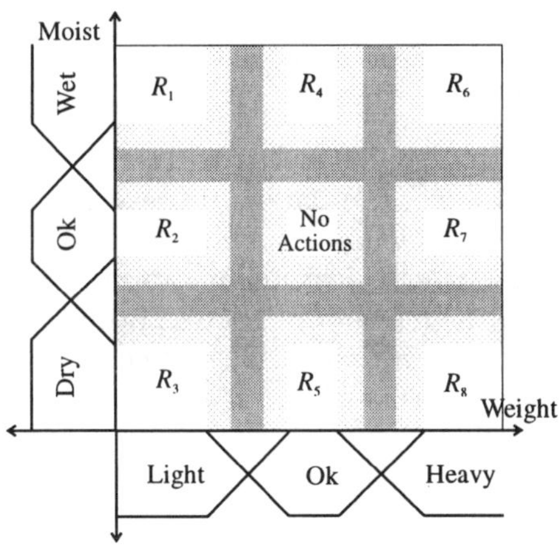

Figure 6: For each error region $i = 1, 2, \ldots, 8$, a set of control actions R_i is defined.

If a quality deviation occurs, the rules with the highest priority, and not prohibited by constraints, will fire with a degree to which the process is considered to be in the respective error states. This selection is implemented in SieFuzzy by considering each of the eight rule sets as separate rule bases and apply mean of maxima inference[4,3]. For the selected rules, max-product inference and center of gravity (COG) defuzzification[4,3] is used to arrive at suitable control actions. Figure 7 illustrates the inference for a case where the powder is too light and partly too dry, resulting in the firing of two rules.

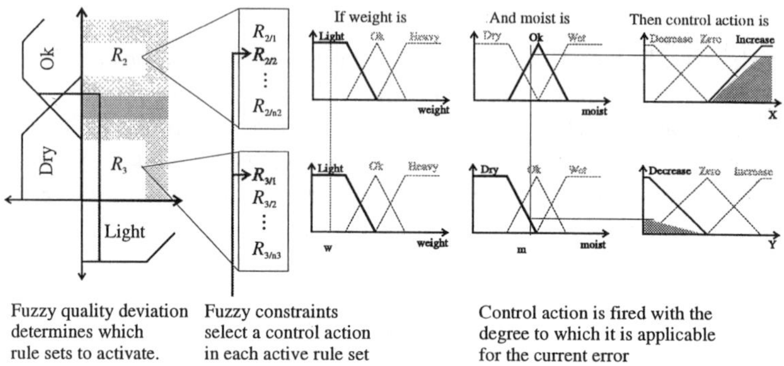

Figure 7: The present quality deviation activates two rule sets in a rule base. From each rule set, one control action is selected based on the present constraints. The selected actions are fired with the degree to which they are considered relevant for correcting the error.

4 Implementation

Process data is read from the DCS bus to the MS-Windows operated PC hosting the fuzzy controller. The data is processed in a spreadsheet, Excel, and stored for possible later analysis. A graphical user interface, see Fig. 9, is implemented in Excel as well. It communicates the control actions inferred by the fuzzy system to the human operator and contains menus for selecting recipes. SieFuzzy, implementing the fuzzy system, is linked with Excel using Dynamic Data Exchange (DDE), giving the data-flow shown in Fig. 8.

5 Tests and results

Knowledge-based systems modeling the behavior of human experts are not straightforward to evaluate. To test the fuzzy system, evaluation has been performed in four stages: In the *first stage* the fuzzy rule-based model was evaluated and tuned using historical data. In the *second stage* the controller was placed on-line. The control actions suggested by the fuzzy controller were judged by the human operators, and the rules were tuned accordingly before the *third stage* testing could take place. Now the actions advised by the fuzzy controller were always carried out. The operators classified the advices into four classes before applying them: Good, Probably Good, Probably Bad, and

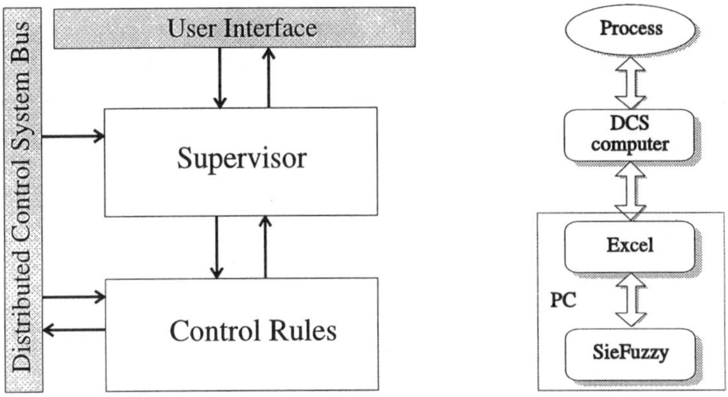

Figure 8: The fuzzy controller (left), and the data transport in the implementation (right).

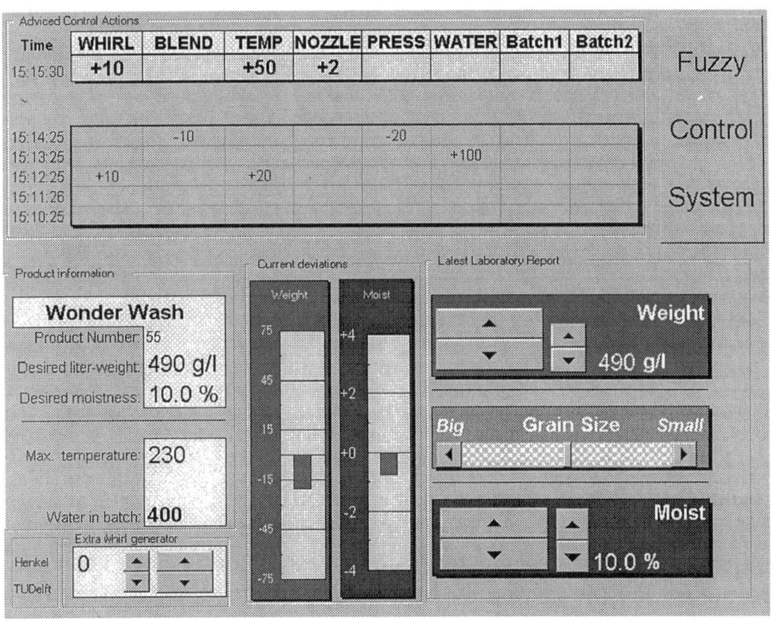

Figure 9: Graphical user interface in Excel.

Bad. Afterwards, the actual effect of the action was classified as: Good, Moderate or Bad. The results from the third stage testing is reported in Table 1. It is seen that the operators judged more than 70% of the advices given as "good". The suggested actions almost always had a good (63%) or moderate (29%) effect, also when they were not judged as "good" by the operators in the first place.

Actual effect	Action classified as: *Good*	*Prob. good*	*Prob. bad*	*Bad*
Good	42 %	13%	8%	-
Moderate	25 %	4 %	-	-
Bad	4 %	-	-	4 %

Table 1: Classification by operators of actions suggested by the fuzzy controller (before application) and their actual effect.

After additional tuning for cases where the controller performed less well, the controller was used as a decision support system during production. In this *fourth stage* testing, the operators sometimes followed the advice from the fuzzy controller fully or partially, and sometimes ignored the advice from the controller. A summary of the results are given in Table 2.

Actual effect	The action was proposed by: *Fuzzy controller*	*Partially by controller*	*Operator alone*
Good	73 %	71 %	67 %
Moderate	19 %	15 %	22 %
Bad	8 %	14 %	11 %

Table 2: Effect of actions proposed fully and partially by the fuzzy controller, and by the operator.

So far, only steady-state operations have been thoroughly tested. It is seen from Table 2 that there is an improvement in the control actions when the operators follows the advice from the fuzzy controller. Typically, the fuzzy controller was found to react faster to quality deviations, using bigger corrective control steps than the operators. In the tests, this provided a faster correction of the quality deviation. Currently, the fuzzy controller is being tested for various products during all production states. The performance is being logged to provide information for extension of the controller before a possible final implementation as a decision support system will be considered.

6 Concluding remarks

The design, implementation and testing of a fuzzy controller for use in the production of washing powders have been described. The controller consists of a supervisor and three fuzzy rule bases modeling the behavior of human operators in different process situations. The rule-based models perform static mapping from the fuzzy error states and constraints of the process into corrective control actions. The dynamics of the process is captured in the supervisor by means of fuzzy rules operating on the error states and the derivative of the error. By letting the supervisor trigger the processing of the rules in the rule-based models, nervous control behavior is avoided without implementing knowledge about systems dynamics into the rule-based models themselves, thereby keeping the rule bases compact and ensuring a high level of transparency.

The controller has been implemented using low cost software and hardware. A MS-Windows operated PC running Excel and SieFuzzy hosts the prototype version of the fuzzy controller. By using the fuzzy controller, Henkel-Ecolab can control the quality of their products within tighter bounds. This will help ensure customer satisfaction.

Acknowledgment

This work was supported in part by the Research Council of Norway and Henkel Ecolab B.V.

References

1. P. Krings, H. Harder, and R. Weber. *Waschen und Waschmittel.* Hüthig Verlag, Düsseldorf, Germany, 1976.
2. Siemens. *SieFuzzy User's Manual version 2.0.* Siemens AG, Industrial and Building Systems Group, Erlangen, Germany, 1995.
3. R.R. Yager and D.P. Filev. *Essentials of Fuzzy Modelling and Control.* John Wiley, New York, 1994.
4. R. Jager. *Fuzzy Logic in Control.* Ph.D. Thesis, Control Laboratory, Delft University of Technology, Delft, 1995.

SEMI-MECHANISTIC MODELING AND ITS APPLICATION TO BIOCHEMICAL PROCESSES

H. A. B. TE BRAAKE
Heineken Technical Services,
Burgemeester Smeetsweg 1,2382 PH,
Zoeterwoude, The Netherlands
E-mail: h.a.b._te_braake@heineken.nl

J. A. ROUBOS, R. BABUŠKA
Department of Information Technology and Systems,
Control Laboratory, Delft University of Technology,
P.O. BOX 5031, 2600GA, Delft, The Netherlands
E-mail: {j.a.roubos, r.babuska}@its.tudelft.nl

The objective of this chapter is to show that neural networks and fuzzy models can be incorporated in a semi-mechanistic modeling environment in a straightforward manner. The procedure is described by the development of a semi-mechanistic model for a real biochemical process, the enzymatic conversion of Penicillin-G. Two different type of black-box model parts: (i) a fuzzy model, and (ii) a neural network, are compared. Finally, the semi-mechanistic model is used to create enough data to make a neural network model for the full process, which fits better in real-time control strategies. Although the results do not carry over directly to other engineering fields, the main ideas, conclusions, and drawbacks will certainly hold for other application areas as well.

1 Introduction

In biochemical industries, most processes are carried out in a batch or fed-batch reactor. Think for example of the production of beer, penicillin or bakers yeast. These types of processes are characterized by their time-varying nature and their recipe-based processing. Furthermore, measuring all important state-variables is either impossible or very expensive. On-line control of these processes therefore is usually limited to the control of temperatures, pH, flows, and easily measurable states. Most of the processes are controlled by standard recipes, stating what to do at a certain time instant. Here, models are usually not constructed for control purposes but for process optimization which leads to optimized recipes or an optimal process lay-out (equipment sizes, etc.).

The modeling of biochemical processes serves in practice only three main purposes:

1. Training of operators,

2. Fault diagnosis,

3. Process optimization.

These three reasons for construction of process models restricts the type of model to be used. In general, black-box models are not easily applicable for process optimization due to the time-varying nature of the underlying process[a]. However, the construction of white-box models (mechanistic models) is also not easy.

White-box models are based on first principles. Besides their prediction properties, they also have the capability to explain the underlying mechanistic relationships of the process. These models are in general to a certain extent applicable independently of the process scale. Furthermore, these models are easy to modify or extend by changing parts of the model. White-box modeling is generally supported by experiments to get some of the parameters. In that case, even after scaling, new experiments are necessary because some of the parameters are difficult to express in a simple way as functions of process characteristics. In a practical situation, the construction of white-box models can be difficult because the underlaying mechanisms may be not completely clear, experimental results obtained in the laboratory do not carry over to practice, or parts of the white-box models are in fact not known.

Due to the given drawbacks, combinations of a-priori knowledge with black-box modeling techniques are gaining considerable interest. Two different approaches can be distinguished: grey-box modeling and semi-mechanistic modeling. In a grey-box model, a-priori knowledge or information enters the black-box model as e.g. constraints on the model parameters or variables, the smoothness of the system behavior, or the open-loop stability [1,2,3]. For example, Lindskog and Ljung [2] tried to find combinations or (nonlinear) transformations of the input signals corresponding to physical variables and used the resulting signals in a black-box model. A major drawback of this approach is that it mainly suffers from the same drawbacks as the black-box model, i.e. no extrapolation is possible and time-varying processes remain a problem.

One can also start by deriving a model based on first-principles and then include black-box elements as parts of the white-box model frame[4,5,6,7,8]. This modeling approach is usually denoted as *hybrid-modeling* or *semi-mechanistic modeling*. The latter term is used in the sequel because the first one is rather confusing with other methods. Johansen [8] describes various techniques to incorporate knowledge into black-box models. He uses a local model structure

[a] Note that process optimization by off-line response surface-like techniques can of course lead to optimized processes. However, this is a static optimization and not an optimization of the dynamic process response.

and formulates the global identification as a nonlinear optimization problem. Thompson and Kramer[6] use the so-called parallel approach of semi-mechanistic modeling. Here, the error between the data and the white-box model is represented by a neural network. They also describe the serial approach where a neural network is used to model unknown parameters.

Several authors applied hybrid models for the modeling biotechnological systems [5,7,9,10]. In [7,9], fed-batch experiments are used to find the parameters of both the mechanistic part and the neural part of their model. Van Can *et al.* [10,11] describes the application of semi-mechanistic modeling with neural networks to describe a pressure process.

The semi-mechanistic modeling strategy is be combined quite naturally with the general structure of white-box models in biochemical processes, since this structure is usually based on macroscopic balances (e.g. mass, energy or momentum). These balances specify the dynamics of the relevant state variables and contain different rate terms. Some of these terms are directly associated with manipulated or measured variables (e.g. in- and out-going flows) and do not have to be modeled any further. In contrast, some rate terms (e.g. reaction rate) have a static mathematical relation with one or more state variables which should be modeled in order to obtain a fully specified model. These terms are then considered as inaccurately known terms. They can be modeled in a white-box way if a static mathematical relation for the rate terms can be based on easy obtainable first principles. If this is not possible, they can be modeled in a black-box way with a nonlinear modeling technique. In the latter case, one obtains the semi-mechanistic model configuration. One of the advantages of the semi-mechanistic modeling strategy is that it seems to be more promising with respect to extrapolation properties[10].

In this chapter, the construction of the semi-mechanistic models is considered. By using black-box parts in the model, one obtains the means for a relatively fast model development since not all relations need to be known in a white-box setting. Several techniques can be used to model the black-box part. An overview is given by Sjöberg[12]. The use of neural networks or fuzzy models provides the model developer with a powerful multi-variable modeling tool, to increase the modeling accuracy and to speed-up the development. The advantage of neural networks is that they are able to model multi-variable relationships in a elegant and simple way. Fuzzy models are more complicated, but give a more transparent model, which means that the user can partly verify the model. Moreover, for these multi-variable settings, a neural network or fuzzy model does not suffer from the curse of dimensionality, which is a well-known drawback of, for example, multi-variable polynomial models.

This chapter is organized as follows: Sec. 2 introduces the semi-mechanistic modeling approach and two methods fuzzy logic and neural networks for modeling the black-box part. In Sec. 3, the semi-mechanistic modeling of a real biochemical process is presented. Further, results and modifications for real-time control purposes are described. Finally, in Sec. 4, conclusions and some comments are given.

2 Construction of Semi-Mechanistic Models

2.1 Semi-Mechanistic Approach

This section describes the semi-mechanistic modeling approach. The obtained model should be able to predict the process states in such a way that it can be used for process optimization, but at the same time, the model should be applicable for various process scales. Further, a transparent model is preferable for process understanding.

Commonly, white-box models of (bio)chemical processes are based on macroscopic balances, e.g. mass, momentum or energy balances. These balances are based on the conservation principle that leads to balances which generally can be written as[13]:

$$\left[\frac{\text{accumulation of } S}{\text{time period}}\right] = \left[\frac{\text{flow of } S}{\text{time period}}\right] - \left[\frac{\text{generation of } S}{\text{time period}}\right] - \left[\frac{\text{consumption of } S}{\text{time period}}\right], \quad (1)$$

where S is a certain quantity, e.g. mass or energy. In general not all of the terms in Eq. (1) are exactly or even partially known. Especially, the modeling of reaction rates (kinetics) and thermal effects is difficult. On the other hand, transport terms can be obtained more easily and accurately. Besides the difficulty of estimating reaction kinetics, in general, some of the parameters will also be time-varying.

Once a white-box modeling structure is known, and it is known which parameters are easy to obtain and which are more laborious to obtain, black-box models can be used to model these otherwise difficult obtainable parameters. The resulting model is a semi-mechanistic model which is defined as a model given by a white-box model structure (e.g. Eq. (1)) where the unknown parts are modeled by black-box models. The modeling procedure is represented in a schematic way by:

1. Obtain a white-box model structure of the process.

2. Estimate the easily obtainable model parameters (e.g. flows, masses, or volumes).

3. Rewrite the model parts that are difficult to obtain in units which are independent of the process geometry. If this is impossible then proceed with the next step, but the model is not scalable any more.

4. Perform batch experiments to obtain data for the modeling of the unknown relations with black-box models (e.g. neural networks, fuzzy models).

5. Use these black-box models as parts of the white-box model structure.

6. Test the semi-mechanistic model. Stop when the model results are satisfying or start again with point 1. 2. or 3. to improve the model

2.2 Fuzzy Model Structure

The use of fuzzy logic and fuzzy sets theory for modeling purposes is explained extensively in other chapters in this book. Therefore, only a brief introduction is given at this point.

In the application as described later on a linguistic fuzzy model [14] is used, which consists of rules in the following form:

$$R_i: \textbf{ If } x_1 \text{ is } A_{i,1} \textbf{ and } \ldots \textbf{ and } x_n \text{ is } A_{i,n} \textbf{ then } y \text{ is } B_i, \quad i = 1, 2, \ldots, K, \tag{2}$$

where R_i denotes the i-th rule, $\mathbf{x} = [x_1, \ldots, x_n]^T$ with $x_j \in X_j \subset \mathbb{R}$ the antecedent (input) variables, and $y_i \in Y \subset \mathbb{R}$ is the consequent (output) variable. $A_{i,1}, \ldots, A_{i,n}$ and B_i are instances of linguistic terms defined in the antecedent and consequent domains, respectively. K denotes the number of rules in the rule base.

The process of deriving an output of a fuzzy model from known values of the inputs is called *fuzzy inference*. In the relational model, the inference proceeds in the following three steps [14]:

1. *Compute the degrees of fulfillment.* Given an input vector $\mathbf{x} = [x_1, \ldots, x_n]$, the degree of fulfillment β_i of the i-th rule Eq. (2) is given by:

$$\beta_i = \mu_{A_{i,1}}(x_1) \ t \ \mu_{A_{i,2}}(x_2) \ t \ \ldots \ t \ \mu_{A_{i,n}}(x_n), \quad i = 1, 2, \ldots, K. \tag{3}$$

   The symbol t denotes a t-norm operator, such as the minimum or product. The fuzzy set (vector) containing the degrees of fulfillment of all the rules is denoted by $\boldsymbol{\beta} = [\beta_1, \ldots, \beta_K]$.

2. *Apply relational composition.* Having computed the degrees of fulfillment $\boldsymbol{\beta}$, which correspond to the given model input, the output fuzzy set

$\boldsymbol{\omega} = [\omega_1, \ldots, \omega_M]$ is derived by means of relational sup-t composition:

$$\boldsymbol{\omega} = \boldsymbol{\beta} \circ_{\mathrm{t}} R \,. \tag{4}$$

The composition operator $\circ_{\mathrm{t}}$ is defined by:

$$\omega_j = \max_{1 \le i \le K} [\beta_i \; t \; r_{i,j}], \quad j = 1, 2, \ldots, M \,. \tag{5}$$

3. *Defuzzify the consequent fuzzy set.* The consequent fuzzy set $\boldsymbol{\omega}$ is defuzzified by using the weighted mean method:

$$y = \frac{\sum_{l=1}^{M} \omega_l \, b_l}{\sum_{l=1}^{M} \omega_l}, \tag{6}$$

where b_l are the centroids of the consequent fuzzy sets B_l. These centroids can be computed, e.g., by the center-of-gravity method or the mean-of-maxima method. In this chapter, the mean-of-maxima mcthod is applied:

$$b_l = \mathrm{centr}(B_l) = \mathrm{mean}\{\mathrm{y} | \mu_{\mathrm{B_l}}(\mathrm{y}) = \max_{\mathrm{y} \in \mathrm{Y}} \mu_{\mathrm{B_l}}(\mathrm{y})\} \,. \tag{7}$$

The construction of the relational model from process measurements proceeds in two steps. First, the membership functions of the antecedent and consequent linguistic terms are derived. Then, the fuzzy relation describing the rule base is estimated. Fuzzy clustering is applied to extract the membership functions from measurements [14].

2.3 Neural Network Model Structure

There are several reasons to choose neural networks to model nonlinear systems. The most important reason is that feedforward neural networks are universal approximators. It is proven that any continuous nonlinear function can be approximated arbitrarily well over a compact interval by an one-hidden layer feedforward neural network. Another important reason to use neural networks is the fact that these can efficiently model multivariable systems. Compared to other universal approximators, for example, polynomial expansions, the *curse of dimensionality* is avoided such that relatively small models are sufficient.

The mathematical structure of neural networks is given by several elements [15]. The basic entity in a static feedforward neural network is a unit, called *neuron*. A neuron calculates a weighted sum of all inputs. To this sum a bias term is added and the result is mapped onto a nonlinear function, which is called the *activation function.* Neurons can be grouped into interconnected layers. A feedforward neural network consists of several uni-directional layers. All outputs of one layer are the inputs for the next layer. This type of network does not contain interconnections between neurons in the same layer. If a certain layer is neither an input layer nor an output layer, then it is called a *hidden layer.* Although several hidden layers can be used, here a three layered network with only one hidden layer is considered. The first layer is the input layer, which contains the source nodes. It directs the input to every neuron in the second layer that is called the hidden layer. This layer contains neurons with a nonlinear activation function. The third layer is the output layer. This one is built in the same way as the hidden layer and may also contain a bounded nonlinear activation function. In this chapter, a linear activation function is used to avoid an unnecessarily bounded output. Every input of a neuron in the hidden layer is multiplied by an activation weight w_{ij}^h. These multiplied inputs and a bias input are added. This is expressed by:

$$z_{j,k} = \sum_{i=1}^{N_i} w_{ij}^h \phi_{i,k} + b_j^h. \tag{8}$$

The output of a neuron of the hidden layer is then given by:

$$v_{j,k} = \sigma(z_{j,k}). \tag{9}$$

The output of the neural network is given by:

$$y_{l,k} = \sum_{j=1}^{N_h} w_{jl}^o v_{j,k} + b_l^o, \tag{10}$$

where j denotes the j-th neuron in the hidden layer, i denotes the i-th network input, l denotes the l-th network output and k denotes the k-th event. N_i the number of inputs and N_h is the number of neurons in the hidden layer. The input of the network is given by ϕ. The weights w_{ij}^h are called the activation weights and the weights w_{jl}^o the output weights. The activation bias is denoted as b_j^h and the output bias as b_l^o. The function σ is the activation function. This function can be any sigmoidal or gaussian function.

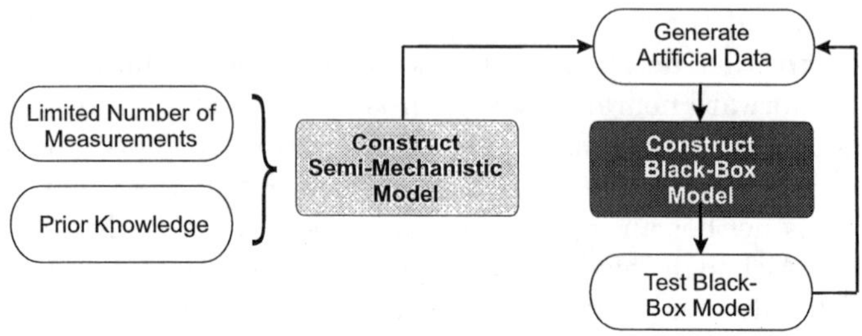

Figure 1: Construction of a black-box model based on data generated by a semi-mechanistic model.

3 Application: Modeling of an Enzymatic Conversion in a Fed-Batch Reactor

3.1 Introduction

This section describes the application of the semi-mechanistic modeling approach to the control of an enzymatic conversion process taking place in a fed-batch reactor. The enzymatic conversion process is based on the conversion of Penicillin G ($PenG$) which is controlled by the pH in the reactor. The enzymatic conversion of $PenG$ was chosen as a test case because:

1. It can very well be performed in a fed-batch mode, which is very common for many (bio)chemical processes.

2. It has complex non-linear kinetics, which is very common for (bio)chemical processes.

3. It allows reproducible experiments, so that the results from testing the predictive properties of a model are not obscured by experimental uncertainties. Unfortunately, this is not always the case for biochemical processes.

4. It has been investigated already by many researchers, so that a comparison with a white-box model is possible.

The enzymatic conversion of $PenG$ is operated in a fed-batch reactor. The model is ultimately used for control purposes. The reason for using the model for control purposes is that by using the model in a controller structure, the model capability of capturing the correct process dynamics can be tested. In a practical situation, the model would mainly be used for recipe optimization.

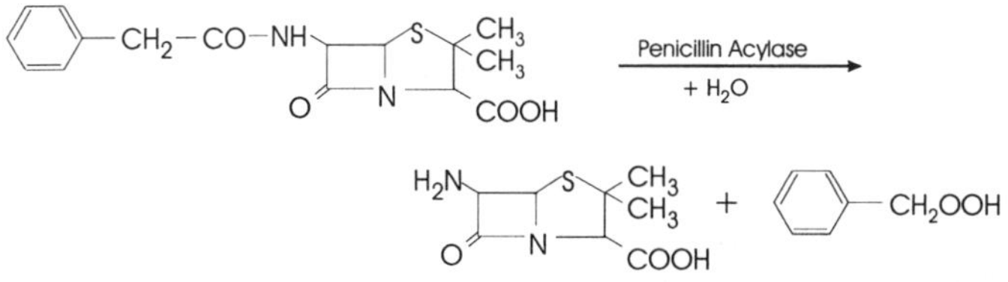

Figure 2: Penicillin-G conversion process.

Usually, a pure black-box model is the easiest way to obtain such an accurate model for control purposes (no extrapolation is necessary). However, it was not possible to measure data sets with enough information to construct a pure neural or fuzzy black-box model. Therefore, a two-step approach is applied: First a semi-mechanistic model is constructed. The resulting model can accurately predict the process outputs with given inputs. Since the obtained model was too complex for use in the controller, a neural or fuzzy black-box model is identified based on data which is created by the semi-mechanistic model. The black-box model is tested on a different test set which is also generated by the semi-mechanistic model. If the tested model is not working properly, new training and test sets can be constructed and the modeling procedure is restarted. This procedure is schematically depicted in Fig. 1.

In this way it is possible to obtain a black-box model for processes were it is very difficult to obtain the necessary dynamics in the input-output data when it is operated in real-time. Using a semi-mechanistic model relaxes the practical constraints, while the obtained black-box model is then used in the controller structure. For optimization purposes the semi-mechanistic model should be used instead of the black-box model, because of its extrapolation properties.

3.2 Description of the Process and Experimental Setup

The modeling strategy is demonstrated using experimental data for the modeling of the enzymatic conversion of Penicillin G ($PenG$) to 6-Amino-penicillanic Acid (APA) and Phenyl Acetic Acid ($PhAH$) by the enzyme Penicillin Acylase (E)(Fig. 2):

$$PenG + H_2O \xrightarrow{E} APA + PhAH. \tag{11}$$

Per component, for example $PhAH$, there is an acid-base equilibrium like:

$$PhAH \rightleftharpoons PhA^- + H^+. \tag{12}$$

These equilibria depend on the pH inside the reactor and can be calculated with the help of titration curves. The pH is defined as:

$$pH = -\log([H^+]), \tag{13}$$

where the square brackets denote a concentration (mol/l). The pH is influenced by the added base (OH^-) because there also exists a well defined equilibrium:

$$H^+ + OH^- \rightleftharpoons H_2O. \tag{14}$$

It is assumed that all these acid-base equilibria are settled much faster than the time constant of the conversion process itself, which means that the dynamics involved with these equilibria can be neglected. Thus there exists a static mapping between the pH and the concentration of the components $PenG$, $PhAH$, APA and buffer:

$$pH = f([PenG], [PhAH], [APA], [Buffer]). \tag{15}$$

Fourteen batch experiments based on different initial substrate and product concentrations were performed to obtain identification data [11]. These experiments are performed at a temperature of 310 K, in the pH range of $5.5 - 8.5$ and at initial $PenG$ concentrations between $10 - 100$ mM. H^+ is released during the conversion due to the acid-base equilibria of the three compounds,. Consequently, the amount of added base (OH^-) that is needed to keep the pH at a set-point during the conversion is a very valuable on-line signal, which can be used to calculate time series of concentrations and conversion rates. Further, 6 fed-batch experiments were done to obtain validation-data which could also be used for extrapolation studies.

The overall experimental setup is presented in Fig. 3. All experiments were performed in a thermostated reactor with a maximum volume of 1500 cm^3, equipped with a stirrer. Solutions with the required concentration $PenG$ were prepared by dissolving known masses in a 50 mM phosphate buffer. To obtain the concentrations of $PenG$, $PhAH$ and APA, the pH was adjusted to $pH = 8$. The reaction was started by adding an accurately known volume of enzyme solution to the reactor. There was no possibility to automatically add acid to the reactor, which puts some constraints on the input signal.

3.3 White-Box Modeling Part of the Semi-Mechanistic Model

The purpose of the models which are to be developed is to predict the pH-profile of a complete conversion of penicillin given only the initial state of the system and the imposed added base (control input) during the experiment.

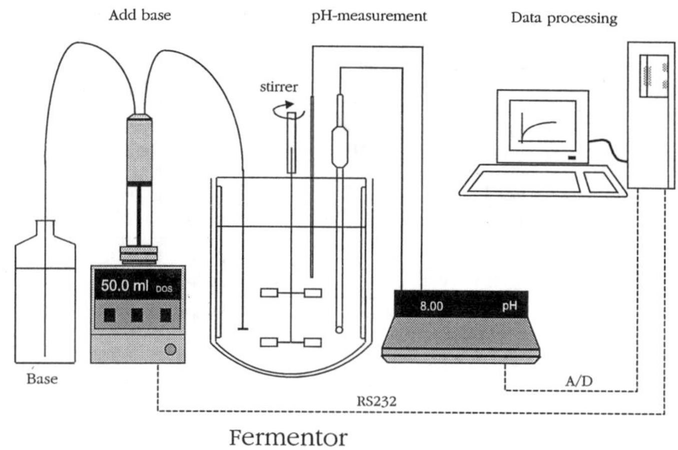

Figure 3: Experimental set-up to perform the enzymatic penicillin conversion process.

The pH can be calculated from the charge balance that can be expressed as:

$$[Charge_{in}]_{k+1} + [Charge_{t0}]_{k+1} + f_{t1}(pH) \cdot [Buf]_{k+1} + f_{t2}(pH) \cdot [PenG]_{k+1} + f_{t3}(pH) \cdot [APA]_{k+1} + f_{t4}(pH) \cdot [PhAH]_{k+1} = 0. \quad (16)$$

in which $[charge_{in}]$ is the concentration of the added ions (charged molecules) due to the addition of base. The initial ions without the three reactants ($PenG$, APA, $PhAH$) and buffer are taken together in the variable $[charge_{t0}]$. The f_{ti} functions describe the pH dependency of the charge of component i (i.e. the equilibria of Eq. (12)). The f_{ti}-functions are assumed to be dependent on the pH only and are modeled in a black-box way. For each component (Buf, $PenG$, APA and $PhAH$), this dependence is measured by recording the titration curves (pH versus added base) of each component. At pH values that are not directly measured, the f_{ti} values are calculated by linear interpolation between f_{ti} values at the nearest higher and lower pH value (i.e. a lookup-table technique is used). The charge balance, Eq. (16), can be solved iteratively to obtain the pH. A detailed description of this rather complex formula can be found in te Braake *et al.* [16].

The $Charge_{in}$-concentration is directly related to the total amount of added base (B in ml):

$$[Charge_{in}]_{k+1} = -\frac{M_B}{V_{k+1}} B_{k+1}, \quad (17)$$

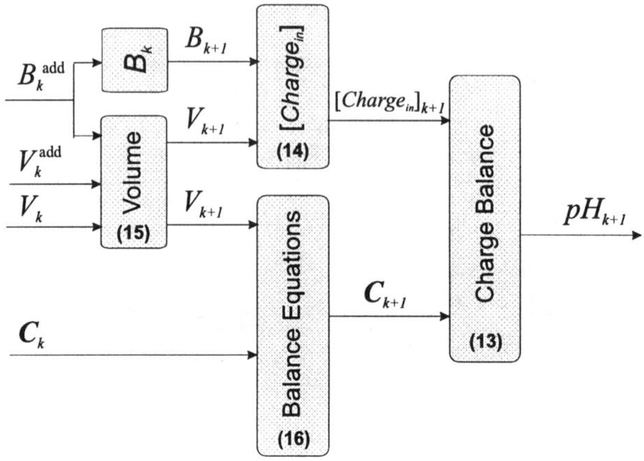

Figure 4: Model of the penicillin conversion process.

in which M_B is the concentration of the added base (mol/l). During the conversion process, the volume changes due to the addition of base or because of added $PenG$ or APA and $PhAH$ ($= V_k^{add}$). The volume effect is expressed in the following equation:

$$V_{k+1} = V_k + V_k^{add} + B_k^{add}. \tag{18}$$

The dynamics of the state variables which are involved in the charge balance can be directly written down in the following discrete balance equations:

$$\begin{aligned} C_{k+1}^i &= \frac{C_k^i V_k - \alpha_i r_k [E]_k V_k \Delta t + V_k^{i,add} C_k^{i,add}}{V_{k+1}}, \\ C &= ([PenG], [APA], [PhAH], [Buf], [Charge_{t0}], [E])^T, \\ \alpha_i &= (-1, 1, 1, 0, 0, 0)^T, \ i \in [1, 6], \end{aligned} \tag{19}$$

where Δt is the sampling time. Now the model contains five unknown terms: the penicillin conversion rate (r in mol/h), which is referred to as the kinetic part of the model, and the four f_t terms (Eq. (16)) which are associated with acid base equilibria of the involved compounds. This part will be referred to as the titration part of the model. In the semi-mechanistic model these five terms will be parameterized by black-box model structures.

Summarizing, to calculate pH_{k+1} based on B_{k+1}^{add}, one first has to calculate $[charge_{in}]$ from Eq. (17). Then one proceeds with the calculation of Eq. (18)

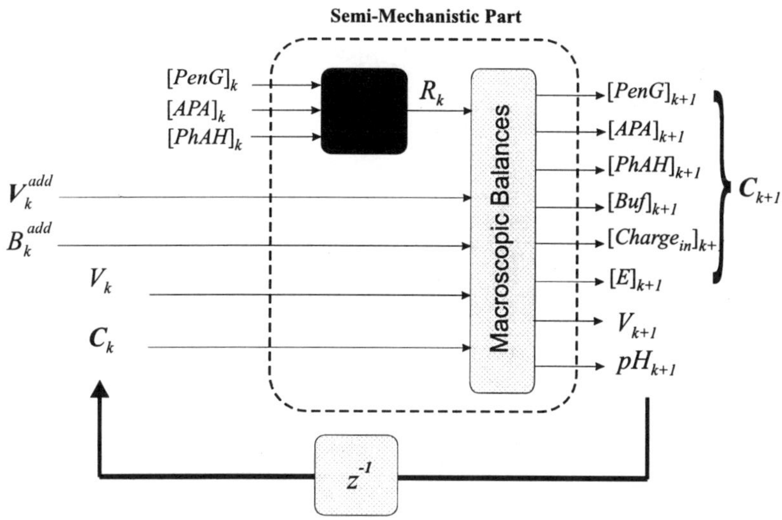

Figure 5: Semi-mechanistic model of the conversion process.

and Eq. (19). The last step is to solve the charge balance given by Eq. (16). By using an iterative search routine, this equation can be solved for pH_{k+1}. Note that the f_{ti}-functions depend on pH_{k+1} only. The complete model is depicted in Fig. 4.

3.4 Black-box Modeling Part in the Semi-Mechanistic Model

For (bio)chemical processes, the proper model structure for the conversion rate term (r) is generally not directly known without separate experiments to reveal the relevant kinetic mechanisms. In the semi-mechanistic model, the conversion rate will be modeled as:

$$r_k = f_{bb}([PenG], [APA], [PhAH], pH)_k, \tag{20}$$

where f_{bb} is a black-box model. In all identification and validation experiments the concentrations of APA and $PhAH$ were equal, thus r can be simplified into:

$$r_k = f_{bb}([PenG], [APA], pH)_k, \tag{21}$$

where f_{bb} is a black-box with three inputs. Fig. 5 gives an impression of the developed semi-mechanistic model.

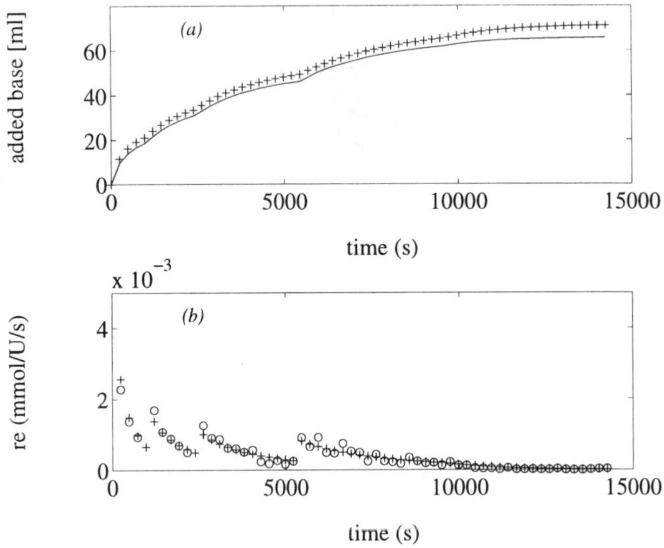

Figure 6: Performance of the semi-mechanistic model with the NN-part on an extrapolation experiment where *PenG* is added to the reactor at various instances. The overall model output is given in the upper plot and the NN-part that estimates the conversion rate r, is given in the lower plot (model output (line), real data (+), NN-output (o))

3.5 Results of the Semi-Mechanistic Model with a NN Part

The number of hidden nodes and related parameters of the neural network are determined in the identification procedure. The identification for the neural network was straightforward. The complexity of the neural network model was increased stepwise by adding an additional neuron to the hidden layer. The parameters of neural networks, varying in number of hidden neurons from 1 to 25, were calculated and the resulting neural network models were tested on the test set (cross validation). The sum of squared errors in this cross validation test no longer decreased when more than 5 neurons in the hidden layer were used. So, a neural network with 5 neurons in the hidden layer was chosen as black-box component to describe the conversion kinetics in the semi-mechanistic model.

The semi-mechanistic model is validated by testing its ability to predict the pH-trajectory of a fed-batch conversion given only the initial state of the system and the base addition in course of time. One result of the semi-mechanistic model is shown in Fig. 6 when the model is used for the prediction of added base. Compared with the measured values, the differences are small. Moreover,from this and other not shown results, it could be concluded that the semi-mechanistic model is also able to extrapolate.

The performance of the semi-mechanistic model is also compared with the performance of a white-box model in which the kinetics are modeled by a Mechaelis-Menten mechanism [17]:

$$r = \frac{r_{\max}[\mathrm{PenG}]}{\mathrm{K_M} + [\mathrm{PenG}] + \frac{\mathrm{K_M}}{\mathrm{K_A}}[\mathrm{APA}] + \frac{\mathrm{K_M}}{\mathrm{K_P}}[\mathrm{PhAH}] + \frac{\mathrm{K_M}}{\mathrm{K_P K_A}}[\mathrm{PhAH}][\mathrm{APA}] + \frac{1}{\mathrm{K_A}}[\mathrm{PenG}][\mathrm{APA}]}, \tag{22}$$

where:

r	:	conversion rate (mol/U/s),
$r_{\max}$	:	maximum conversion rate (mol/U/s),
K_M	:	Michaelis–Menten coefficient (mol/l),
K_{A}	:	inhibition coefficient for APA (mol/l),
K_{P}	:	inhibition coefficient for PhAH (mol/l).

It should be noted that even if the structure of Eq. (22) is known, the estimation of the above parameters from data is a nontrivial optimization problem, which requires good initial values for convergence[17].

With respect to interpolation properties, the semi-mechanistic model was more accurate than the white-box model based on the Michaelis–Menten kinetics and with respect to extrapolation properties a more or less comparable performance was obtained. Both models were identified with the same data, but for the white-box model significantly more knowledge, such as complicated kinetics and equilibrium thermodynamics, is necessary to construct the model. [17].

3.6 Results of the Semi-Mechanistic Model with a Fuzzy Model Part

In this section, the structure of the fuzzy model and the identification technique are explained. It is expected that the specific conversion rate r depends on the concentrations of $PenG$, APA and $PhAH$ in a nonlinear way, but the exact form of this dependence is considered unknown:

$$r = f_{bb}([PenG], [APA], [PhAH]) . \tag{23}$$

A *fuzzy linguistic model* [14] is used to approximate the unknown function f_{bb}. The model consists of if–then rules, such as:

If [PenG] is LOW **and** [APA] is LOW **and** [PhAH] is LOW **then** r is FAST,

If [PenG] is HIGH **and** [APA] is LOW **and** [PhAH] is LOW **then** r is MODERATE,

etc.

The rule base of the model contains rules for all possible combinations of the antecedent linguistic terms (LOW, MEDIUM, HIGH, etc.). The membership functions have been extracted from the process measurements by *fuzzy clustering*, and the consequent linguistic terms have been identified by using *fuzzy relational techniques* [18,19]. Next, the structure of the model and the identification method are described in more detail. The data sequences

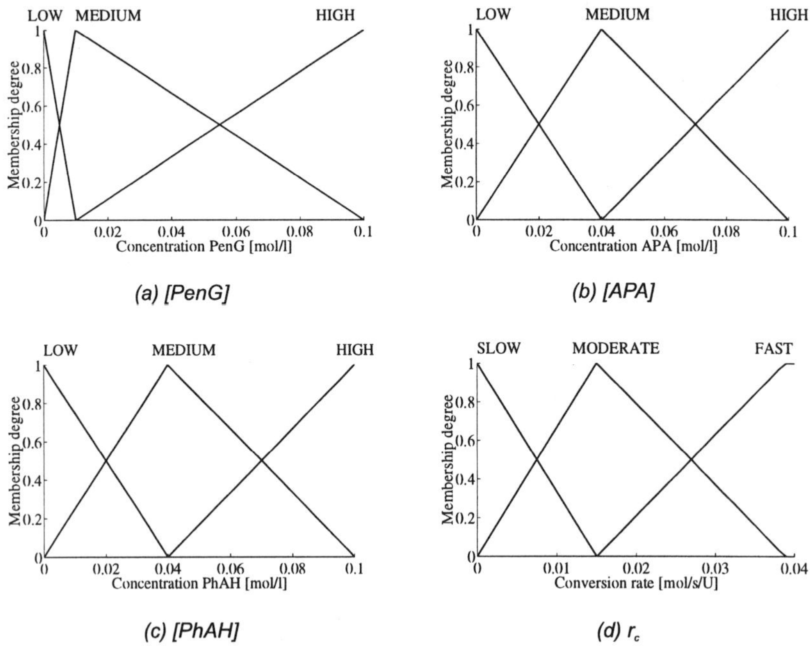

(a) [PenG] (b) [APA] (c) [PhAH] (d) r_c

Figure 7: Membership function for the fuzzy model.

($PenG$,APA,$PhAH$,r)are clustered by the GK algorithm [14]. The number of clusters is set to three, since the number and information content of the data do not allow for more clusters. As shown later, the model derived from the three clusters is sufficiently accurate. If this was not the case, more measurements would be needed, based on experiments at different initial conditions.

The total number of fuzzy sets was reduced, since some of the clusters project into similar membership functions [20]. As a result, three triangular membership functions per variable are obtained (Fig. 7). The rule base contains rules for all possible combinations of the antecedent linguistic terms, and this is given in Table 1. The consequent of the i-th rule is the linguistic term

Table 1: Rule base of the fuzzy model.

#	CF	[PenG]	[6-APA]	[PhAH]	conversion rate
1.	(0.48)	LOW	LOW	LOW	FAST
2.	**(0.55)**	LOW	LOW	MEDIUM	SLOW/MOD
3.	(0.11)	LOW	LOW	HIGH	SLOW
4.	**(0.54)**	LOW	MEDIUM	LOW	MODERATE
5.	**(0.74)**	LOW	MEDIUM	MEDIUM	SLOW
6.	(0.25)	LOW	MEDIUM	HIGH	SLOW
7.	(0.10)	LOW	HIGH	LOW	SLOW
8.	**(0.76)**	LOW	HIGH	MEDIUM	SLOW
9.	**(0.71)**	LOW	HIGH	HIGH	SLOW
10.	**(0.93)**	MEDIUM	LOW	LOW	FAST
11.	**(0.63)**	MEDIUM	LOW	MEDIUM	FAST
12.	(0.04)	MEDIUM	LOW	HIGH	MOD/FAST
13.	**(0.75)**	MEDIUM	MEDIUM	LOW	FAST
14.	**(0.82)**	MEDIUM	MEDIUM	MEDIUM	MODERATE
15.	(0.22)	MEDIUM	MEDIUM	HIGH	MODERATE
16.	(0.19)	MEDIUM	HIGH	LOW	MOD/FAST
17.	**(0.72)**	MEDIUM	HIGH	MEDIUM	MODERATE
18.	(0.47)	MEDIUM	HIGH	HIGH	MODERATE
19.	**(0.90)**	HIGH	LOW	LOW	FAST
20.	(0.19)	HIGH	LOW	MEDIUM	FAST
21.	(0.00)	HIGH	LOW	HIGH	—
22.	(0.37)	HIGH	MEDIUM	LOW	FAST
23.	**(0.50)**	HIGH	MEDIUM	MEDIUM	FAST
24.	(0.06)	HIGH	MEDIUM	HIGH	MOD/FAST
25.	(0.08)	HIGH	HIGH	LOW	MOD/FAST
26.	(0.06)	HIGH	HIGH	MEDIUM	MOD/FAST
27.	(0.04)	HIGH	HIGH	HIGH	MOD/FAST

corresponding to the largest element $R_{i,j}$ in the i-th row of the fuzzy relation. If two terms have equal $R_{i,j}$, both are given. The numbers in brackets preceding the rules are certainty factors (CF), obtained as the difference between the maximum and minimum $R_{i,j}$ in each row of the fuzzy relational matrix. The certainty factors indicate to what degree each particular rule is capable of describing the relationship between the antecedent and the consequent variables. Rules with low certainty factors are less reliable than rules with high values. A low CF indicates insufficient or contradictory data in the particular operating region defined by the respective membership functions. Certainty factors of 0.5 and above are printed in bold. Certainty factors close to zero indicate that no, or very little data were available to establish the particular rules. The rules (without the certainty factors) and the membership functions were presented to an expert, who confirmed the overall correctness of most of the rules and also the relevance of the membership functions, except for rule 2. As was expected, the major disagreement concerned the rules with low certainty factors. This is *PenG* concentrations. Additional experiments could be designed in order to obtain more data in this region. The expert was not able to assess the numerical values of the membership function parameters directly, but he confirmed that the distribution of the membership functions over the domains with respect to the relative influence of APA and $PhAH$ on the conversion kinetics is realistic.

The fuzzy model obtained here is used as a predictor of the conversion rate, and is incorporated into the macroscopic balance Fig. 5. The upper plot in Fig. 8 compares the conversion rate estimated by the fuzzy model, to the rate computed from the experimental data for two different validation data sets, that had not been used for the identification. The lower plot in Fig. 8 compares the base added, as simulated by the model and measured for the same two experiments. It should be stressed that the values in Fig. 8 are calculated recursively, with only the initial concentrations given. One can see that the model can predict the entire batch very accurately.

The numerical accuracy of the semi-mechanistic fuzzy model is comparable with that obtained with a similar hybrid approach using a neural network (Sec. 3.5). A drawback of the latter approach is that the neural network remains a black-box model, and does not provide any additional information about the process. The validity of the neural network must be assessed on the basis of numerical simulations only. Neural networks, however, are quite easy to train, even for non-experts, provided a sufficiently rich data set is available. The construction of the relational fuzzy model, on the other hand, is generally more complicated, and requires good knowledge of the identification methods.

3.7 Black-box modeling by use of the semi-mechanistic model

A black-box neural model of the process is attractive for control purposes, because this model can be used straightforward in controllers and usually is very fast to simulate. However, as already mentioned, the Penicillin-G conversion process can not easily be handled by black-box models due to the limited number of real data. Therefore, the slow semi-mechanistic model is used to generate data to be used as the training and test-set for a neural black-box model. The black-box model must predict the pH (model output) based on the base addition (model input).

As could be expected, the search for a valuable black-box model was not straightforward. Several configurations have been tested and most of them were not able to model the process accurately enough. This was mainly due to the generated training sets. Identification of a fixed neural model configuration

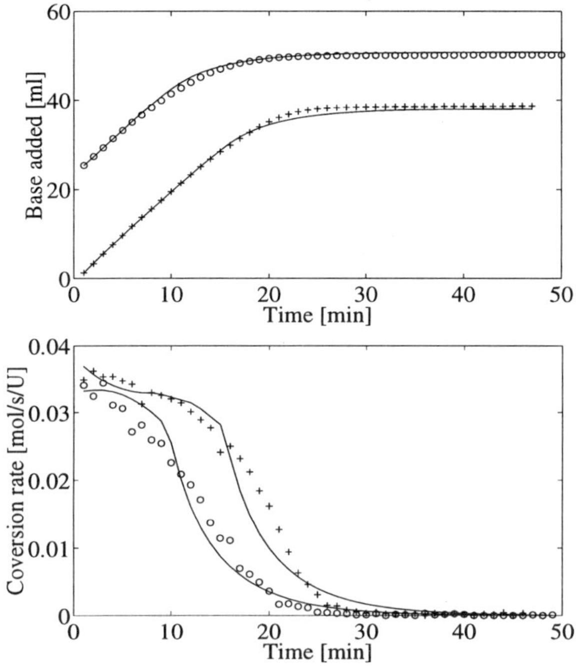

Figure 8: Performance of semi-mechanistic model with the fuzzy submodel. The overall model output is given in the upper plot and the fuzzy model part that estimates the conversion rate r, is given in the lower plot (model output (line), real data (+,o)).

for changing training sets results in a completely different performance. After testing several configurations and training sets, the best performing configuration is:

$$y_{k+1} = f_{NN}(u_k, uk - 1, [PenG]_0, k, y_k, y_{k-1}, y_{k-2}), \tag{24}$$

with $y_k = 10^6 \cdot 10^{-(pH)_k}$ and $u_k = B_k^{add}$. Note that the network output is not chosen to be the pH. By choosing this ad-hoc network output, which relates to the absolute amount of acid, better modeling properties of the neural model are obtained. The discrete-time k, is added because of the time-varying nature of the process. The network contains six hidden neurons. The training set is constructed by performing twenty simulation experiments with the slow semi-mechanistic model. The training results on a one step ahead prediction are very good, however, free run results are not always satisfactory. Fig. 9 shows the free run result on a simulated experiment. The neural model, given by the dashed line, is not very accurate, but it is stable. Moreover, in te

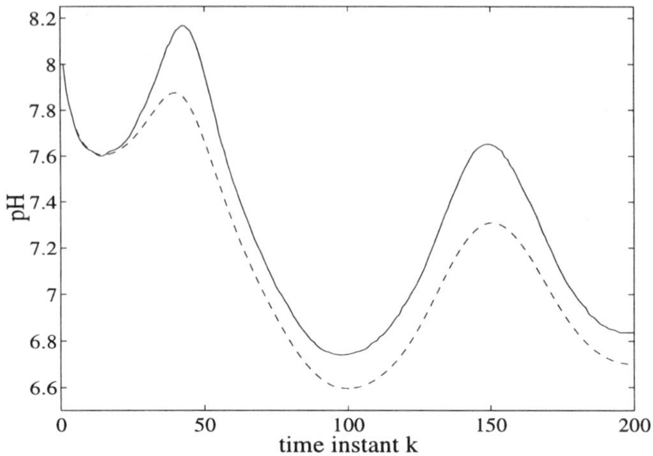

Figure 9: Free run results of the black-box neural model (process output (—), Neural model (- - -).

Braake *et al.* [16] it is shown that the model performs well enough to be used in both a standard nonlinear predictive controller based on a nonlinear optimization algorithm, as well as in a linear MBPC controller based on input/output feedback linearization.

4 Conclusions

A semi-mechanistic modeling strategy is outlined based on white-box modeling combined with black-box parts. This method can be used to model systems where a priori knowledge is available and experimental data is insufficient to make a black-box model. The semi-mechanistic modeling is applied to a biochemical application in which Penicillin–G is converted. First, an easy to obtain neural network is used for the prediction of the kinetics in the Penicillin–G conversion in the semi-mechanistic model. Second, a linguistic fuzzy model for the kinetics in the Penicillin–G conversion was developed from the experimental data. A posteriori analysis of the model indicates that most of the obtained rules are in good agreement with the expert's knowledge. Both models provide good numerical predictions of the conversion kinetics. When combined with the white-box model of the process (macroscopic balances), accurate prediction of the overall process behavior is obtained as well. Finally the semi-mechanistic model is used to create enough data to identify a full black-box model for real-time control purposes. The model prediction is worse than the semi-mechanistic one, but it is stable and performs well enough to be used in a controller.

The use of dynamic black-box models for process optimization is usually not appropriate. The focus of modeling (bio)chemical batch or fed-batch processes should be the development of mechanistic models. This type of model, however, needs long development times and is therefore expensive. The use of semi-mechanistic models give the practitioner the possibility to increase the efficiency of the modeling process easily and to reduce the development costs.

References

1. H.J.A.F. Tulleken. Grey-box modeling and identification using physical knowledge and bayesian techniques. *Automatica*, 29:285–308, 1993.
2. P. Lindskog and L. Ljung. Tools for semi-physical modeling. In *Proceedings IFAC SYSID*, volume 3, pages 237–242, Kopenhagen, Danmark, 1994.
3. T. Bohlin and S.F. Graebe. Issues in nonlinear stochastic grey-box identification. In *Proceedings IFAC SYSID*, volume 3, pages 213–218, 1994.
4. H.A.B. te Braake. *Neural Control of Biotechnological Processes.* PhD thesis, Delft University of Technology, Department of Electrical Engineering, Delft, The Netherlands, 1997.
5. R. Simutis and A. Lübbert. Exploratory analysis of bioprocesses using artificial neural network-based methods. *Biotechnology Progress*, 13:479–487, 1997.
6. M.L. Thompson and M.A. Kramer. Modelling chemical processes using prior knowledge and neural networks. *AIChE Journal*, 40:1328–1340, 1994.
7. D.C. Psichogios and L.H. Ungar. A hybrid neural network - first principles approach to process modeling. *AIChE Journal*, 38(10):1499–1511, 1992.
8. T.A. Johansen. *Operating regime based process modeling and identification.* PhD thesis, Department of Engineering Cybernetics, Norwegian Institute of Technolgy, University of Trondheim, Norway, 1994.
9. A. Tholudur and W.F. Ramirez. Optimization of fed-batch bioreactors using neural network parameter function models. *Biotechnology Progress*, 12:302–309, 1996.
10. H.J.L. van Can, H.A.B. te Braake, C. Hellinga, K.Ch.A.M. Luyben, and J.J. Heijnen. Strategy for dynamic process modeling based on neural networks and macroscopic balances. *AIChE Journal*, 42:3403–3418,

1996.

11. H.J.L. van Can. *Efficient Mathematical Modeling for Bioprocesses - based on Macroscopic Balances and Neural Networks.* PhD thesis, Delft University of Technology, Delft, The Netherlands, 1997.
12. J. Sjöberg, Q. Zhang, L. Ljung, A. Benveniste, B. Delyon, P.-Y. Glorennec, H. Hjalmarsson, and A. Juditsky. Nonlinear black-box modeling in system identification: a unified overview. *Automatica*, 31:1691–1724, 1995.
13. G. Stephanopoulos. *Chemical Process Control: An introduction to theory and practice.* Prentice Hall International, 1994.
14. R. Babuška. *Fuzzy Modeling for Control.* Kluwer Academic Publishers, Boston, 1998.
15. S. Haykin. *Neural Networks; A Comprehensive Foundation.* Macmillan College Publishing Comp., New York, USA, 1994.
16. H.A.B. te Braake, E.J.L. van Can, R. Babuška, and H.B. Verbruggen. Predictive control of the pH in a penicillin conversion process. In *AIRTC*, pages 475–480, Kuala Lumpur, Malaysia, 1997. IFAC.
17. H.J.L. van Can, H.A.B. te Braake, S. Dubbelman, C. Hellinga, K.Ch.A.M. Luyben, and J.J. Heijnen. Understanding and applying the extrapolation properties of models based on neural networks in macroscopic balances. *AIChE Journal*, 44:1071–1089, 1998.
18. W. Pedrycs. An identification algorithm in fuzzy relational systems. *Fuzzy Sets and Systems*, 13:153–167, 1984.
19. W. Pedrycs. *Fuzzy Control and Fuzzy Systems.* Wiley and Sons, New York, 2nd, extended edition, 1993.
20. M. Setnes, R. Babuška, U. Kaymak, and H. R. van Nauta Lemke. Similarity measures in fuzzy rule base simplification. *IEEE Transactions on Systems, Man and Cybernetics - Part B: Cybernetics*, 28(3):376–386, 1998.

FUZZY CONTROL OF A FILLING SYSTEM IN PACKAGING MACHINERY

C. FANTUZZI
Università di Ferrara, Dipartimento di Ingegneria
V. Saragat, 1, 44100 Ferrara - Italy.
E-mail: cfantuzzi@ing.unife.it

F. BASISSI
Tetra Brik Research and Development
v. Delfini 1, P.O. Box n. 636 C.P.O. I-41100 Modena - Italy
E-mail: fabio.bassissi@tetrapak.com

This chapter describes an experimental application of fuzzy logic control to the filling system of a packaging machinery. The filling system is a computer controlled device that sets accurately the quantity of product in each package produced by the machinery. The filling process behavior may change rapidly because of variations in product properties (e.g. viscosity) and disturbances (mainly inlet product flow), making this process non-linear and difficult to be controlled. The chapter details also some experimental tests performed at Tetra Brick laboratory in Modena. Results from those tests show good performance of the fuzzy controlled filling systems.

1 Introduction

Fuzzy control theory has some specific features with respect to other methodologies applied in system control. In particular the fuzzy controller is (in general) a static non-linear function which parameters can be design either using human heuristics or a mathematical model of the controlled system. A quick introduction to fuzzy control theory may be found in [1] and [2], while a deeper introspection may be found in some textbooks [3,4,5]

In the context of industrial control these aspects, non linearity and intuitive feeling with the design procedure, are very important mainly because of two reasons:

- Most industrial processes are non-linear and often require non-linear controllers.
- Usually mathematical models are not easy or economically convenient to use in industrial plants. In some cases heuristic knowledge or trial-and-error procedure, even less formal, may be more practical and convenient for system analysis and control design.

This chapter describes an application of fuzzy control methodology in a real system, in which several aspects have been considered, like control performances, time consumed for parameter tuning and acceptance of the technology by field technicians.

The chapter is organized as it follows: Section 2 concerns with the overall description of the packaging machinery, while Section 3 refers specifically to the filling system, with particular emphasis on the technological aspects that pose some challenging problems to the control system design. Section 4 describes the methodology used in the design of the fuzzy controller, while Section 5 reports results form experimental tests. The chapter ends with some final conclusions reported in Section 6.

2 The Tetra Brik packaging machine

The Tetra Pak^{TM}Company is a market-leader in the field of machinery for packaging of liquid foodstuff like milk, tomato sauce, fruit juices, etc. Among various packaging system, the Tetra Brik system is one of the most famous, as it is widespread in about 150 countries all over the world.

Tetra Brik machinery, developed and produced by Tetra Brick Packaging System^{TM}, are used to package fresh products (milk) as well as long storage products (long-life milk, fruit juice), in various package format (form 100 *ml* to over two liters).

Figure 1 shows the TBA 19©machine, which produces small size packages (100 to 300 *ml*).

Differently from Tetra Pak^{TM}competitors, the package is made simultaneously with the product filling. This assures a better package sterility and improves the final product quality.

In Figure 2 the packaging process is depicted. The paper, stored in rolls, is subjected to various processing steps before the filling and package forming, namely product information stamping, sterilization process, etc.

The package forming is obtained through a mechanical device which bends the paper to obtain a cylinder which ends are subsequently soldered and cut. The last processing phase consist in folding the two ends of the package to give the well-known parallelpiped shape to the final product.

As shown in Figure 2, the product is conveyed into the paper cylinder through a pipe. The filling system regulates the level of the product inside the paper tube, aiming at controlling the amount of product inside each package. Characteristics and problems of this system are the subjects of the following section.

Figure 1: The Tetra Brik^{TM}TBA 19©packaging machine.

3 The filling system

The product level in the packages is regulated through a pneumo-hydraulic valve which is activated by means of a current-pressure converter. The control software is run by a proprietary board called TMCC (Tetra Pak Multi Purpose Compact Controller), with a 8051 micro-controller programmable in "C" language.

The measure of the product level inside the paper tube is obtained by means of an array of Hall sensors which are coupled with a set of magnets placed on an ball-cock inside the paper tube. A scheme of the filling system is depicted in Figure 3.

The product level measure is very poor, as it is obtained as an array of sensors, each of them having a strong non-linear behavior.

For this reason the sensor output is quite close to the digital output of a Analog-to-Digital converter with a small number of bit of resolution. The range of measure is about 16 *cm* obtained with a sixteen Hall sensor array (each sensor covers about one centimeter).

The disturbances acting on the filling system are of various types, for example the pressure of the incoming product may change because is greatly

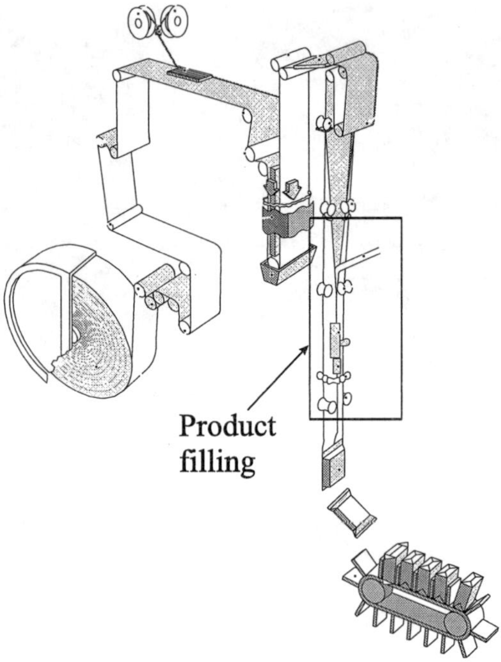

Figure 2: Production path of the packages

influenced by the start and the stop of other packaging machines, that modifies the gain of the actuation system (in particular the pneumo-hydraulic valve). An other important source of disturbance is the soldering and cutting phase, during which the tube of paper is compressed by handling jaws, and, therefore, the level of the product rises inside the tube.

The controller algorithm of standard machines is based on a PID, which has two sets of parameters, one for the starting phase, and one other for the steady state phase. This double configuration of the PID is needed to speed-up the starting phase of the machine, ensuring also good performance in steady state.

4 Fuzzy Control of the filling system

Fuzzy control has been considered in this case because of several reasons, namely:

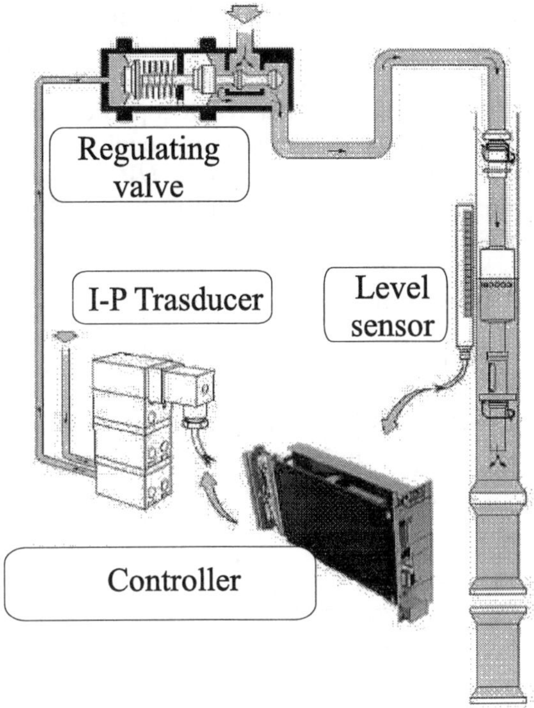

Figure 3: Control loop of the filling system.

- The process is inherently non-linear and, in particular, the process regulation has different requirements during the machine start-up and steady state.
- The level sensors give a poor signal, quantized with a low resolution. Therefore a control look-up table is easy to define using fuzzy rules.
- The PID is difficult to set up correctly, and requires an high effort during the machine initial adjustment. Fuzzy rules seemed more simply and intuitive to be set up.

The fuzzy controller used in this application is an improved version of the standard fuzzy-PD controller[3] with two inputs (level tracking error and tracking error increment) and one output (the current set-point for the I/P transducer).

The main improvement with respect to the standard structure is a self-adjustment mechanism that tunes some controller parameters on line. As a matter of fact, the actuation system (actually the product flow control, see Table 1) has a non linear behavior that requires particular care during controller design. Moreover, during normal working condition, such behavior may change (e.g. because different inlet pressure or I/P transducer setting) and therefore an adapting mechanism is necessary to maintain good performance of the actuation system.

Current [mA]	valve shift [mm]	flow [l/s]
7.62	0	0
8.43	1.88	0.0014
9.03	4.58	0.0014
9.53	6.36	0.381
9.82	7.20	0.592
10..9	8.11	0.712
10.55	10.10	0.786
10.98	11.68	0.798
10.85	11.58	0.792

Table 1: Static behavior of the control actuation system.

The proposed fuzzy controller (see Figure 4) has a double adaptation mechanism:

- The first is the controller output offset trimming, that is necessary because the active working range of the I/P transducer is in [7,11] mA. Moreover this range may change slightly because component factory setting or temporal shift.
- The second is self-tuning of the output gain of the fuzzy controller, which shapes the controller input-output map accordingly to the error level.

The global scheme of the core of the filling control system is depicted in Figure 4. The fuzzy controller has the sampled level error $e(k)$ and the discrete derivative of the sampled level error $de(k) = \frac{e(k)-e(k-1)}{T}$, where the sampling period T is 90 mS. The output of the controller $v(k)$ is multiplied by a variable gain, which is a function of the steady state level error and added to an offset. The offset is computed as the value of the output for which the steady state error is lower than a set value.

The controller has been tuned using a simulation model of the filling machine under Matlab-Simulink©software environment. The last part of the work

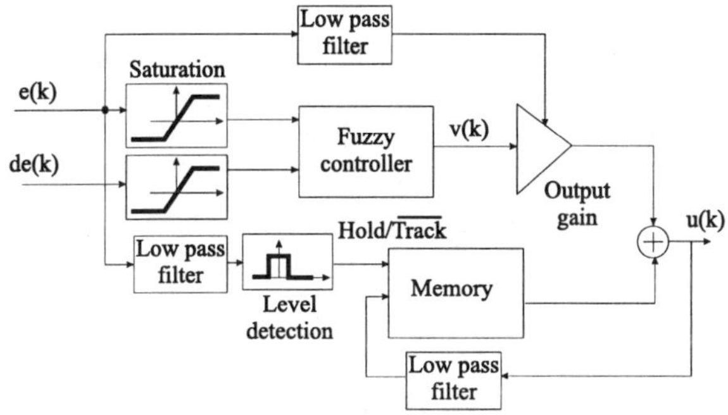

Figure 4: The block scheme of the filling control system.

regarded the experimental results of the application of the fuzzy controller to a real machine. These results are discussed in the next Section.

5 Experimental results

The proposed controller has been tested on a machine in the Tetra Pak research and development laboratory, aiming to make a comparison with respect to standard PID control methodology currently used in the filling machinery.

We analyzed both the technician comments on the general aspect of the new methodology as well as the global filling system performances.

The comments by Tetra Pak technicians may be summarized as it follows:

> The Fuzzy controller has an intuitive feeling which allowed a quick understanding of the basic steps of the controller set-up. This aspect made very easy and fast the system adjustment.
>
> Despite many methods for PID tuning[6], the controlled process has such a great complexity, both because technological and physical aspects, that the PID tuning was possible only by means of simple trial-and-error process. Therefore, regarding this aspect, no difference has been experienced between the two control methodologies.
>
> Moreover, the non-linearity of the fuzzy controller allowed the use of a single structure instead the two sets of parameters needed for the PID, to cope with the start and steady state phases.
>
> Fuzzy controller has much more parameters to adjust than PID controller and the adoption of a standard structure (i.e. the fuzzy-PD rules sets) has helped very much in the initial design phase.

To compare the performance of the fuzzy and PID control system, we made several experiments. Among then we choose a significant case which is described in the following. The results gained in the whole test set (see for detail [7]) are consistent to the result presented as follows.

In Figure 5 data acquired from a run test of the machine are shown. The test regarded a cold start (filling pipe completely empty) as well as two warm starts (filling pipe partially full).

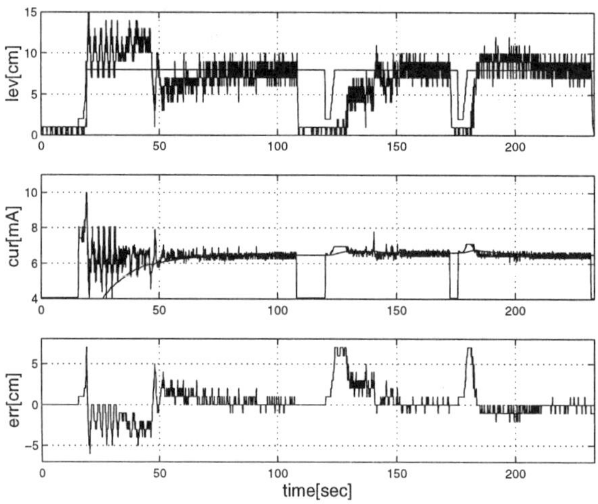

Figure 5: Experimental results on the application of the Fuzzy control. The graphics show (top) the product level (middle) the controller output and (bottom)the tracking error.

The top graphic shows the actual measure level of the product in the forming tube. The reference level is about 8 cm. At first sight the control performance seems very bad. However, it should be kept in mind that the resolution of the sensor is about one centimeter, therefore an evident oscillation is unavoidable. From the quality point of view, the performance of the filling system controlled by a fuzzy system is completely acceptable.

The middle graphics shows the output of the controller. In the first seconds of running there is a appreciable stress of the actuation system, largely due to the initial gain and offset of the controller. In fact, those values are defined on the basis of standard machine parameters, which may differs from the actual values.

In a later stage the adaptation mechanism, both for gain and offset, leads to a lower chattering in the control variable, and, therefore, in lower stress in the whole actuation system.

The bottom graphic shows the level error, which is inside one centimeters, which is actually the lower limits because of technological aspects.

In Figure 6 the experimental results of the application of the PID controller to the filling system are shown. The test regarded a cold start with two successive warm starts (160 and 220 seconds).

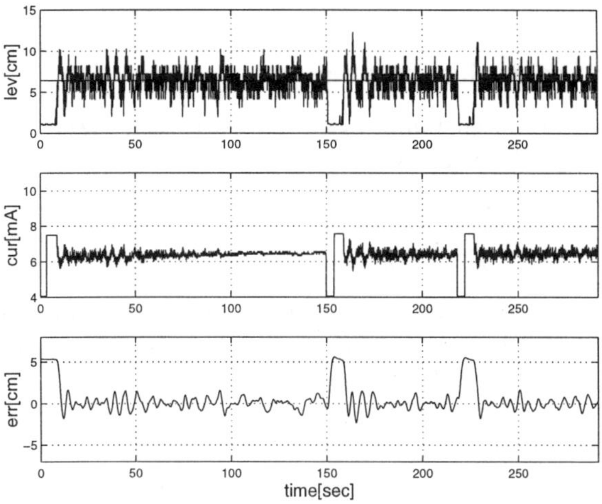

Figure 6: Experimental results on the application of the PID control. The graphics show the product level (top) the controller output and (middle) the tracking error (bottom).

In this case the reference level is 6.4 centimeter. The PID has two set of parameters, The first is used in the cold start, in which the filling pipe is completely empty, when the second is used in the steady state condition.

The steady state error (bottom graphic) is about one centimeter (in average) with some large peaks (5 centimeters).

To comment these performance results, we can conclude that the fuzzy controller, which has a more complex structure than PID, performs quite better, both in term of actuators stress and level of error.

6 Conclusions

In this chapter the fuzzy control of a filling system for packaging machinery is analyzed and discussed. This industrial process is quite complex to be controlled because of several aspects, both physical and technological.

Tests on the application of fuzzy-PD controller to this process, with some improvement developed for this particular application, show good performance in terms of tracking error and low actuator stress.

7 Acknowledgements

The Authors wish to thank Prof. C. Bonivento (DEIS, University of Bologna) for his interesting comments and suggestions which helped very much the developing of this work.

A The fuzzy controller parameters

The core of the proposed filling system consist of a fuzzy-PD control. The input-output map of the controller is:

$$\frac{\sum_{k=1}^{n} \min[\mu_{E_{i(k)}}(e), \mu_{DE_{j(k)}}(de)]Y_k}{\sum_{k=1}^{n} \min[\mu_{E_{i(k)}}(e), \mu_{DE_{j(k)}}(de)]} \tag{1}$$

where μ_E and μ_{DE} are, respectively, the membership functions on the error e and discrete derivative of the error de input variables and values Y_k are the output singleton (see also Table 4). The subscripts $i(k)$ and $j(k)$ mean that the actual membership function in the formula are picked up form the rule data base accordingly to the k-th rule evaluated in the sum.

The membership function are described by the tables 2 and 3, while the fuzzy rules are shown in Table 4.

References

1. C.C. Lee. Fuzzy logic in control systems: Fuzzy logic controller, part I and II. *IEEE Trans. Sys. Man, Cybern.*, 20(2):404–435, 1990.
2. C. Fantuzzi. Basis of fuzzy control. In C. Bonivento, C. Fantuzzi, and R. Rovatti, editors, *Fuzzy Logic Control - Advances in Methodology.* World Scientific, 1998.
3. D. Driankov, H. Hellendoorn, and M. Reinfrank. *An Introduction to Fuzzy Control.* Springer Verlag, 1993.
4. Li-Xin Wang. *A Course in Fuzzy Systems and Control.* Prentice Hall, 1997.
5. Fuzzy logic control - advances in methodology, 1998. Bonivento, C. and Fantuzzi, C. and Rovatti, R. Editors.
6. Karl J. Astrom and Tore Hagglund. *Pid Controllers.* Instrument Society of America, 1995.
7. Bernardetta Macchini. Fuzzy control of a filling machine. Master's thesis, University of Bologna, 1998. in Italian.

NB	NS	ZE	PS	PB
10	0	0	0	0
10	0	0	0	3
7	3	0	0	7
3	7	0	0	10
0	10	0	0	7
0	7	3	0	3
0	3	7	0	0
0	0	10	0	0
0	0	7	3	0
0	0	3	7	0
0	0	0	10	0
0	0	0	7	3
0	0	0	3	7
0	0	0	0	10
0	0	0	0	10
0	0	0	0	10

Table 2: The table shows the values of the membership functions $\mu_{E_i}(x)$, where x is the range of the sensor input (16 centimeters, 1 centimeters each step).

NB	NS	ZE	PS	PB
10	0	0	0	0
10	0	0	0	0
10	0	0	0	0
10	0	0	0	0
5	5	0	0	0
0	10	0	0	0
0	5	5	0	0
0	0	10	0	0
0	0	5	5	0
0	0	0	10	0
0	0	0	5	5
0	0	0	0	10
0	0	0	0	10
0	0	0	0	10
0	0	0	0	10
0	0	0	0	10

Table 3: The table shows the values of the membership functions $\mu_{DE_i}(x)$, where x is the range of the sensor input (16 centimeters, 1 centimeters each step).

DE/E	NB	NS	ZE	PS	PB
PB	0	-1	-9	-9	-9
PS	1	0	-1	-1	-9
ZE	1	1	0	1	1
NS	10	1	1	0	-1
NB	10	10	10	10	0

Table 4: The fuzzy rules data base.

FAULT DETECTION AND DIAGNOSIS OF ELECTRO-MECHANICAL DRIVES AND ACTUATORS

ROLF ISERMANN AND DOMINIK FUESSEL

Institute of Automatic Control,
Darmstadt University of Technology
Landgraf-Georg-Str. 4, D-64283 Darmstadt
email: {RIsermann,DFuessel}@iat.tu-darmstadt.de

Fault detection and diagnosis can be substantially improved by the application of fuzzy logic. Fault detection can benefit from non-linear fuzzy models and fault diagnosis can profit from a transparent system which can be learned from experimental or simulated data. A fuzzy logic-based diagnosis is advantageous since it allows the easy incorporation of a-priori known rules and lets the user understand the inference of the system. In this contribution, the successful use of a dynamic fuzzy model for the fault detection of an actuator is presented. Furthermore, some methods of learning fuzzy rules for diagnosis are reviewed, and a neuro-fuzzy approach that is especially well suited for the needs of a fault diagnosis is presented. Two examples illustrate the approach.

1 Introduction

Within the automatic control of technical systems supervisory functions serve to indicate undesired or unpermitted process states, and to take appropriate actions in order to maintain the operation and to avoid damage or accidents. The following functions can be distinguished:

1. *monitoring*: measurable variables are checked with regard to tolerances, and alarms are generated for the operator.

2. *automatic protection*: in the case of a dangerous process state, the monitoring function automatically initiates an appropriate counteraction.

3. *supervision with fault diagnosis*: based on measured variables, features are calculated, symptoms are generated via change detection, a fault diagnosis is performed and decisions for counteractions are made.

The classical methods 1. and 2. are suitable for the overall supervision of the processes. To set the tolerances, compromises have to be made between the detection size of abnormal deviations and unnecessary alarms because of normal fluctuations of the variables. Most frequently simple, limit value checking is applied, which works especially well if the process operates approximately in a steady-state. The big advantage of the classical limit-value-based supervision methods is their simplicity and reliability. However, they are only able to react after a relatively large change of a feature, i.e. after either a large sudden fault or a long-lasting gradually increasing fault. In addition, an in-depth fault diagnosis is usually not possible. Therefore 3. *advanced methods of supervision and fault diagnosis* are needed, which satisfy the following requirements:

- Early detection of small faults with abrupt or incipient time behaviour.
- Diagnosis of faults in the actuator, process components or sensors.
- Detection of faults in closed loops and supervision of processes in transient states.

The goal for the early detection and diagnosis is to have enough time for counteractions such as other operations, reconfiguration, maintenance or repair. The earlier detection can be achieved by gathering more information, especially by using the relationship between the measurable quantities in the form of mathematical models. For fault diagnosis, the knowledge of cause-effect relations has to be used. A general scheme for all the supervisory functions and resulting actions is given in Fig. 1.

Fault Detection and Fault Diagnosis

The main tasks of supervision and fault diagnosis can be subdivided into *fault detection* by analytic and heuristic symptom generation, and *fault diagnosis*. These tasks will be briefly explained below.

1.1 Symptom Generation

The analytical knowledge about the process is used to produce quantifiable, analytical information. To do this, data processing based on measured process variables has to be performed, to generate first the characteristic values by

- Limit value checking of direct measurable signals. The characteristic values are the exceeded signal tolerances.
- Signal analysis of directly measurable signals by the use of signal models like correlation functions, frequency spectra, or the autoregressive moving average (ARMA). The characteristic values are, e.g., variances, amplitudes, frequencies or model parameters.

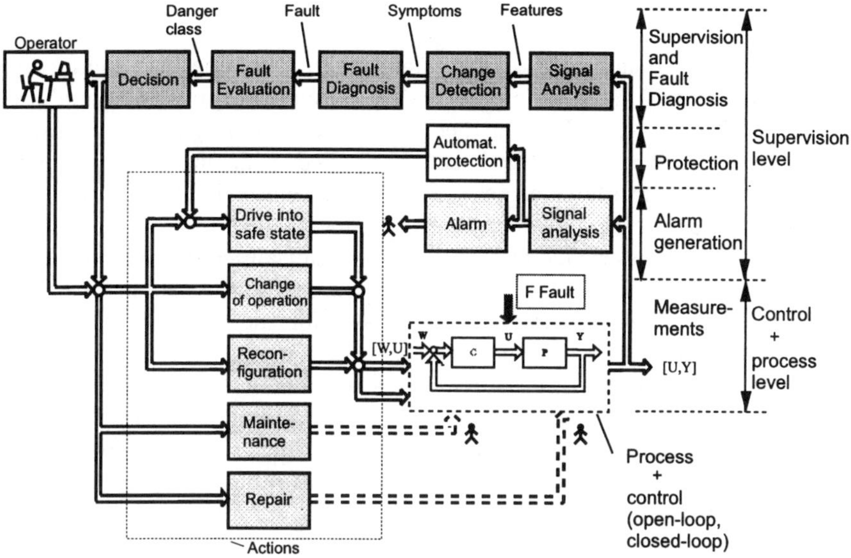

Figure 1. General scheme of supervision methods.

- Process analysis by using mathematical process models together with parameter estimation, state estimation and parity equation methods. The characteristic values are parameters, state-variables or residuals.

Modern control theory has given additional tools for the employment of model-based symptom generation. While most methods were originally developed for linear process models, they now become available for non-linear models like for instance Takagi-Sugeno fuzzy models. These allow the precise description of non-linear plants such as actuators that typically have saturation behavior, direction dependent hystereses effects, non-linear friction and alike.

In addition to the symptom generation using quantifiable information, heuristic symptoms can be produced by using qualitative information from human operators. Through human observation and inspection, heuristic characteristic values in the form of special noises, colours, smells, vibration or the process history are obtained. In this way heuristic symptoms are generated, which can be represented as linguistic fuzzy variables (e.g., small, medium, large) or as vague numbers (e.g., around a certain value).

1.2 Fault Diagnosis

The task of fault diagnosis consists of determining the type, size and location of the fault, as well as its time of detection, based on the observed analytical and heuristic symptoms. If no further knowledge of fault-symptom causalities is available, classification methods can be applied, which allow a mapping of symptom vectors into fault vectors. To this end, methods like statistical and geometrical classification or neural nets and fuzzy clustering can be used. If, however, a-priori knowledge of fault-symptom causalities is available, e.g. in the form of causal networks, diagnostic reasoning strategies should be applied. Many authors report the use fuzzy rule bases for the diagnosis. These can conveniently implement the operator knowledge who is trained in performing the fault diagnosis. If measurements are available, these can be used to learn the fuzzy system from the data. Such an approach is straightforward and facilitates problems that can arise if conflicting or incomplete operator knowledge is to be used in a fuzzy system. Desirable, however, is also an automatic derivation of diagnosis tools like fault trees that have been used for many years. Those graphical structures are very intuitive and well-known among operators.

In the following, the use of fuzzy logic for the fault detection and diagnosis of electro-mechanical actuators will be shown. In the next section, a non-linear fuzzy model will be employed for the fault detection of an electro-pneumatic valve. Then, the use of a neuro fuzzy decision tree will be examined as a diagnosis tool for the valve and an electrical d.c. motor. Finally, some conclusions will be given.

2 Fuzzy Model-Based Symptom Generation

For the diagnosis of an electro-pneumatic valve, an approach was developed that utilizes a non-linear fuzzy model. The features of this model structure make it particularly suitable for the fault detection task. The approach heavily depends on a model for the non-linear process which represents the static and dynamic behavior and is transparent enough so that characteristic values can be extracted. An ideal model to meet the requirements is the local linear model tree (LOLIMOT) algorithm which has been developed by Nelles[1]. The algorithm is based on the idea of approximating a non-linear dynamic function by piece-wise linear models. The physical inputs, their time delayed values and the time delayed values of the process outputs form the model input space. Each local model is valid only on a certain subregion of the input space. The partitions where the local models are valid are not crisp but fuzzy. Therefore, the validity of one local model i for a certain point in the input space is expressed by a weighting function Φ_i continuously defined over the whole input space and scaled from zero (not valid) to one (fully valid). The output of the overall model is calculated as the weighted sum of all M local-linear-model parameters, multiplied with the according inputs x_i.

$$\begin{aligned} y = & \left(\sum_{i=1}^{M} w_{0i}\Phi_i\right) + \left(\sum_{i=1}^{M} w_{1i}\Phi_i\right) x_i + \ldots + \left(\sum_{i=1}^{M} w_{ni}\Phi_i\right) x_n \\ = & w_0 + w_1 x_1 + \ldots + w_n x_n. \end{aligned} \tag{1}$$

where $w_{0i} \cdots w_{ni}$ are the parameters of the i-th linear regression model, $x_1 \cdots x_n$ are the inputs and Φ_i is the normalized Gaussian weighting function for the i-th model. The offset parameter w_{0i} defines the operating point.The superimposed parameters w_i define a linear model in the actual set point. Fischer and Nelles[2] refer to this as dynamic linearization.

Symptom Generation

As mentioned above, the LOLIMOT process model is well suited for generation of symptoms which can be used for fault detection purposes. One category of symptoms are signal based ones. They can be derived in closed loop operation and normally do not require any additional process excitation. The simplest symptom is the output error between model and process, which is named residual symptom S_r in the following. The performance of a controller in constant operation regions and during set point changes also characterizes the process and symptoms can be derived by defining different control performance indices (CPI). One rates the difference between control reference signal W and the controlled variable y in constant operating regions:

$$S_{CPI1} = \sum_{i=1}^{l} (W(k-i) - y(k-i))^2 \tag{2}$$

Another possibility is the use of the time interval needed to enter a predefined region around the reference signal. In the following, the region is defined by ±2% of the new set point. The symptom is named S_{t98}.

A similar symptom can be derived from the control command signal u:

$$S_{CPI2} = \sum_{i=1}^{l} u^2(k-i) \tag{3}$$

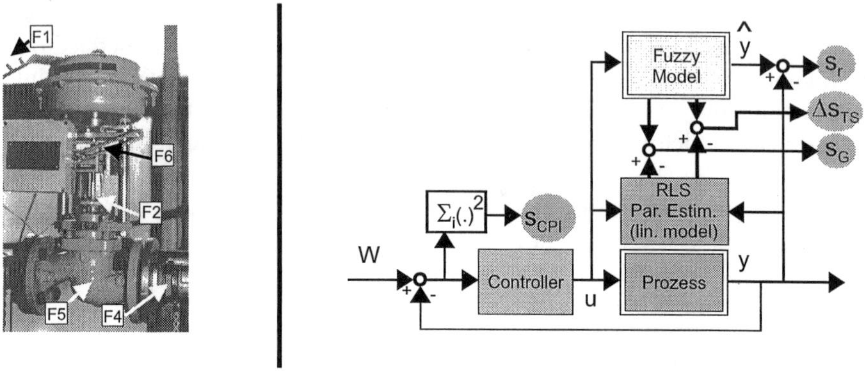

Figure 2. a. and b.: Electro-pneumatic valve (left) and general scheme for the symptom generation involving a Takagi-Sugeno type fuzzy model (right).

For the diagnosis the symptoms derived from the process have to be compared with the nominal ones. For this purpose, a simulation with the model in closed loop control (MCL) can be run in parallel to the process in closed loop (PCL) using the same reference signal W. Fig. 3 depicts this approach and the symptom generation.

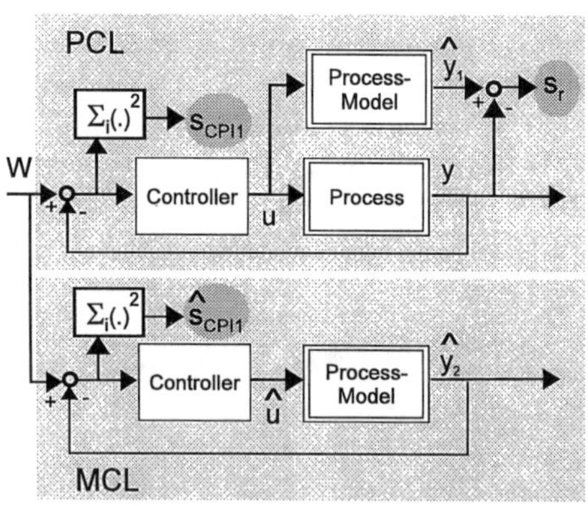

Figure 3. Symptom generation using the process model.

The third class of symptoms are characteristic values which define the process behavior $y = f(u)$. A common way is to identify a linear model (Hoefling et. al.[3]) in a small region around the actual set point. This requires

sufficient process excitation which is often not fulfilled during normal operation. In order to meet this requirement, excitation signals can be added to the command signal u leading to a temporary degraded control performance but enabling a parameter estimation. The nominal values can be derived from the model by dynamic linearization. Here the special structure of the LOLIMOT model is very helpful. Important symptoms are time constants S_{Ts}, gain S_G, offset S_O, and dead times S_{Td}.
With the help of the dynamic model (see Fig. 2.b), a set of symptoms can be constructed:

- Deviation of the time constant $\left(\frac{S_{Ts}}{S_{TSnom}}\right)$
- Change of the control performance in the closed-loop situation (ΔS_{CPI1})
- Deviation of the rise time ΔS_{T98}
- Change of the offset of the internal position controller of the valve (ΔS_O)
- Change of the model gain $\left(\frac{S_G}{S_{Gnorm}}\right)$
- Change of measurement and mode prediction (S_r)

The diagnosis based in these symptoms will be shown in a later section.

3 Fuzzy Logic-Based Diagnosis Methods

All of the methods dealt with in the following combine the symptom information in a diagnostic system and make decisions about most probable fault causes and give hints for a quick counteraction. This step is generally referred to as the fault *diagnosis*.

As symptom generation and systems handled become more complex and the allowable time for creating and maintaining a diagnostic system is limited, there is a need for self-learning approaches which can support the designer in creating a knowledge base. This leads to the field of *fuzzy rule generation* from experimental data as an alternative.

In the following, some established concepts of the automatic generation of static, Mamdani-type fuzzy systems are reviewed. It will be explained why an adaptive component is essential for those systems. After a few words about the creation of membership functions, some results from the learning of logic operators will be presented. Finally, a self-learning structure which is especially suited for the needs of fault diagnosis will be presented and examples will be given.

3.1 Creating Fuzzy Rules from Data

There are many ways of extracting fuzzy rules from experimental data. Some of those can be categorized into

- Statistical approaches
- Evolutionary methods
- Neuro-fuzzy systems
- Other methods

Among the other methods are in particular techniques from *machine learning* leading to so-called *decision trees*. Their theory is quite evolved and they are very powerful tools for diagnostic tasks (for a good reference, see Michie[4] and Breiman[5]). The result is not generically a fuzzy system, but the tree can be converted into fuzzy rules.

Statistical approaches compare the appearance of a fuzzy rule premise term with possible conclusions. They build rules whenever there is a statistically significant correlation between two terms. One method is the so-called ROSA approach (Krone et al.[6]).

Membership functions that are created from a grid are also the basis of *evolutionary algorithms*, especially genetic approaches. They benefit from a simple coding scheme of the rules when the membership functions are fixed. It is straightforward to formulate a binary coding for the membership functions and possible conclusions in order to seek for a well performing rule base. Examples are given by Nelles[7] and also Ishibushi[8].

The main drawback of these algorithms is the pre-defined location and size of the membership functions. If they are not located such that a separation of the different classes is possible, any combination thereof will fail to deliver an optimal classification system. The selection of different rule tables can give some relief to that problem but it seems very simple to overcome the problem with such a "brute force" approach.

Therefore, some sort of *adaptive component* needs to be incorporated in the diagnosis system so that it can construct the optimal decision boundary between the classes. That leads to the concepts of neural networks. The prototype solution for that task is a simple multilayer perceptron (MLP). Given enough data points, it can fit the exact classification boundary.

Unfortunately, MLPs also lack any interpretability. Other concepts have, therefore, been presented based on a combination of neural networks and fuzzy systems. That general description can be further be divided into fuzzy systems that have parameters being adjusted by neural nets and *hybrid* approaches with an inherent equivalence between the neural network and the fuzzy system, see Fig. 4. The latter have been proposed for example by Ayoubi[9], Caminhas[10], Nauck[11] and others.

While these concepts seem to be appealing in the first place, one has to take care, however, that the interpretability they claim to have, is really given. An *interpretable* system usually requires a *constrained* system. One has to make sure, for instance, that the ordering of the membership functions is not lost, that they do not overlap too strongly or that the number of rules is not too large. On the other hand, these constraints might also hinder the system from being flexible enough to appropriately solve the given problem.

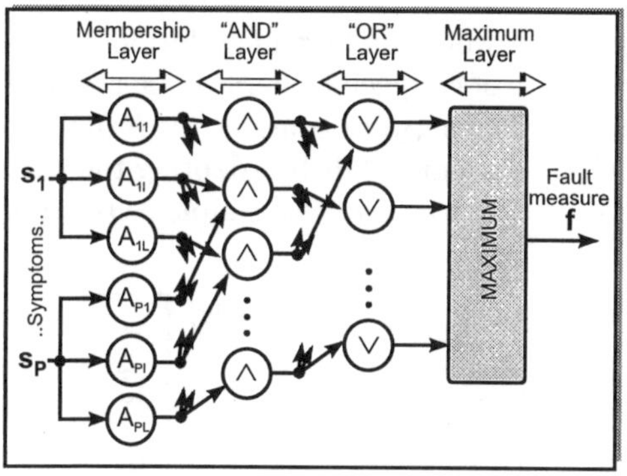

Figure 4. Typical hybrid neuro fuzzy approach.

Transparency

Why is interpretation important? Consider using a multilayer perceptron network for the diagnosis. Supposed, the network is trained and performs well on the data. If, however, data outside the training range is used, it will produce unreliable results. An alternative would be to use radial basis function networks. Due to their local approximation they will probably reject new data points and prevent a wrong result by delivering very small outputs.

Unfortunately, neither of the two approaches fits the need of a fault diagnosis problem. There, one very often has to deal with a lack of appropriate data points since data from faulty processes is often hard to get. Sometimes, these faulty processes are simply not available or the artificial introduction of the fault is dangerous or costly. Simulated data can only partially solve that problem. Hence, one must acknowledge that a good generalization capability outside the training set is essential since the data is never completely satisfactory. Only a transparent system where the user can follow the decision making process can help, because the generalization can be judged and influenced appropriately.

It has to be mentioned, furthermore, that a transparent system must be as simple as possible. Clearly, a fuzzy system with many rules is not transparent. Each individual rule is transparent, but the sheer number hinders a real understanding of the system.

Learning Logic Operators

But can hybrid neural networks find simple solutions? In order to investigate the learning capabilities of a special operator, a network with two hidden layers was constructed. The network consists of ANDOR operators which can be adapted from AND to a logical OR by adapting the parameter α (compare Fig. 5.b). The data set that was used is a simple XOR-problem (see Fig. 5.a).

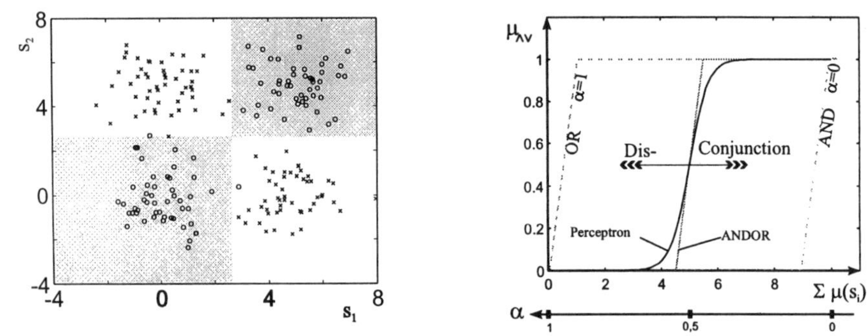

Figure 5. a. and b.: Data set from the XOR-problem (left) and continuous approximation of the ANDOR operator by a sigmoidal function (right).

In this example, the membership functions of the hybrid network are fixed and only the characteristics of the operators and the connection weights are optimized with a non-linear optimization algorithm. The objective function is given by a simple least squares approach. The weights as well as the parameters α were bounded between 0 and 1. The optimization was performed with sequential quadratic programming. The resulting network's first layer ends up being an AND and the second one transforms into OR functions as expected. It codes the following rule set:

$$\begin{array}{l}\text{IF } x_1 \text{ small AND } x_2 \text{ small OR}\\ \text{IF } x_1 \text{ large AND } x_2 \text{ large THEN class "o"}\\ \text{IF } x_1 \text{ large AND } x_2 \text{ small OR}\\ \text{IF } x_1 \text{ small AND } x_2 \text{ large THEN class "x"}\end{array} \tag{4}$$

That is exactly what was expected.

When creating a system, however, the correct size of the network is normally not known beforehand (as was the case here). To simulate that situation, the same network is confronted with a simpler data set. That data can be described with rules as

$$\begin{array}{l}\text{IF } x_1 \text{ large OR } x_2 \text{ large THEN class "x"}\\ \text{IF } x_1 \text{ small AND } x_2 \text{ small THEN class "o"}\end{array} \tag{5}$$

The results can be seen in the first half of Fig. 6. It is interesting to note that the network learned the rules from (5). That is apparent after the connections are pruned (with a weight of zero) and the corresponding neurons removed. The network even collapses to one hidden layer since some of the neurons are clearly redundant. When the input membership functions are also learned and were randomly initialized, the result is not satisfactory. While the network still performs a decent job (but not as good as before), an interpretation can hardly be given. In this simple example, one can still explain the results, but they are not similar to the rules from (5). Only when additional constraints are introduced, the picture changes. Here, Gaussian membership functions characterized by a center c_j and a standard deviation σ_j were used. Demanding

$$\begin{aligned} c_j &< c_{j+1} \\ 0.5(\sigma_{j+1} + \sigma j) &< (c_{j+1} - c_j) \end{aligned} \tag{6}$$

ensures sensible membership functions. Now, the optimization again yields similar results as before, with the same membership functions. These constraints obviously have a positive effect on the optimization: They narrow the search space significantly and remove many local minima. While they ensure the transparency of the system, they also act as a *regularization* for the optimization.

Creating Membership Functions

In the previous example, membership functions were learned by the optimization. Sometimes, however, one already needs those to select rules or as a good initialization for an optimization. If one does not want to use a fixed grid, the only possibility is to create the functions using a clustering algorithm. This

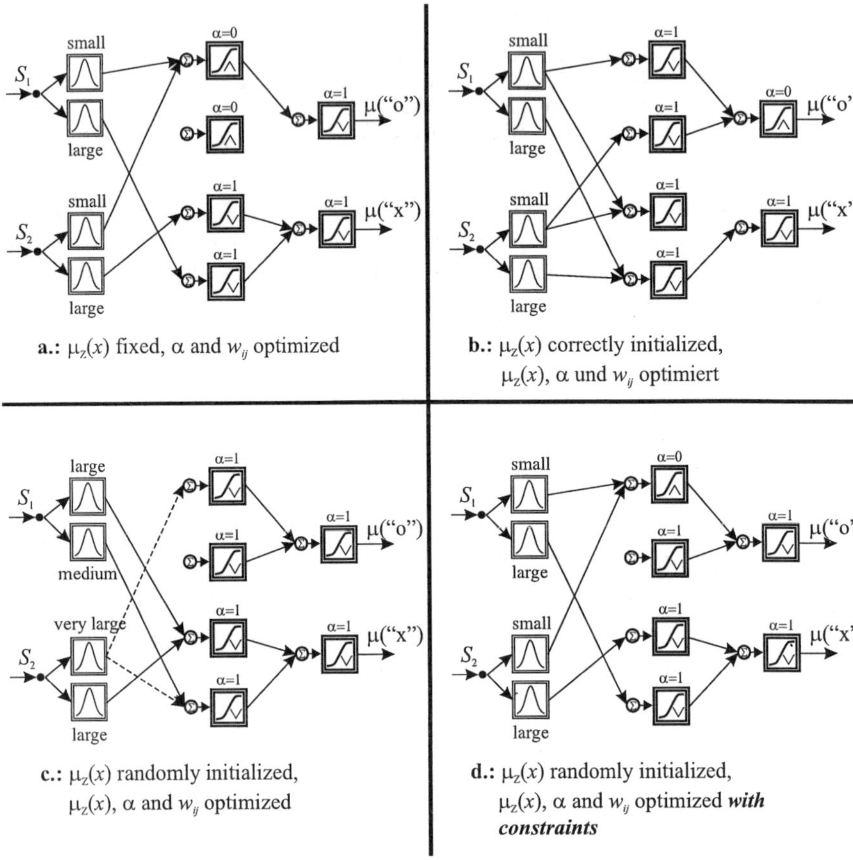

Figure 6. Results of a network with two layers of ANDOR neurons. It is investigated whether the operators move towards the correct logic element (AND, symbolized by $\wedge$ and OR, symbolized by $\vee$), by adjusting the parameter α, the weights w_{ij} and the membership functions with a least-squares approach. The results in a. and b. are as expected. The network can be interpreted. Connections yielding a zero weight have already been removed in the figure. When all parameters are learned and the membership functions are allowed to freely develop, the result is not good. Figure c. shows the resulting network. It has connections different from 0 or 1 and the membership functions are shaped in an unexpected way. The resulting mean square error is also higher than in the other cases. Only after the introduction of constraints (see the text) a satisfactory result can be achieved.

is very frequently done either by unsupervised Fuzzy C-Means clustering or Kohonen Maps or similar approaches. A better approach would be to use a supervised clustering or - if the physical insight is given - create them by hand.

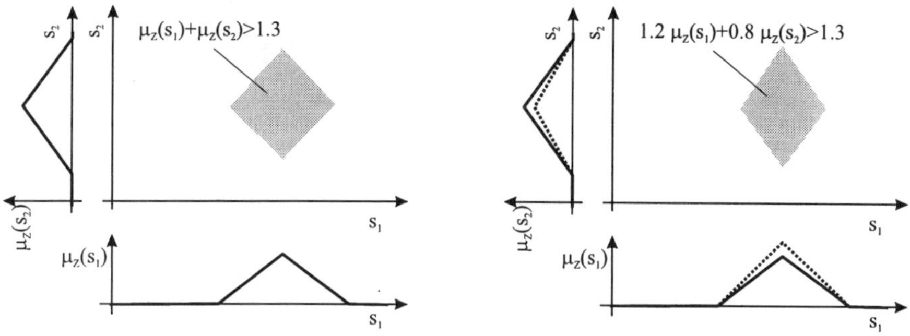

Figure 7. Effect of a weighting of the membership values in a neuro-fuzzy system. When a weighted sum of membership values (the μ_Z) is subject to a thresholding, the effect of a weight variation is similar to that of differently wide membership functions.

But even if sub-optimal membership functions are used, sometimes the results are still satisfactory. That can be explained by the effects of other parts of a network. Consider a simple weighted sum of the membership values and a threshold being applied to it. It can then be shown that this weighting effectively acts like a variation of the membership functions would (see Fig. 7). The variation of the membership functions is "hidden" in the weight pattern.

3.2 SELECT - Self Learning Classification Tree

In this section, a methodology is suggested that tries to overcome some of the problems mentioned above. It consists of a series of very simple fuzzy rules or - equivalently - decision neurons, that create a diagnostic tree. Fig. 8 shows its structure. The nodes of the tree are decisions that deliver a membership value or belief (sometimes even a probability as will be shown below) of an event E_j. That event can either already be a particular fault event or a part of it. The neurons use a continuous form of an accumulative AND operator approximated by a sigmoidal.

$$\mu_{\wedge,j}(\underline{s}(k)) = \frac{1}{1+e^{-\lambda\left(\left(\frac{2}{P}\sum_{i=1}^{P} w_{ij}\mu_Z(s_i(k))\right)-(0.5+(P-1))\right)}} \tag{7}$$

The neurons, therefore, perform a *linear* classification in the membership space.

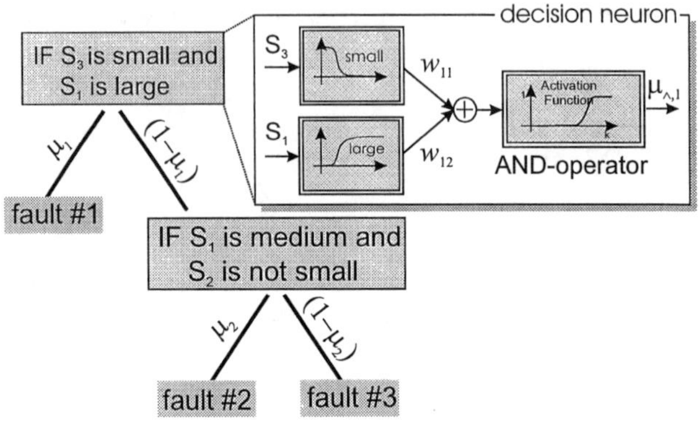

Figure 8. Example of a SELECT tree.

Following Bishop[12] one can show that the output of such a function can be seen as a-posteriori probabilities if one uses triangular membership functions and the input patterns have a Gaussian distribution. In that case, also the membership function values are Gaussian.

Consider the simple case with two classes C_1 and C_2 with equal covariance matrices Σ

$$p(\underline{\mu}_Z \mid C_k) = \frac{1}{(2\pi)^{d/2} |\Sigma|^{1/2}} e^{\left(-\frac{1}{2}(\underline{\mu}_Z - \underline{c}_k)^T \Sigma^{-1} (\underline{\mu}_Z - \underline{c}_k)\right)} \tag{8}$$

Hereby μ_Z is the output of the membership layer. Following Bayes' theorem gives the a-posteriori probability of data belonging to class 1

$$\begin{aligned} p(C_1 \mid \underline{\mu}_Z) &= \frac{p(\underline{\mu}_Z \mid C_1)P(C_1)}{p(\underline{\mu}_Z \mid C_1)P(C_1) + p(\underline{\mu}_Z \mid C_2)P(C_2)} \\ &= \frac{1}{1 + e^{(-a)}} \end{aligned} \tag{9}$$

with

$$a = \ln \frac{p(\underline{\mu}_Z \mid C_1)P(C_1)}{p(\underline{\mu}_Z \mid C_2)P(C_2)} \tag{10}$$

Using the conditional probabilities of (10), leads to

$$a = \underline{w}\underline{\mu}_Z + w_0 \tag{11}$$

with

$$\underline{w} = \Sigma^{-1}(\underline{c}_1 - \underline{c}_2) \tag{12}$$

$$w_0 = -\frac{1}{2}\underline{c}_1^T \Sigma^{-1} \underline{c}_1 + \frac{1}{2}\underline{c}_2^T \Sigma^{-1} \underline{c}_2 + \ln \frac{P(C_1)}{P(C_2)} \tag{13}$$

Hence, the results can be interpreted in a statistical sense.

Tree Building Procedure

In order to create a structure similar to that in Fig. 8, one has to find the series of rules or neurons. That is done by a procedure shown in Fig. 9. After the creation of membership functions the procedure searches the best rule to identify a certain event E_j. The rule should fire strongly when data from that event is presented and reject all other data. During that process, the candidate fuzzy rules "grow" by incorporating more and more premise terms into them. If a given complexity is reached or an event is sufficiently described by one rule, this rule becomes part of the tree by forming a new decision neuron. The data set is applied to it and all data points that are selected by the rule are removed from the training set. Then, the rule growing starts recursively.

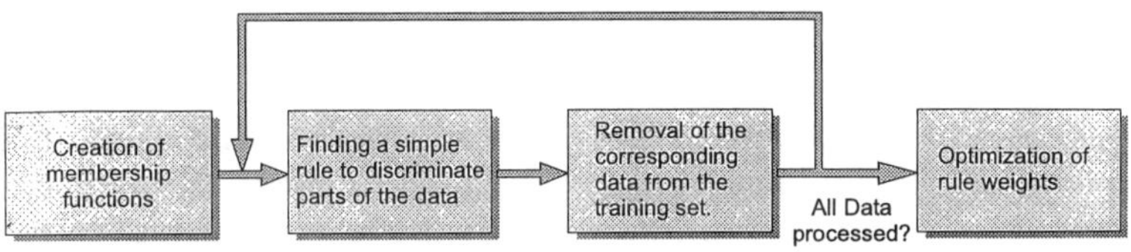

Figure 9. Self-Learning Classification Tree - Procedure.

The described approach has a number of advantages:

- As the data set decreases in size, the process speeds up.
- The problem of finding rules becomes easier as the data from the different events are removed. That way, rules for easily detectable classes and events are created first.
- The rules tend to have a minimum size. In fault diagnosis examples, the observed rules typically work well with 1-3 premise terms, even if more features are available. This can be explained because not all of the features are necessary for the isolation of every class.

Optimization

In order to improve the classification accuracy of the structure, the weights w_{ij} in the tree are optimized with a least-squares approach. This is done *after* the tree structure has been constructed. In order to keep the interpretation of the neurons as rules, the weights are constrained from 0 to 1. That allows to see them as a measure of the importance of a symptom for that rule.

The optimization is performed using the Levenberg-Marquardt Algorithm and implementing the constraints with the aid of auxiliary variables γ_{ij}

$$w_{ij} = \frac{Pe^{\gamma_{ij}}}{\sum_{i=1}^{P} e^{\gamma_{ij}}} \tag{14}$$

This ensures not only $0 < w_{ij} < 1$ but also $\sum_i w_ij = const.$ Hence, the importance of the symptoms is always *relative* to each other.

This optimization can again be performed sequentially, one node at a time. This leads to a very fast procedure, because if n nodes each having c weights are to be optimized, that will yield n optimizations in a c-dimensional space. The computational complexity is of the order $c \cdot \mathcal{O}(n^2)$ instead of a parallel optimization of the order $\mathcal{O}((cn)^2)$. In addition, the problems associated with the "curse of dimensionality" are greatly reduced.

Incorporation of Structural Knowledge

In most applications, a certain amount of knowledge is present. It should be exploited when creating a diagnostic system. Even if exact numbers for thresholds etc. are not known, there usually is some insight into the process like physical understanding about similar faults or similar effects of faults on certain symptoms. The benefits from such information are sometimes neglected, if a diagnosis system is designed with the aim to be solely learned from measured data.

Furthermore, the selection of the symptoms for the diagnosis becomes a matter of *robustness*. Some symptoms are affected by faults for which they are not an appropriate indicator. In an experimental environment, it is virtually impossible to get enough measurements to adequately reflect such an influence. This leads to diagnosis systems which work well only under the experimental conditions but fail otherwise. A diagnosis of a fault should therefore be based on the *appropriate subset* of all available symptoms.

Often, different faults can be categorized into larger groups when their effects on the process are similar. It is then advantageous to find a classification system for the larger groups first and later separate within them. This leads to the concept of a *hierarchical* diagnosis system.

It is therefore suggested to use a-priori knowledge to *structure* the diagnosis system. The designer builds groups of faults and identifies the corresponding relevant symptoms to first differentiate between and later within them. The exact decisions can be found automatically if enough measured data is available. This procedure can be implemented with the SELECT approach. Additionally, one can incorporate more detailed knowledge into the individual rules if desired.

3.3 Application Results

The SELECT procedure has successfully been applied for the diagnosis of different example processes.

Valve Diagnosis

One is the electro-pneumatic valve from Fig. 2.a. Different faults can be introduced into the system among which the following were considered (compare also Fig. 2.a):
Supply pressure leak to membrane chamber (F1), increased friction of the valve shaft (F2), fault at the flow rate sensor (F3), increased flow resistance of the valve (partial clocking) (F4), erosion of the valve (simulated by bypass) (F5), wrongly adjusted position controller (F6).

Using those, a SELECT Tree evolved with the following rules:

IF $\left(\frac{S_G}{S_{Gnorm}}\right)$ not large AND S_τ large AND ΔS_{T98} changed THEN F4
ELSE IF ΔS_O changed AND ΔS_{T98} unchanged AND ΔS_{CPI1} changed THEN F6
ELSE IF $\left(\frac{S_{Ts}}{S_{Tsnom}}\right)$ large AND ΔS_{T98} unchanged AND ΔS_{CPI1} unchanged THEN F2
ELSE IF $\left(\frac{S_G}{S_{Gnorm}}\right)$ large AND ΔS_O unchanged AND ΔS_{CPI1} unchanged THEN F3
ELSE IF $\left(\frac{S_G}{S_{Gnorm}}\right)$ not reduced AND ΔS_{CPI1} unchanged THEN F5
ELSE IF ΔS_{CPI1} changedTHEN F1
ELSE REJECTION

(15)

The achieved classification rate was 81% when all six faults were considered. A diagnosis without the erosion fault resulted in a classification rate of 84% (always computed with a 3-fold cross-validation scheme). The complete rule set is very compact and small. One can easily oversee its generalization behavior.

More importantly, the rule set reveals some insight to the user. The friction fault, for instance, is diagnosed by a higher time constant of the model which makes a lot of sense, since a rusty valve shaft will be sticky and slower than a good one.

D.C. Motor Diagnosis

The SELECT procedure was also applied to the diagnosis of a d.c. motor that can be examined at an experimental test bench. A fault detection based on parity equations and parameter estimation is implemented. For details refer to Fuessel[13]. It allows the detection of 14 different fault situations. A hierarchical diagnosis system was learned using the data from 140 experimental runs. The resulting decision tree given in Fig. 10 proved to be highly suitable for the diagnosis. It achieved 100% classification rate in a cross-validation scheme. Details have been omitted to visualize the concept only. Every branching of

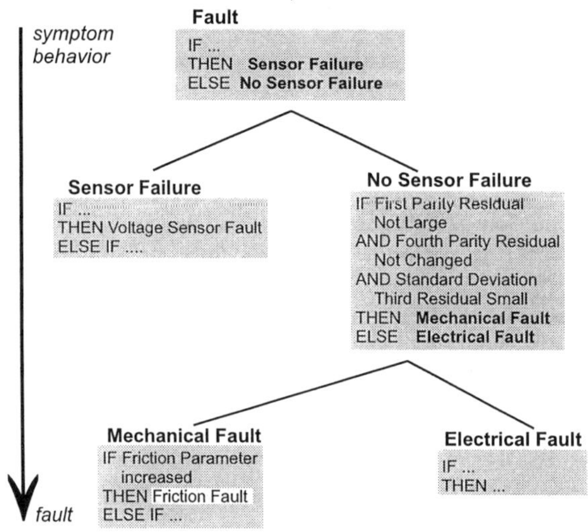

Figure 10. Hierarchical fault diagnosis system. Each block comprises a fuzzy classification tree.

the tree is connected to a decision learned with the SELECT approach, i.e. it contains a fuzzy rule. In every block of the lowest level, a classification tree decides which individual fault has occurred.

The groups of faults have been selected following basic understanding of the d.c. motor supervision concept. Firstly, the three total sensor breakdowns are different from other faults due to their strong effects on all symptoms. Since the motor can be understood as a combination of an electrical and a mechanical component, faults on these two parts were again treated separately. Accordingly, the appropriate subsets of symptoms for the diagnosis were selected. For the diagnosis of mechanical faults, for instance, the estimates of the electrical parameters are unreliable indicators, although some mechanical faults can influence them.

Relation to Fault Trees

The resulting hierarchical classifier can also be interpreted as a set of fuzzy fault trees. If one *reverses* the order of the structure and traces the decisions leading to a particular fault through the tree, it is possible to explicitly draw a fault tree for each individual fault. Fig. 11 shows one fault situation (increased friction in the motor) as an example. The intermediate steps like "Mechanical Fault" from Fig. 10 become *events* of the fault tree.

Similar fault trees can be pictured for the other faults. They are therefore created through a mixture of quantitative (measured data) and qualitative (structural knowledge) information.

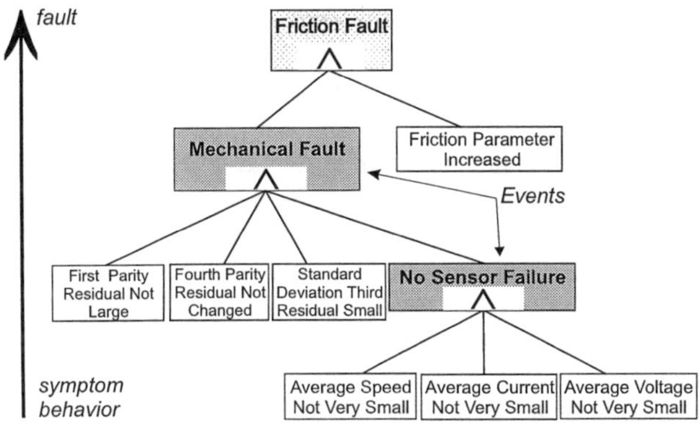

Figure 11. Fault symptom tree for one particular fault extracted from the diagnostic system tree in Fig. 10.

4 Conclusions

In this contribution, the use of fuzzy logic for the fault detection and diagnosis of electro-mechanical systems and actuators has been shown. Methods of learning fuzzy rules were reviewed. It has been emphasized that an interpretable system is essential for diagnostic tasks. A neuro-fuzzy structure was suggested and its application shown. Two examples illustrated the useful application of fuzzy logic both as a tool for building non-linear models and as a mean of creating diagnosis systems that designers can interact with.

References

1. O. Nelles and R. Isermann. Basis function networks for interpolation of local linear models. In *Basis Function Networks for Interpolation of Local Linear Models*, 1995.
2. M. Fischer and O. Nelles. Fuzzy model-based predictive control of processes with fast dynamics. In *2nd International Symposium on Fuzzy Logic and Applications, Zürich, Switzerland*, 1997.
3. T. Höfling and R. Isermann. Parameter estimation triggered by continuous-time parity equations. In *Proceedings of the ACC 95*, 1995.
4. D. Michie, D.J. Spiegelhalter, and C.C. Taylor. *Machine Learning, Neural and Statistical Classification.* Ellis Horwood, 1994.
5. L. Breimann and J.H. Friedmann. *Classification and Regression Trees.* Wadsworth and Brooks, Monterey, Ca, 1984.
6. A. Krone and H. Kiendl. Rule-based decision analysis with fuzzy-ROSA method. In *EFDAN 96*, pages 109–114, 1996.
7. O. Nelles. Furega - fuzzy rule extraction by a genetic algorithm. In *EUFIT Aachen 1996*, 1996.
8. H. Ishibuchi, K. Nozaki, N. Yamamoto, and H. Tanaka. Selecting fuzzy if-then rules for classification problems using genetic algorithms. *IEEE Transactions on Fuzzy Systems*, 3(3):260–270, 1995.
9. M. Ayoubi. Fault diagnosis with dynamic neural structure and application to a turbocharger. In *Proceedings of the SAFEPROCESS'94, IFAC Symposium on Fault Detection, Supervision and Safety for Technical Processes*, 1994.
10. W.M. Caminhas, H. Tavares, and F. Gomide. A neurofuzzy approach for fault diagnosis in dynamic systems. In *Proceedings of the Fifth IEEE International Conference on Fuzzy Systems, FUZZ-IEE '96*, pages 2032–7, 1996.
11. D. Nauck and R. Kruse. Nefclass - a neuro-fuzzy approach for the classification of data. In *Proceedings of the 1995 ACM Symposium on Applied Computing*, pages 461–465, 1995.
12. C. M. Bishop. *Neural Networks for Pattern Recognition.* Oxford University Press, 1995.
13. D. Füssel and R. Isermann. Hierarchical motor diagnosis utilizing structural knowledge and a self-learning neuro-fuzzy scheme. In *Proceedings of the IECON 98*, pages 1883–1888, 1998.
14. D. Füssel, P. Ballé, and R. Isermann. Closed-loop fault diagnosis based on a nonlinear process model and automatic fuzzy rule generation. In *Proceedings of the IFAC SAFEPROCESS 97*, pages 359–364, 1997.

FUZZY LOGIC FOR TRAFFIC MANAGEMENT AND CONTROL

HANS HELLENDOORN
Siemens Nederland N.V., P.O. Box 16068, NL-2500 BB, The Hague The Netherlands, e-mail: Hans.Hellendoorn@siemens.nl
Control Engineering Laboratory, Delft University of Technology P.O. Box 5031, NL-2600 GA Delft, The Netherlands

Fuzzy systems are very appropriate to represent human knowledge. In traffic technology this property of fuzzy logic is important because it can be hard to make mathematical models due to human influences and complex connections between input parameters. One example of the use of fuzzy logic in traffic control is in highway speed control systems. The goal is to optimally use the highway. We describe a fuzzy system for traffic flow control and incident recognition that is in use since some time. Another example that we describe is fuzzy logic in forecasting whether a particular parking garage is full or not. We describe the input parameters and the structure of the fuzzy system. Furthermore, we present an overview of a forecasting application for a down town city area.

1 Introduction

Fuzzy logic is an intelligent technique that can be used in a wide area of applications: in nonlinear control of electric motors, in diagnosing nuclear power plants, in the control of washing machines, in automotive industry, etc. One area where fuzzy logic has turned out to be very useful is *traffic technology.* This is due to a number of reasons. The first one is the general demand for more *intelligence* in every single aspect of life. This holds for domestic appliances, for industrial installations, and, in particular, for traffic technology. Everyone who drives a car knows the frustrations when one has to wait for a red traffic light although no car is coming from other directions. Modern traffic lights with sensors and built-in chips we expect to be intelligent enough to avoid a situation described above. In the same way we expect that highways can recognize the number and kind of cars that drive on it and can react correspondingly.

The second reason is the increased expectation of service. Suppose someone living outside who wants to go shopping downtown. This person may want to call the traffic information service line or watch teletext to obtain a forecast of the traffic situation for the next few hours, such that she can decide to go downtown by car or proceed to some Park and Ride station in the neighborhood.

A third reason has to do with *modelling.* It is very hard to make a model of the traffic situation because many factors play a role and it is not always clear which role they play. If someone wants to give a prognostication of the traffic situation in the next hour the best he can do is give an estimation based on a number of vague expectations. Fuzzy logic is very appropriate to express these vague expectations. One can design fuzzy rules that exactly represent the vague knowledge of the traffic expert. Furthermore, linguistic variables like "Time of Day" can be very well expressed by a set of fuzzy sets.

A fourth reason is connected with *subjectivity.* This argument is highly interwoven with the former argument. One can imagine that one advises people not to go to a certain parking garage because it is almost sure it will be full within half an hour or there is congestion on the way towards it. The quality of the forecast highly depends of the behavior of the people. If they trust the system and act according to it the traffic situation in particular around other parking garages will be different than in case everyone neglects the signs. The same holds for traffic regulation systems on highways that should be adaptive with respect to the behavior of the motorists.

In this chapter we present a number of examples of applications in traffic control. The first application is in traffic regulation on highways. The idea is that there are signs on the highway like given in Fig. 1.

Figure 1: An example of a sign above the highway. One can see a warning for fog and a speed limit of sixty kilometers per hour.

The problem is to have an optimal highway utilization rate. In case there is much traffic on the highway, i.e., one needs more capacity on the road, one has to have a speed limit that harmonizes the velocities of the cars and decreases the distances between cars. We will describe a fuzzy system that performs this.

We will also describe a system that forecasts the utilization rate of a parking garage that stands in the neighborhood of a shopping center. The problem here is that many parameters have to be taken into account, where some are difficult to describe and where it is hard to estimate how their values influence the behavior of people like, for example, certain weather conditions.

Furthermore, we will describe how neural networks can be used to improve the forecasting quality of such a fuzzy system when enough data are available. We could improve an initial fuzzy system from 78% to 86–88% forecasting quality using neural networks. However, the use of expert know-how and detailed comparison of forecasting figures with the actual situation resulted in approximately 93%. One needs such a high percentage, or an even higher one, to make people belief in the system.

Another example concerns a part of a large city with some 120 measuring points for traffic flow and additional information from several places with large parking places. The question here is to describe the traffic flow out of the individual data points, i.e., to extract the information out of the data. A model of the traffic flow can be used to forecast the traffic flow in the future, an important new item in traffic technology.

2 The Fundamental Diagram

Siemens has used fuzzy logic in traffic flow regulation and traffic light control. Both applications are based on the so called fundamental diagram, see Fig. 2.

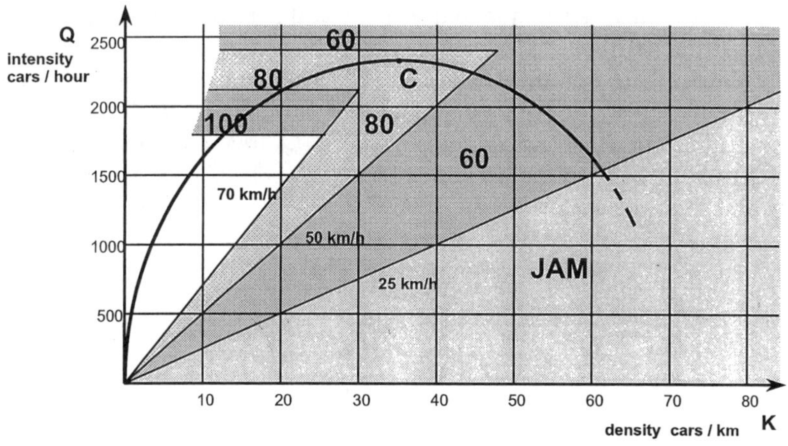

Figure 2: The fundamental diagram.

On the horizontal axis there is K, denoting the number of cars per kilometer, on the vertical axis there is Q, denoting the traffic volume, the number of cars per hour. There are several diagonal lines labeled with velocities. The line "25 km/h," for example, describes the situation where every car drives 25 km/h. In this case, for this particular highway, there are 60 cars on one kilometer and 1500 cars passing in one hour. Here it is assumed that the cars are standard, i.e., a fixed ratio of trucks, vans, limousines, etc. The parabola in

Fig. 2 describes the behavior of uncontrolled traffic. If Q, the ratio of cars per hour increases, K, the number of cars per kilometer, will also increase, until a certain point C, depending on the kind of road we are describing and the allowed maximal velocity. In our figure this point is reached at approximately 35 vehicles per kilometer and 2300 vehicles per hour. From this point on the road becomes overloaded and the average velocity decreases, which means that Q decreases. When all traffic stands, we have K at its maximum and Q equal to zero.

Note that there are a large number of different presentations of this fundamental diagram depending of country, application type, and sensor system. In some books, for example, the traffic volume Q is given on the horizontal axis, and the distance between the cars (slightly different from K) is presented on the vertical axis.

3 Fuzzy Traffic Flow Control

The goal of the fuzzy system is to direct the traffic to or keep the traffic at the left part of the parabola. This can only be achieved by decreasing the velocity of all the cars if the traffic reaches the critical point C. The fuzzy system forecasts how the traffic is moving along the parabola or, with the usual very chaotic traffic situation, in any point of the K-Q-area, and fires fuzzy rules to bring the traffic back to the left part of the parabola. An example of how this works is shown in Fig. 3.

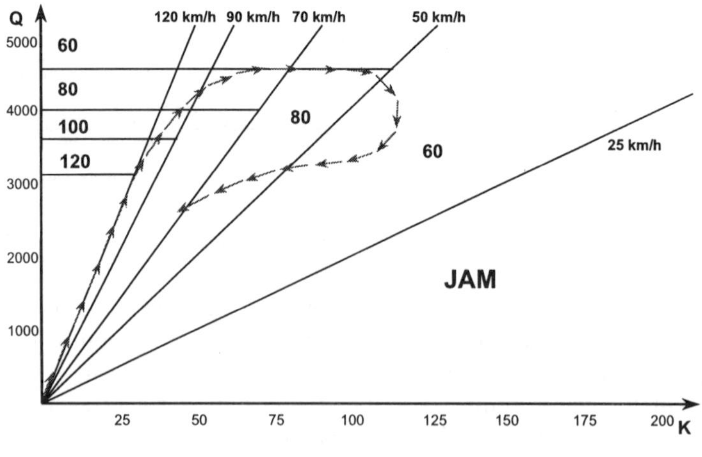

Figure 3: The dynamic fundamental diagram.

Each part of the field is described by fuzzy sets, i.e., the direction of each arrow in Fig. 3 is determined by one or more fuzzy rules of the fuzzy system.

It has turned out that the fuzzy system could very well recognize the occurrence of accidents (see Fig. 4).

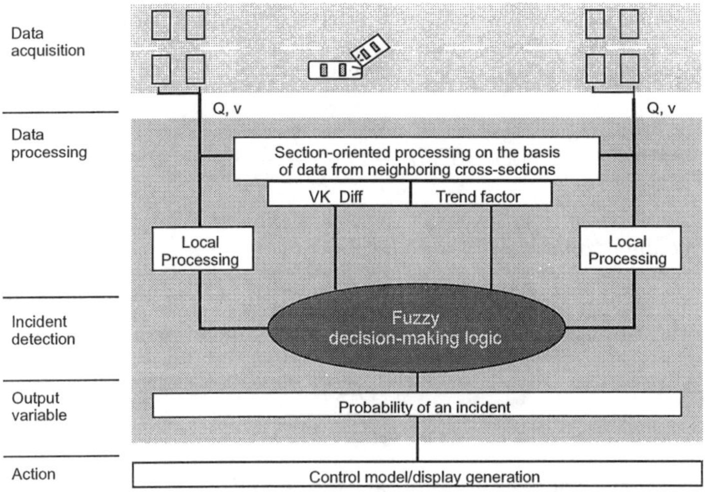

Figure 4: The recognition procedure of accidents (from [1]).

Frequent drivers will know that this is not so easy. On a full highway it often happens that the traffic flow suddenly stops, and then, after a few minutes, starts again and arrives high velocities. One of the tasks of the fuzzy system is to discriminate between this effect and an accident. The problem is to detect platoons, to use the platoon probability to make an incident pre-analysis, and from there to come up with an incident probability. These tasks can very well be performed with fuzzy logic. As described in [1] the following indicators play a crucial role in the incident detection part of the system:

3.1 Speed density difference VK_Diff

This indicator is defined as

$$\sqrt{\frac{(V_{\text{free}} - V_i)^2}{V_{\text{free}}^2} + \frac{D_i^2}{D_{\max}^2}} - \sqrt{\frac{(V_{\text{free}} - V_{i+1})^2}{V_{\text{free}}^2} + \frac{D_{i+1}^2}{D_{\max}^2}} \qquad (1)$$

This formula is used to compare the speed and the local density $D = Q/V$ (V is the mean speed) at point i with point $i+1$, see Fig. 4. An incident between measured cross-section will cause lower velocities and higher densities at point i, causing an increment in VK_Diff.

3.2 Trend factor

The trend factor measures how many cars drive into and out of an interval. If more vehicles enter the interval than depart from it, an incident is probable. The trend factor $F_T(k)$ is defined by

$$F_T(k) = F_T(k-1) \cdot \frac{Q_i(k)}{Q_{i+1}(k)} \tag{2}$$

3.3 Traffic volume trend

This Q_{trend} is calculated at both i and $i+1$. If it decreases at point $i+1$ and remains the same at point i an incident might have been occurred. It is calculated by

$$Q_{\text{trend}} = \frac{Q(k)}{\tilde{Q}(k-1)}, \tag{3}$$

where $\tilde{Q}(k-1)$ is the convex sum of Q at time k and $k-1$.

These three indicators are used in the incident detection part of the fuzzy system.

The whole fuzzy system was implemented on the road from Munich to Nuremberg and was tested in parallel to the conventional system. The result was that the fuzzy system could better forecast critical situations. In particular, accidents where one or two lanes were blocked could be detected three minutes earlier with the fuzzy system than with the conventional system. Within these three minutes 125 to 160 additional vehicles are warned. One of the reasons is that the fuzzy system can better differentiate between several kinds of traffic, e.g., normal cars and small and large trucks. The system is fully implemented now and guarantees a gain in safety.

4 The Parking Problem

In many large cities systems are implemented that tell a driver when he enters the city about the situation of the parking garages downtown. Usually this is done as follows: there are signs with, for instance, the following information:

P Marienplatz	361
P Stachus	120

The meaning of these signs is that the actual situation in the parking garage "Marienplatz" is 361 free places and the actual situation in the parking garage "Stachus" is 120 free places. When it is still a long way to the parking garage it often happens that people drive in the direction of one of these parking garages and the closer they come to the parking place, the less free places appear on new signs. When they finally reach the parking garage it is full. A better concept here would be to have signs that forecast the situation of a parking garage after the estimated driving time towards the garage, e.g. like

P Marienplatz	full
P Stachus	free

To make such a prognosis one needs a lot of knowledge about the estimated traffic situation in the city and about the expected number of other cars also driving to the parking garage. This last number depends on the general interest to go downtown, which again depends on a large number of influence parameters like time of the day, season, special events, weather conditions, etc. It is hard to say how each parameter influences the traffic and why people are going downtown during the one hour and not during the other one. Nevertheless, traffic experts know how to deal with these parameters. Furthermore, graphics of parking garages utilization rates show a clear dependence on, say, time of the day and day of the week. Fuzzy logic is optimally appropriate to, on the one hand, represent the know-how of the expert and, on the other hand, to use the data available to improve itself. Fig. 5 shows an example of the utilization rate of one of the parking garages. One can see the number of occupied places in the parking garage for one whole day. Furthermore, one sees how many cars are driving in and out.

Hence, the goal of a fuzzy system is to decrease

- the traffic searching for free parking lots,
- the time to look for free parking lots,
- the number of accidents,
- the environmental pollution, and
- the gas consumption.

Furthermore, such a system can be built-in in a car, like the RDS system, such that people can obtain information about the actual traffic situation already before entering the city. This can help them to decide to drive to a P+R station and take the underground. Alternatively, people can call a special

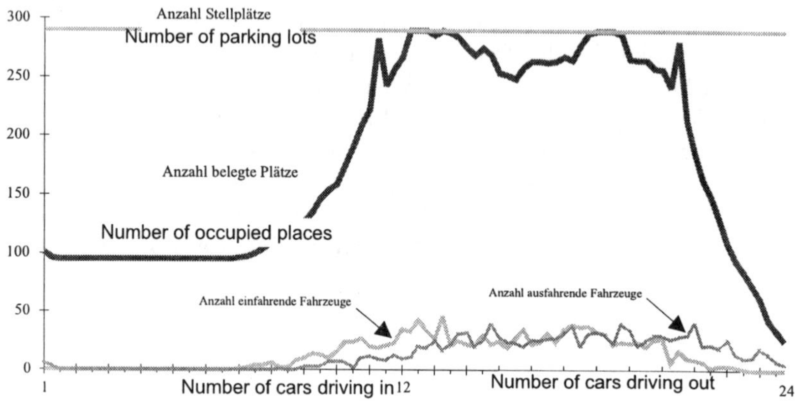

Figure 5: An example of the utilization rate of a parking garage from 0.00h to 24.00h. The curve shows clearly that the data refer to a Thursday when the shops are opened until 20.30h.

phone number or watch teletext to learn about the actual traffic situation before deciding what to do.

We have mentioned before that there are many parameters that influence the traffic downtown. The task of the designer of a fuzzy system is to talk with the traffic expert to get information about which parameters are important and which can be neglected. In our system, after discussions with the traffic expert, we have taken into account the following parameters:

Number of cars driving into and out of the parking garage. This is not a fuzzy variable but a crisp number. For short-time prognoses this is the most important parameter.

Traffic density. In the system we have been using four sensors on different places downtown that together give a global estimation of the momentary traffic density.

Time. There are several kinds of time: time of the day (during the early afternoon many people go shopping), day of the week (on Monday the situation is different from Saturday), season.

Weather conditions. It is difficult to describe the influence of the weather on the behavior of the people.

Special traffic situations. Examples are roadworks, holidays, etc.

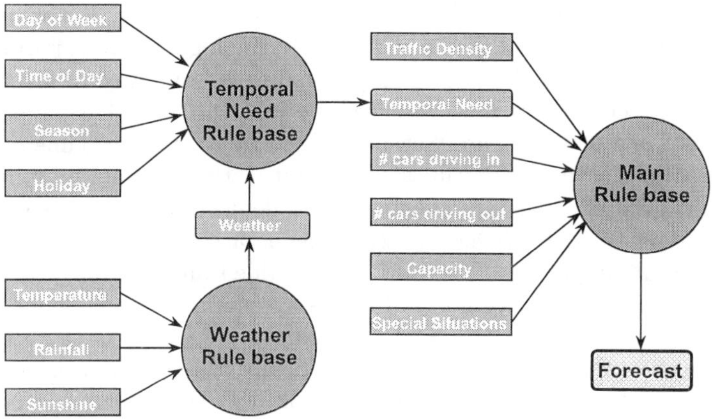

Figure 6: The structure of the fuzzy system.

We have used a special kind of parking garage, viz., a parking garage

- that is situated downtown, near the shopping center, and is mainly used by people that go shopping,
- where most people during the day only park several hours, i.e., there is a relatively low percentage of long-term parkers,
- that, when empty, cannot be filled within one hour, due to the structure of the parking garage.

These assumptions can be changed for different fuzzy systems.

For this application we have used data from ten parking garages in a large city in North Rhine Westphalia in Germany. We also used weather data, that we got from a special institute in Offenbach.

5 The Fuzzy System

In Fig. 6 we have given an overview of the fuzzy system. If we consider all the inputs we will find that "Special Situations," "Day of Week," and "Holiday" cannot be considered as fuzzy variables. Nevertheless, "Special Situations" can be considered fuzzy, because, although every special situation is different, they can be subdivided into fuzzy classes. "Holiday" is a Boolean variable, although also here one can imagine some scaling or classes of holidays, e.g., how many schools are closed. "Day of Week" is definitely crisp.

The other variables are all considered as fuzzy variables. It was important to talk with experts from Siemens Traffic Technology about partitioning the domains. Several variables can be partitioned with, say, three to seven equally divided fuzzy sets, others, however, need special treatment. This, in particular, holds for "Time of Day" and "Season." The fuzzy variable "Time of Day" is subdivided into seven fuzzy values, viz., night, early morning, morning, noon, afternoon, evening, and late evening. The determination of each fuzzy set corresponding to each of these fuzzy values is a complex task, that returns during the tuning process. "Season" is subdivided into five fuzzy values, viz., the four seasons and December. Here, e.g., summer is not defined by the position of the sun (June 21 – Sept. 21), but by the feeling of the people that are going to buy clothes, etc. More specifically, this means that spring and autumn are defined by relatively steep, small fuzzy sets, whereas summer and winter have large support interval.

Each rulebase consists of a number of rules. The weather rulebase determines the influence of the weather on the traffic situation. Rules are, e.g., as follows:

```
RULE SUB_Rule_I_good weather_1
  IF (Temperature IS high)
    AND (Rainfall IS dry)
    AND (Sunshine IS sunny)
    THEN (Weather IS small_traffic)
END
```

This means that when it is nice weather (hot, not raining, sunny) it will be relatively quiet downtown in comparison to other weather conditions.

A rule in the Temporal Need rulebase is for example

```
RULE SUB_Rule_II_Day_of_week_5
  IF (Holiday IS false)
    AND (Time_of_day IS evening)
    AND (Day_of_Week IS NOT weekend)
    AND (Day_of_Week IS NOT Thursday)
    THEN (Temporal_need IS low)
END
```

This means that on a normal day (Thursday evening shops are opened, not a holiday, not during the weekend) it is relatively quiet during the evening. This is independent from the weather situation.

In the main rulebases there are rules that express the prognosis for the time after one hour. An example of such a rule is

```
RULE Rule_II__8
  IF (Capacity IS critical)
```

AND ((Temporal_need IS high)
OR (Temporal_need IS very_high))
THEN (Forecast IS parking_garage_full)
END

So, if the parking garage is almost full and it is to be expected that many vehicles are driving towards the parking garage then the prognosis is that it will be full within one hour.

The problem of chaining these rulebases has extensively been described in
6

6 Tuning the System with Neural Networks

In the beginning of the project we started with first building an initial rule base for six parking garages in the same city that all had approximately the same conditions, namely, they were all close to shopping centers downtown. We also had available data and information about other parking garages, but these were so different in character that we did not use them anymore (e.g., one was close to the opera). This initial system was for all parking garages the same and had a prediction quality of 78%. This is definitely too low for such a system to be installed, but a good result for a first prototype. We then had many discussions with traffic and parking garage experts and could highly improve the system. We have taken test data from two weeks, one in June and one in August where we came to the following results:

Parking garage	Prediction quality
PG1 J	0.9548
PG1 A	0.9239
PG2 J	0.9147
PG2 A	0.9086
PG3 J	0.9298
PG3 A	0.9272
PG4 J	0.9415
PG4 A	0.9051
PG5 J	0.9147
PG5 A	0.8849
PG6 J	0.9415
PG6 A	0.9374

Better results can be obtained when one tunes each parking garage separately. Nevertheless, the average of the above given values is 0.9237 which is a really good result.

We also used a neural networks approach based on fuzzy associated maps (FAM) from Kosko [8]. FAM is nothing else than a clustering method based on neural networks. In systems with chained rule bases like we have here it is usually rather complicated to use neural network because the back propagation becomes very complicated, i.e., it is difficult to find the place where the error was made that influenced mostly the prediction quality. Using FAM we were able to improve the quality from 78% to 86–88% which is a real good result although it is less than the achieved 92% we got by manual tuning. We also used FAM to tune each parking garage separately. Here we could achieve results of approximately 93% average prediction quality which is an excellent result for a system without human know-how about the traffic situation. If one can use neural networks plus data to tune each system this makes sense. We expect to obtain improved results with the neural-fuzzy approach described in [3] in combination with the ADALINE (adaptive linear element) or LLM (Local Linear Map) approach.

7 Traffic flow modeling

Another application in traffic control is taking place in Cologne, Germany. From 120 induction loops we gather information about the traffic flow in a part of the city. These data including information about special events like fairs, sporting events, etc. are analyzed and used to deliver a daily prognosis message in the newspaper and an hourly traffic report on the local television.

The situation is as follows. A part of the city consists of two bridges, a number of large roads with crossroads, etc. Everywhere in this area there are induction loops in the road that keep track of the number of cars and the kind of cars passing along. All this information is gathered in a central computer and put into a large database. The data is just raw material containing the desired information. There are a number of problems that have to be tackled to extract this information out of the data.

First, the information coming from the induction loop contains errors. Some of the induction loops may be defect, others now and then deliver strange, unexplainable data. Therefore it is necessary to build a graph of all measuring points and to demonstrate the traffic flow through the graph. This graph is used to perform a consistency check between measuring points and to demonstrate the traffic flow through the graph. If one node in the graph produces data that cannot be explained by neighboring nodes, these neighboring nodes are used to adapt the data. The graph also has knowledge about the usual traffic division on the lanes. This information is used to check if a sensor on a lane is error prone.

Secondly, there may be special situations like the closing of lanes or complete roads for work on the road. It is impossible to expect that the system will always be informed about such situations. Therefore the above mentioned graph is not only designed for defective induction loops but also for these situations. Furthermore, there can be an accident, the system is able to detect this.

Thirdly, it is too expensive to regularly control the system by people on location. The presence of people who tune the system could greatly improve it. They would be able to see circumstances that the system cannot 'see.' For cost reasons, the system has to do without this additional information.

There is also information that greatly improves the quality of the system. The most important source of information is the presence of 'sinks' and 'sources' within the graph that can be described surprisingly well. Two main examples in the part of Cologne that we considered were the fair and the sporting stadium. The boards of these institutes deliver information about the number of people they expect to come to their events. This information is of a high quality. Information from past events plus the corresponding traffic data tells us how many cars are to be expected and the time of the day these cars will come. This information is fed into the graph. A *sink* is a node in the graph where cars 'disappear,' e.g., on a parking place of the fair, the stadium, or a company. An example of a source is the closing of the fair or the end of a sporting event, when all cars return to the roads to take part of the traffic. These sinks and sources are well described and contain valuable input information for the graph.

We also use another source of information connected with different groups of people taking part of the daily traffic, e.g., people that go to work, people that go shopping, etc. Each of these groups can well be represented by a Gauss-function

$$f_i(x) = e^{-\frac{(x-\mu)^2}{2\sigma^2}} \tag{4}$$

where μ is the center and σ is the spread. Each Gauss-function can be multiplied by a factor α_i to express its relative weight. The daily curve $d(x)$ can then be approached by

$$d(x) \approx \sum_i \alpha_i f_i(x) \tag{5}$$

There is, for example, a 'blue collar' Gauss-function with mean value 7.30h and a spread of, say, one hour, a 'white collar' Gauss-function with mean value e.g., 8.30h, a 'morning shopping' function with mean value 10.30h, etc. Furthermore, there is a very broad Gauss-function with a spread of over twenty hours and a center a 13.00h representing traffic that can hard be grouped into one of the standard categories. Depending on the day these functions can be scaled using experience and traffic data. Using this experience together with, e.g., weather forecast data enable us to present a good forecast of the traffic flow. A neural network can be used to learn the α's, μ's and σ's.

8 Conclusions

We have shown some examples of the use of fuzzy logic in traffic control. The first approach dealt with traffic flow control on highways and accident recognition systems. We have shown that fuzzy logic led to improved prognosis qualities and better flow control. In the second example we have shown that fuzzy logic can be used to estimate the number of cars driving to a number of parking garages. This can be used to put signs that forecast the situation of a parking garage when someone drives in that direction. Siemens now uses fuzzy logic to build larger and more complicated traffic control systems that integrate a number of local systems.

Some open questions have also theoretical importance:

1. Our system predicts for the time one hour from now. It would be interesting to change the system so, that it predicts for 90 minutes, two hours, etc. The difficulty here is that one should not have to change the whole rulebase or numerous fuzzy sets when one wants to have another prediction scope.

2. There should be an extra input parameter that deals with the type of parking garages, so that the system can be easily applied to all types of parking garages.

3. The most difficult question is how to integrate human and social aspects into the system. The prognosis quality is completely different for one and the same system when people trust the system or not. The system expects that people believe what it is saying but this will normally not be the case. In this respect the system is completely different from similar systems in telecommunication technology.

References

1. F. Busch and A. Ghio, Automatic Incident Detection on Motorways by Fuzzy Logic, *Int. Conf. Traffic and Transport Solutions*, Amsterdam, 1994.

2. D. Driankov, H. Hellendoorn and M. Reinfrank (1993), *An Introduction to Fuzzy Control*, Heidelberg, Springer Verlag.
3. W. Hauptmann and K. Heesche, A Prototype of an Integrated Fuzzy-Neural System, *Proc. of the EUFIT'94,* 6achen, 1994.
4. H. Hellendoorn and C. Thomas, On Defuzzification in Fuzzy Controllers, *Journal of Intelligent and Fuzzy Systems*, **1**(2)(1993)109–123.
5. H. Hellendoorn and C. Thomas, The ξ-Quality Defuzzification Method, *Proceedings of the fifth IFSA World Congress*, Seoul, 1993, pp. 1159–1162.
6. Driankov, D. & Hellendoorn, H., "Chaining of Fuzzy IF-THEN Rules: Some Problems," *Proceedings of the FUZZ-IEEE/IFES International Conference*, Yokohama, 1995, pp. 103–108.
7. J. Konrad, *Entwurf und Programmierung einer Fuzzy-Steuerung für eine Verkehrsbeeinflussungsanlage auf BAB*, Siemens internal memorandum, 1992.
8. B. Kosko, *Neural Networks and Fuzzy Systems,* Englewood Cliffs, N.J., Prentice-Hall, 1991.

FUZZY MODELING AND CONTROL IN AVIONICS

GERARD SCHRAM
SKF Engineering and Research Centre B.V.
Kelvinbaan 16, 3439 MT Nieuwegein, The Netherlands

MARCEL OOSTEROM, HENK VERBRUGGEN
Faculty of Information Technology and Systems
Delft University of Technology
Mekelweg 4, 2628 CD Delft, The Netherlands

Improving the flight safety and increasing the efficiency of designing flight control laws are two key issues in modern flight control. For flight safety, we need flight control systems that have good performance under nominal as well as under exceptional, hazardous conditions. On the other hand, the design of a flight control system is very time-consuming and hence very expensive. In this perspective, the application of fuzzy modeling and control is reviewed and discussed.

1 Introduction

An increasing amount of attention has been paid in recent years to intelligent control of aeronautical systems [1,2]. The integration of qualitative knowledge (heuristics) and quantitative knowledge (analytical models) through fuzzy modeling and control offers interesting new possibilities. For example, the aggregation of human pilot heuristics and analytical aircraft models has yielded simple and transparent flight control laws [3,4]. Besides for aircraft, fuzzy control has been suggested for helicopters [5,6,7] and spacecraft as well [8,9]. Fuzzy modeling on the other hand has been applied to the identification of control system failures [10,11] and to the modeling of the human pilot [12].

From a pragmatic point-of-view, however, it may be wondered why fuzzy logic could play a role in the control of aeronautical systems. The (aero)dynamics are well described by first principles modeling and, although of nonlinear nature, the adaptation of controllers to varying dynamics has been successfully applied in practice. Therefore classical control structures are found in all present-day, fly-by-wire control systems [13,14]. This is however only possible when *extensive design efforts* are executed for the various circumstances to be expected. Moreover, *exceptional* situations such as control system failures need advanced detection mechanisms and control law reconfiguration such that performance and hence flight safety is maintained. For these reasons, the application of fuzzy modeling and control can be justified as will be explored.

The first part is devoted to the question whether fuzzy modeling and control can reduce the design effort that is necessary for the development of flight control systems. The design is mainly focussed on linear single-input-single-output (SISO) structures and only valid for limited regions of the flight envelope. Removing the interaction between the SISO loops and including extensive gain-scheduling procedures requires much additional design effort. Results of previous work based on the Mamdani approach of fuzzy control are reviewed. Recent European initiatives towards an integrated classical-fuzzy approach with the Takagi-Sugeno type of fuzzy modeling and control are described as well.

The second part focuses on the flight safety issue. To ensure future safety for the increasing number of flights, more stringent safety requirements are demanded for future control systems [15,16]. Therefore, robustness against exceptional situations such as system failures is required by applying advanced fault detection, diagnosis and reconfiguration of the flight control laws. The role of fuzzy modeling and control is again reviewed and discussed.

2 Towards a more efficient design of flight control laws

2.1 Overview of previous work

The first application of fuzzy logic in flight control has been introduced by Larkin [3]. The basic goal of this fuzzy controller is to formalize the human pilot's control strategy by a collection of **if–then** rules. The control task involves the landing of a simulated aircraft in the vertical plane. Elevator deflection and engine speed commands are formulated in the form of fuzzy rules as a function of glide slope and airspeed deviations. In this way, a PD-like fuzzy controller has been derived.

A more elaborated simulation study into fuzzy flight control has been performed by Steinberg [4]. Following a three dimensional flight path, the task involves the control of speed, climb rate, and attitude in order to allow a safe landing. The strategy of the pilot consists of several subtasks, e.g. roll angle control. If the desired roll angle is positive large (roll angle "error" positive large), then the pilot imposes a positive lateral displacement on the stick:

> **If** roll angle error is *positive large*
> **then** lateral stick position is *positive large.*

The rule describes a proportional relation between roll angle error and lateral stick position. The rules are usually a combination of proportional as well as derivative action. A typical rule from the climb rate rule base is:

> **If** climb rate error is *near zero*
> **and** climb acceleration is *positive large*
> **then** longitudinal stick position is *positive medium.*

The rules are part of a set of six rule-based fuzzy controllers. The results show significant *performance* benefits with respect to the conventional controller in both guidance and control tasks but the design effort is not reduced[4]. Basically similar results are obtained in comparable studies[17,18,19,20,21,22].

2.2 Fuzzy control results of the RCAM design challenge

More recently, the Group for Aeronautical Research and Technology in Europe (GARTEUR) formulated the so-called Research Civil Aircraft Model (RCAM) design challenge[23]. Major European airframe manufacturers, relevant research institutes, and universities were involved. The goal of the project was to investigate the efficiency of advanced control techniques for the design of flight control systems, in order to reduce the development cost. The problem involved the control of a two-engined, civil aircraft during the landing phase.

A *simplified* scheme of the aircraft model in the vertical plane is shown in Figure 1. The aircraft moves with an inertial velocity V at an altitude h along the inertial flight path γ. The angle between the horizontal plane and the aircraft body is the pitch angle θ. Due to wind velocities, the air-relative velocity differs from the inertial velocity and is given by the airspeed V_A. The angle between the airspeed direction and the aircraft body is called the angle of attack α.

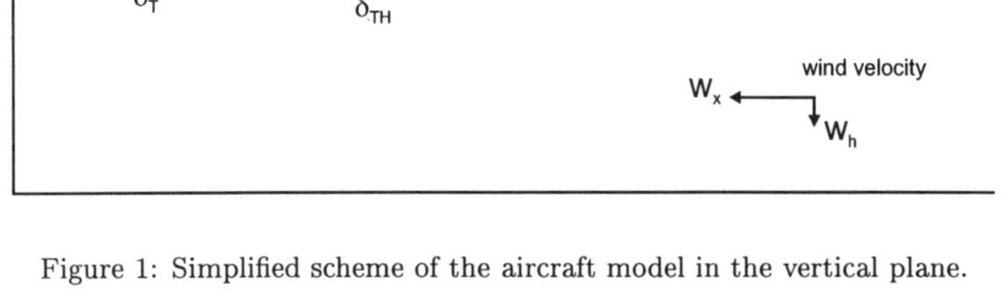

Figure 1: Simplified scheme of the aircraft model in the vertical plane.

The actuators to control the longitudinal aircraft motion are the two engines and the tail-plane. The throttle setting $\delta_{TH} = \delta_{TH1} = \delta_{TH2}$ determines the thrust of both engines which initiates a deceleration or an acceleration. The deflection of the tail-plane δ_T introduces a pitch angular velocity, called the pitch rate q. The objective of the longitudinal controller is to regulate altitude and airspeed by throttle setting and tail-plane deflection.

Figure 2 shows a simplified scheme of the aircraft model in the horizontal plane. The deviation from the desired flight path is indicated by the lateral deviation y_{lat}. The inertial direction and velocity of the aircraft are defined by the inertial track angle χ and the inertial velocity V. The angular position of the aircraft body with respect to the inertial reference frame is described by the heading (or yaw) angle ψ and the roll angle ϕ. Due to the wind velocity, the inertial velocity differs from the air-relative velocity V_A. The angle between the aircraft body and the airspeed direction is defined as the sideslip angle β.

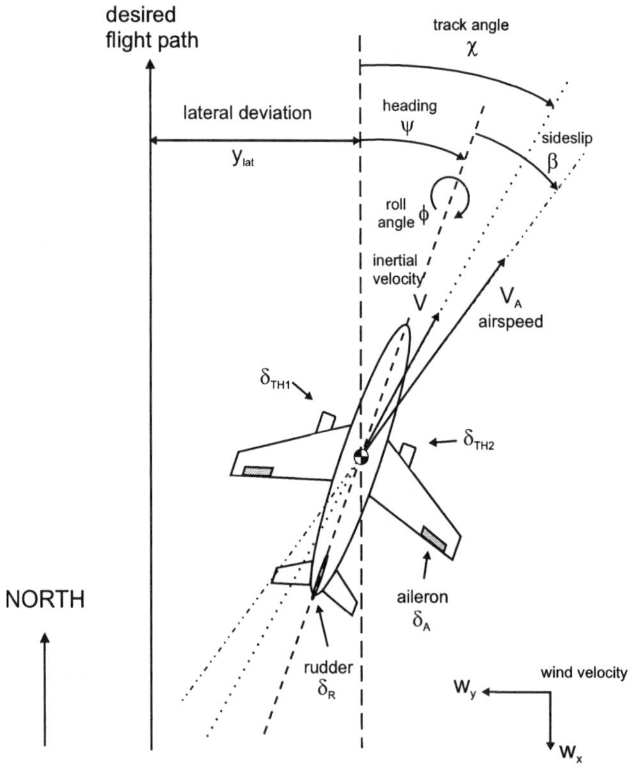

Figure 2: Simplified scheme of the aircraft model in the horizontal plane.

The actuators to control the lateral aircraft motion are the ailerons and the rudder. The (opposite) aileron deflections δ_a introduce a roll angular velocity, called the roll rate p, while the rudder deflection δ_r causes a yaw angular velocity, called the yaw rate r. The objective of the lateral controller is to track the desired flight path with minimal sideslip angle by using ailerons and rudder deflections.

The dynamic behavior of the aircraft is nonlinear, of high-order, and strongly coupled. Both horizontal and vertical flight paths had to be followed accurately, under various conditions. The performance specifications included rise time, settling time, overshoot, and steady state deviation requirements. Robustness was tested for mass, center-of-gravity, and airspeed variations. Passenger comfort and safety indexes included acceleration limits, damping specifications, and bounds on certain variables.

Twelve controller designs were extensively evaluated and compared within the project. One of the controller designs was based on fuzzy control and is reviewed here [24,25]. Since the objective was to show how the human pilot's heuristics on controlling an aircraft can be incorporated, the Mamdani type of fuzzy control was selected. Experienced human pilots know how to handle an aircraft satisfactorily, i.e., they know how to combine competing criteria such as performance, robustness, safety, and passenger comfort without mathematical formulations. The design procedure consisted of three steps. In the first step, simple SISO classical control laws for stability augmentation (attitude control) were defined and tuned. The second step used the experience of the human pilot to define multi-input-multi-output (MIMO) fuzzy control laws. The fuzzy control laws were hereby initialized as a linear function between the input variables and the control outputs. In the final step the controller was fine-tuned by modifying the rule consequents based on the time responses in the design specifications. Additional rules for low airspeed and engine failure were included as well showing that gain scheduling and handling of exceptional situations can be readily incorporated.

In Figure 3, the test scenario used in the RCAM controller evaluation is shown. The flight mission is to follow a horizontal flight path with strong lateral wind and engine failure (0-1), followed by a sharp right turn (1-2) and a descending flight path (2-4). The desired flight path and the simulated trajectory of the aircraft with fuzzy controller can be hardly distinguished.

The fuzzy controller was compared with respect to the other control methodologies. Most of the controllers fulfilled the criteria and showed comparable results. This was exactly the objective of the design challenge. The aim was not to give the answer to the question "which control method is best", but rather to show, step by step, how modern control could be applied. The ultimate goal was to reduce the necessary design effort. All methods demonstrated improvements in specific areas compared to the classical approach. The main results concerning the fuzzy logic controller were:

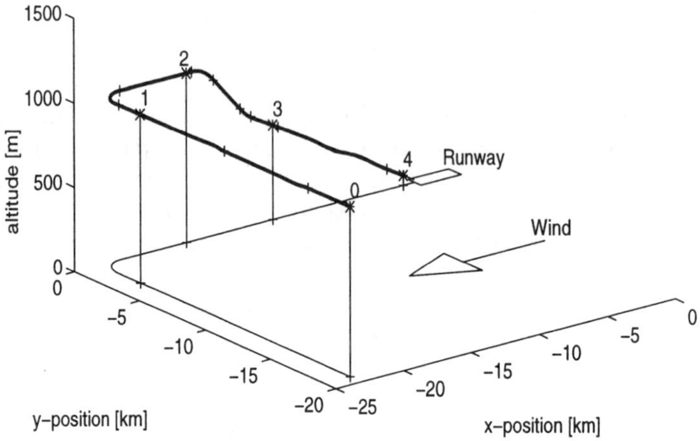

Figure 3: RCAM test scenario: following a three-dimensional flight path.

- The translation of human heuristics into the fuzzy controller is a *systematic* procedure. The heuristics can directly be used to define MIMO control laws with good decoupling. Furthermore, the use of linguistic terms makes the interpretation of (tuning) the controller straightforward. However, tuning the fuzzy controller must still be performed by trial-and-error (in fact for all control methods). Linear control tools, e.g., in the frequency domain, can hardly be used to synthesize and to analyze the fuzzy controller. The final verification of the controller has therefore to be done by extensive (randomized) simulations.

- The approach yields intuitive, simple rule-based control laws of low-order. This is very important, since black-box controllers like high-order state feedback or neural networks are not acceptable in the aeronautical community. The *transparency* is of main importance in the validation of flight control systems. Therefore, no problems are expected concerning validation and certification of the fuzzy control laws.

- Robustness is achieved *in*directly since the (nonlinear) controllers are of low gain and through the adaptation by gain-scheduling, e.g., as a function of airspeed. This is exactly the way human pilots achieve robustness.

In conclusion, the Mamdani approach of fuzzy control contributed to a *small* degree to the efficiency improvement of the design of flight control laws. The main problem was that a trial-and-error tuning procedure cannot be completely avoided. In fact, the problem asks for an automated design procedure regardless of the control methodology that is used. Furthermore, analytical tools to verify the performance and the robustness of this nonlinear fuzzy controller are still lacking. Based on these findings, future applications of the Mamdani approach are expected to be only useful in that area of flight control and guidance where pilot techniques can be copied and where the precision requirements are not too stringent.

One has to realize that the most profound and most concise knowledge of the aircraft is found in the model of the aircraft dynamics based on first principles. Although the modeling requires much effort, such an aircraft model is developed anyway for simulation purposes. As an alternative design approach, a (fuzzy) local-linear model could be derived from the nonlinear aircraft model. Based on the fuzzy model, local-linear controllers can be designed using linear control techniques, e.g., in the form of classical control laws. The obtained fuzzy model-based controller is a transparent gain-scheduled linear controller [26,27]. Research initiatives into this direction are discussed below.

2.3 Gain-scheduling based on the Takagi-Sugeno approach

The Delft University of Technology is currently involved in a Brite Euram project which aims at developing a cost effective fly-by-wire control system for the category of small commercial aircraft. The participants of the project are Israel Aircraft Industries, Alenia Aerospazio Aeronautica, Centro Italiano Richerche Aerospaziali (CIRA), Marconi Electronic Systems, National Aerospace Laboratory (NLR), and the Israel Institute of Technology (Technion). An important issue is the application of fuzzy logic (FL) in the gain-scheduling procedure in both the design and implementation phase. From the pilot's point-of-view it is desirable that the response of the aircraft is nearly invariant throughout its flight envelope. Consequently, the use of fixed gains in the control laws is unlikely to satisfy this preference. Gain-scheduling is the general method to overcome this deficiency.

The *conventional* approach is to define a grid of operating points covering the flight envelope and to define the feedback gains of the control laws such that the closed-loop aircraft dynamics are satisfactory in all these operating points. The feedback gains are put in a look-up table and shaping functions are used to define the feedback gains for flight conditions between the operating points. The common design procedure is to start with two extreme operating points in the flight envelope, to determine the feedback gains in order to obtain the required aircraft behavior in those flight conditions, and to interpolate between them. When the aircraft behavior in flight conditions between these operating points is not satisfactory, another operating points is added to the grid. In this way the complete flight envelope is covered. This trail-and-error approach is very time-consuming. Moreover, the expertise of the flight control engineer plays an important role.

Instead, a partially automated procedure of designing the gain-scheduled control laws is proposed, based on the *Takagi-Sugeno* approach. Based on input-output data that is obtained from the nonlinear aircraft model, the operating regime is automatically partitioned in local areas through fuzzy clustering[28]. Fuzzy clustering will result in a number of subsets covering the initial dataset. To what extend a certain point belongs to a certain subset is defined by the membership functions that are generated automatically. The subsets identify the areas where local linear models are valid. Each cluster prototype (i.e. the center of each subset) will serve as an operating point for the gain-scheduling procedure. Due to the inherent smooth transients between the local models and the "optimal" grid of operating points the number of local models is probably lower than in the case of following the conventional procedure.

In each operating point, the feedback gains will be calculated using the local linear model of that particular subset, based on conventional control techniques. The gains are used in a scheduling mechanism that determines the control law gains over the whole flight envelope. The fuzzy logic controller can be thought of as a replacement of the "conventional" look-up tables and interpolation functions. This is implemented in the following way:

$$K = \sum_{i=1}^{n} K_i \cdot w_i, \text{ with } \sum_{i=1}^{n} w_i = 1 \qquad (1)$$

where w_i is the degree-of-fulfillment of the flight condition with respect to subset i and K_i the feedback matrix based on the linear model of subset i. The computation of the control gains in each operating point can also partly be performed automatically. Moreover, identifying the operating points in the conventional approach requires experience of the flight control engineer in contrast to the fuzzy clustering technique.

2.4 Control law scheduling of spacecraft

Together with the European Space Research and Technology Centre (ESTEC), the Delft University of Technology also investigates whether fuzzy logic techniques could be used in the guidance, navigation and control system of the

X-38 CRV. The X-38 is a spacecraft attached to the International Space Station (ISS) and will serve as a re-entry vehicle for astronauts on-board. The X-38 will departure from its docking port of the ISS and should reach a particular landing site. The system will have autonomous control law reconfiguration features for changes in mission scope, changes in mission environment, and failure tolerance characteristics.

The short flight of the X-38 is cataloged in three basic phases: the de-orbiting, the re-entry and the landing. For each of these phases the aircraft behaves differently, which requires scheduling between different control laws. During the de-orbiting phase the control of the spacecraft is carried out with small thrusters. During the re-entry, the density of the atmosphere is increasing which enables the employment of the aerodynamic control surfaces. Therefore, the control power will be transferred from the small thrusters to the control surfaces which requires a structural change of the control laws. The environmental conditions will change gradually. For this purpose, some control law scheduling needs to be introduced, e.g., as a function of dynamic pressure. Once the dynamic pressure reaches a certain value, the spacecraft can be controlled by the control surfaces only. From this point on, the gain-scheduling as described in the previous section can be employed. For the final descent to landing a para-foil parachute is deployed.

Due to the inherent smooth transition and gradual interpolation between the control modes, it is expected that only two control modes are required for the transition from the de-orbiting to re-entry phase when fuzzy logic is employed as the scheduling mechanism. This reduces the control system complexity and hence the design effort while providing sufficient control performance.

3 Towards fault-tolerant control

The first step towards fault-tolerant control is fault detection and diagnosis (FDD). Fault detection refers to the detection of off-nominal operation. The determination of the type, the size and the location of the fault is called fault diagnosis. The improved awareness and a better understanding of the degraded performance caused by control system anomalies could help the pilot in recovering the aircraft and eventually reconfiguring the control system. Among many possible faults, sensor and actuator faults can have a serious impact on the performance of the control system [29,30,31] and are considered here.

In order to achieve fault detection and diagnosis, some kind of system redundancy is in principle required. *Direct* or *parallel* redundancy means that multiple independent components (e.g. triplex or quadruplex replication) with a majority vote selection mechanism are used. The non-impaired identical component is brought into service to replace an impaired component when a fault occurs. The alternative form is *functional* or *analytical* redundancy. In this case, the redundancy results from flexible logic that synthesizes missing measurements or control forces using other types of sensors or actuators that are not impaired. It requires fewer additional components at the cost of increased computation. In aeronautical applications, there is a tendency not to substitute direct redundancy entirely by the analytical alternative, but to suppress the (physical) redundancy to some extent. For example, the output of a sensor could be reconstructed from other types of sensors and used in a voting scheme, hereby reducing the number of necessary sensors.

Traditional FDD methods are based on the monitoring of signal trends. The signals are either directly used or after some signal processing. Typical quantities that could be used are the magnitudes of the signals, mean values, limit values, trends, correlation coefficients, etc. The fault is detected and identified when the signal or a combination of signals exceed a predefined threshold. Fuzzy logic could play a role in the implementation of common sense logic in the arbitration of signal trends. An example is the (self-)monitoring of two parallel "channels", from control stick position via control laws to elevator position, by checking the following fuzzy rule:

> **If** control stick *back* (pitch up)
> **and** elevator in *upward* position
> **and** pitch rate sensor signal is *increasing positive*
> **then** the channel is *correct*

In the definition of the antecedent and consequents conditions, some basic understanding of the aircraft dynamics is of course necessary. Assume that a discrepancy between the two (parallel) control stick sensors signals arises. From this discrepancy, no conclusions can be drawn concerning which one has failed. Applying the above rule in both channels provides additional information to select the correct control stick deviation signal. If reliable, this duplex system with common sense logic could provide the same safety level as a conventional triplex system, and hence reduces the number of hardware components.

Since accurate models of aeronautical systems are generally available, *model-based* approaches of FDD can in principle provide more accurate information about the fault status. The general model-based approach of fault detection is to compare measured and reconstructed outputs of the system. Using special types of observers, robust failure indicators (residuals) can be defined such that they are non-zero when faults occur, while being as insensitive as possible to disturbances, model uncertainties, and measurement noise [32,33,34]. In order to isolate the fault, a specific test is usually formulated such as the multiple-model hypothesis test [32]. Each fault hypothesis is modeled separately; the most likely hypothesis indicates the fault [35,36]. However, hypothesis tests get complex and computationally intensive when increasing the number of possible faults.

In comparison to FDD schemes that are based on linear process models, nonlinear approaches are not yet well-established. Recent developments in "soft-computing" technologies suggest nonlinear model alternatives in the form of artificial neural networks and fuzzy systems [37,38,39,10,11,40,41,42,43]. For safety critical systems like aircraft, the FDD scheme must have a transparent structure. Therefore, the *combination* of fuzzy models and linear residual generators is proposed in [25,44,45] and is shortly described in the remaining part of this section.

Linear residual generators, e.g., based on parity equations, are applicable only at one operating point. In order to make the residuals robust to varying operating conditions, the parameters in the residual generators are smoothly scheduled with respect to the relevant physical system parameters using fuzzy rules. The operating conditions under which a local linear (Takagi-Sugeno) system hold are described by the antecedent part of the rule. The local linear residuals form the consequent part. By using fuzzy antecedent propositions, a gradual, smooth interpolation between the operating points is achieved. The benefits from both the well-established linear residual generating technologies and from the nonlinear, rule-based modeling methodologies are retained.

Assuming that each fault has a unique direction in the nonlinear residual space, the type (scaling, bias) and magnitude of the fault can then be identified from the residual vector. For this purpose, a recursive estimation procedure is proposed. In the recursive estimation, the magnitude of the failure is estimated. Given the estimate, the fault is classified into a set of hazard and non-hazard fault classes. For each class, a corresponding set of control laws can be defined. The classes are represented by fuzzy sets which results in a smooth transition from normal to "fault" control modes.

Extensive simulations are conducted on the nonlinear RCAM model under realistic assumptions about actuator dynamics, measurement noise and turbulence [23]. Depending on the cargo of the aircraft, the mass of the aircraft varies between 100,000 kg and 150,000 kg. The airspeed of the aircraft varies between 54 and 90 m/s. The five control effectors of the aircraft model are the ailerons, tail-plane, rudder, and two engines. The 11 given measurements are pitch rate, roll rate, yaw rate, roll angle, sideslip angle, forward load factor, vertical load factor, airspeed, and velocities of the aircraft body in the inertial reference frame. Based on the aircraft model, parity equations are derived in 15 operating points and integrated in a Takagi-Sugeno system with airspeed and mass as antecedent variables. Five membership functions are defined for airspeed between 54-90 m/s, and three membership functions are defined for mass between 100,000-150,000 kg, respectively. The combinations of airspeed and mass describe the flight conditions which span the operating range. The nonlinear residuals are used to estimate the effectiveness (between -1 and 1) and bias of the control surfaces and engines.

One of the investigated scenarios includes a simultaneous engine failure and 50% degradation of the effectiveness of the ailerons, e.g., caused by a defect hydraulic actuator that drives the control surfaces, at t = 10 seconds. The control surface deflections and throttle setting are shown in Figure 4. The flight conditions are airspeed 85 m/s and mass 110,000 kg. Although this fault scenario is quite serious, the original control laws can still compensate automatically for the faults. First of all, the loss of thrust of engine no. 1 is compensated by an increase of the thrust of engine no. 2. Because of the yawing moment that results from this asymmetric thrust, non-zero deflections of the rudder and ailerons are necessary. Secondly, because the effctiveness of the ailerons is decreased, the variance of the aileron deflections is twice as large.

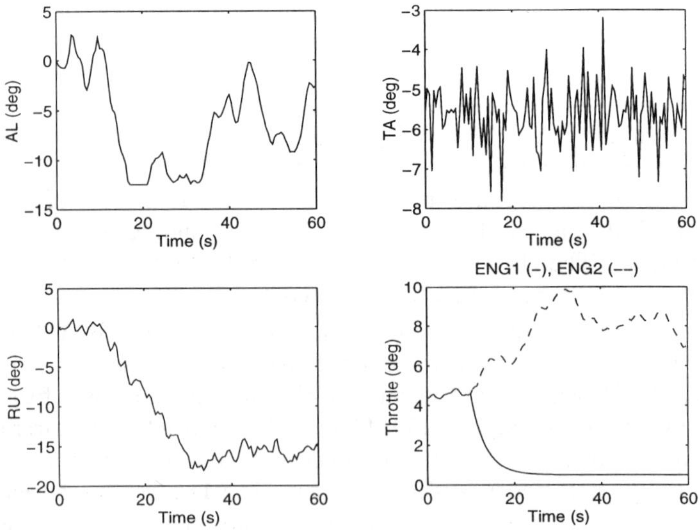

Figure 4: Control surface deflections and throttle settings in case of engine no. 1 failure and 50 % degradation of ailerons (AL = ailerons, RU = rudder, TA = tail-plane, ENG = engine).

The estimated scaling and bias factors are shown in Figure 5. The results show that both the scaling factors of the ailerons and engine no. 1 are correctly estimated (0.5 and 0, respectively). The bias estimates also show an error for engine no. 1. The estimated value corresponds exactly to the negative of the equilibrium (trimmed) throttle position, see Figure 4. Notice that the other parameters are correctly estimated as well (no faults) despite of the large deflections.

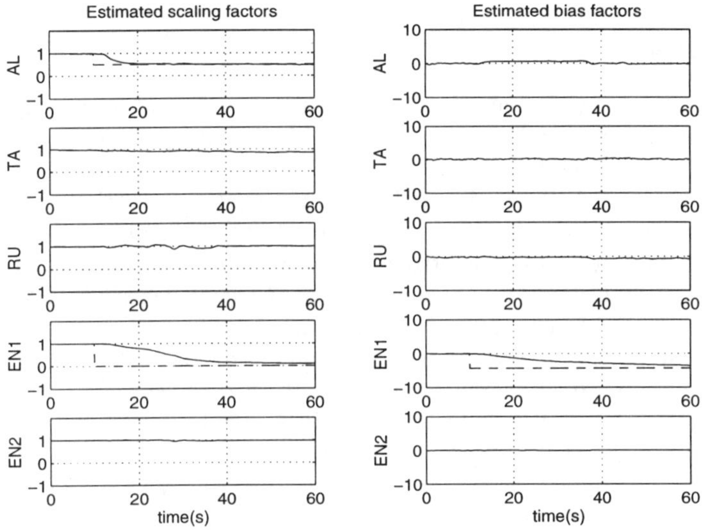

Figure 5: Estimates in case of engine no. 1 failure and 50% degradation of ailerons. The dashed lines indicate the faults.

After fault detection and diagnosis, the control laws can be reconfigured such that the performance remains as much as possible. For this purpose; the Takagi-Sugeno approach of fuzzy control can be applied as well; each local linear controller represent the control laws for the specific hazard and non-hazard fault situations. Since the fault classes are modeled by fuzzy sets, smooth transitions and gradual interpolation between the control modes are automatically achieved [25,46].

4 Conclusions

Reduced design effort and increased safety have been emphasized as the main motivation to introduce new methodologies such as fuzzy modeling and control in avionics. Since the (aero)dynamics of aeronautical systems are well-understood, approaches that combine classical and fuzzy modeling and control techniques are suggested. Gain-scheduling and control law scheduling in a Takagi-Sugeno framework can offer a reduction of the design effort and increases the transparency with respect to the conventional approaches. Monitoring with (common sense) fuzzy logic, nonlinear residual generation, and smooth reconfiguration with fuzzy logic are examples of pragmatic contributions to fault-tolerant control which ultimately improves flight safety.

References

1. M. Steinberg. Potential role of neural networks and fuzzy logic in flight control design and development. In *AIAA Aerospace Design Conference*, AIAA-92-0999, Irvine, CA, USA, 1992.
2. R. F. Stengel. Towards intelligent flight control. *IEEE Transactions on Systems, Man, and Cybernetics*, 23(6):1699–1717, 1993.
3. L. I. Larkin. A fuzzy controller for aircraft flight control. In M. Sugeno, editor, *Industrial Applications of Fuzzy Control*, pages 87 – 103. Elsevier Science Publishers B.V. (North-Holland), 1985.
4. M. Steinberg. Development and simulation of an F/A-18 fuzzy logic automatic carrier landing system. In *IEEE conference on Fuzzy Systems*, pages 797–802, San Fransisco, CA, USA, 1993.
5. M. Sugeno, M. F. Griffin, and A. Bastian. A fuzzy hierarchical control of an unmanned helicopter. In *IEEE 3rd Conference of Fuzzy Systems*, Korea, 1993.
6. T. Y. Jiang, J. V. R. Prasad, and A. J. Calise. Adaptive fuzzy logic flight controllers for rotorcraft. In *AIAA Guidance Navigation and Control Conference*, AIAA-96-3850, San Diego, CA, USA, 1996.
7. C. Phillips, C. L. Karr, and G. Walker. Helicopter flight control with fuzzy logic and genetic algorithms. *Engineering Applications of Artificial Intelligence*, 9(2):175–184, 1996.
8. H. R. Berenji, R. N. Lea, Y. Jani, P. Khedar, A. Malkani, and J. Hoblit. Space shuttle attitude control by reinforcement learning and fuzzy logic. In *IEEE International Conference on Fuzzy Systems*, pages 1396 – 1401, San Fransisco, CA, USA, 1993.
9. G. Schram, W. M. van Buijtenen, R. Babuška, and H. B. Verbruggen. Neuro-fuzzy control of satellite by reinforcement learning. In *Computational Engineering in Systems Applications MultiConference (CESA-96)*, pages 1056 – 1060, Lille, France, 1996.
10. E. G. Laukonen and K. M. Passino. Fuzzy systems for function approximation with application to failure estimation. In *IEEE International*

Symposium on Intelligent Control, pages 184 – 189, Columbus, Ohio, USA, 1994.

11. W. A. Kwong, K. M. Passino, and S. Yurkovich. Fuzzy learning systems for aircraft control law reconfiguration. In *IEEE International Symposium on Intelligent Control*, pages 333 – 338, Columbus, Ohio, USA, 1994.
12. K. KrishnaKumar, P. Gonsalves, A. Satyadas, and G. Zacharias. Hybrid fuzzy logic flight controller synthesis via pilot modeling. *Journal of Guidance, Control, and Dynamics*, 18(5):1098 – 1105, 1995.
13. D. T. McRuer, I. Ashkenas, and D. C. Graham. *Aircraft Dynamics and Automotic Control.* Princeton University Press, New Jersey, 1973.
14. M. B. Tischler. *Advances in Aircraft Flight Control.* Taylor & Francis, 1996.
15. C. A. Shifrin. Aviation safety takes center stage worldwide. *Aviation Week & Space Technology*, November 4:46 –48, 1996.
16. NASA. *Minutes of the NASA Safety Workshop on Control in Adverse Conditions.* St. Louis, MO, USA, November 19-21 1997.
17. S. Chui and S. Chand. Fuzzy controller design and stability analysis for an aircraft model. In *American Control Conference*, pages 821– 826, 1992.
18. N. Imbert and A. Piquereau. Evaluation des techniques de contrôle flou pour le pilotage d'avion. In *Flight Mechanics Panel Symposium*, pages 8.1 – 8.10, Turin, Italy, 1995. AGARD.
19. H. D. Joos, M. Schlothane, and G. Grübel. Multi-objective design of fuzzy controllers using ANDECS. In *IEEE/IFAC Symposium on Computer-Aided Control System Design*, pages 75 – 82, Tucson, AZ, USA, 1994.
20. M. Livchitz, A. Abershitz, U. Soudak, and A. Kandel. On the role of applied fuzzy logic in automated landing of unmanned aircraft. In *IFSA World Congress*, pages 129 – 131, Sao Paulo, Brazil, 1995.
21. Z. Zhou and G. Reichert. Application of fuzzy logic control theory to the design of aircraft flight control systems. In *Congress of the International Council of the Aeronautical Sciences*, pages 351 – 358, Sorrento, Italy, 1996.
22. S. R. Swanson. Fuzzy control of the shuttle training aircraft. In M. Jamshidi, L. Zadeh, A. Titli, and S. Boverie, editors, *Applications of Fuzzy Logic*, pages 387 – 395. Prentice Hall, 1997.
23. GARTEUR. *Robust Flight Control, A Design Challenge.* R. F. Magni, S. Bennani, J. Terlouw (eds). Lecture Notes in Control and Information Sciences 224. Springer Verlag, 1997.

24. G. Schram and H. B. Verbruggen. RCAM: A fuzzy control approach. In J. F. Magni, S. Bennani, and J. Terlouw, editors, *Robust Flight Control, A Design Challenge*, Lecture Notes in Control and Information Sciences 224, pages 397 – 416. Springer Verlag, 1997.
25. G. Schram. *Intelligent Flight Control – A Fuzzy Logic Approach.* PhD thesis, Delft University of Technology, Delft, The Netherlands, 1998.
26. R. Palm, D. Driankov, and H. Hellendoorn. *Model Based Fuzzy Control.* Springer Verlag, 1997.
27. D. Driankov. Fuzzy gain scheduling. In C. Bonivento, C. Fantuzzi, and Rovatti R., editors, *International Summer School in Fuzzy Logic Control - Advances in Methodology*, pages 111 – 148, Ferrara, Italy, 1998. World Scientific.
28. R. Babuška. *Fuzzy Modeling and Identification.* Kluwer Academic Publishers, 1998.
29. NTSB. Final report AAR-79/17 on American Airlines, Inc., DC-10-10 accident, N110AA, Chicago-O'Hare International Airport, Chicago, Illinois, May 25, 1979, 1979. National Transportation Safety Board, USA.
30. NTSB. Final report AAR-92/06 on United Airlines Flight 585, Boeing 737-291 accident., 1992. National Transportation Safety Board, USA.
31. J. T. McKenna. Blocked static ports eyed in Aeroperu 757 crash. *Aviation Week & Space Technology*, page 76, November 11 1996.
32. A. S. Willsky. A survey of design methods for failure detection in dynamic systems. *Automatica*, 12(6):601 – 611, 1976.
33. R. Isermann. Process fault detection based on modeling and estimation methods - A survey. *Automatica*, 20(4):387 – 404, 1984.
34. P. M. Frank. Fault diagnosis in dynamic systems using analytical and knowledge-based redundancy - A survey and some new results. *Automatica*, 26(3):459 – 474, 1990.
35. T. E. Menke and P. S. Maybeck. System/actuator failure detection in the Vista F-16 by multiple model adaptive estimation. *IEEE Transactions on Aerospace and Electronic Systems*, 31(4):1218 – 1228, 1995.
36. P. Eide and P. S. Maybeck. An MMAE failure detection system for the F-16. *IEEE Transactions on Aerospace and Electronic Systems*, 32(3):1125 – 1135, 1996.
37. D. A. Handelman and R. F. Stengel. Combining expert system and analytical redundancy concepts for fault-tolerant flight control. *Journal of Guidance*, 12(1):39 – 45, 1989.
38. R. F. Stengel. Intelligent failure-tolerant control. *IEEE Control Systems*, 11(3):14 – 23, 1991.

39. M. R. Napolitano, C. I. Chen, and S. Naylor. Aircraft failure detection and identification using neural networks. *Journal of Guidance, Control, and Dynamics*, 16(6):999 – 1009, 1993.
40. R. Isermann and P. Ballé. Trends in the application of model-based fault detection and diagnosis of technical processes. In *IFAC World Congress*, San Francisco, CA, USA, pages 1 – 12 (Volume N), 1996.
41. P. M. Frank and B. Köppen-Selinger. New developments using AI in fault diagnosis. *Engineering Applications of Artificial Intelligence*, 10(1):3 – 14, 1997.
42. P. Ballé and R. Isermann. Fault detection and isolation for nonlinear processes based on local linear fuzzy models and parameter estimation. In *American Control Conference*, pages 1605 – 1609, Philadelphia, USA, 1998.
43. P. M. Frank and T. Marcu. Fuzzy techniques in fault detection, isolation, and diagnosis. In *Chapter in this book*, 1998.
44. G. Schram, S. M. Gopisetty, and R. F. Stengel. A fuzzy logic – parity space approach to actuator failure detection and identification. In *AIAA Aerospace Sciences Meeting and Exhibit*, AIAA 98-1014, Reno, NV, USA, 1998.
45. G. Schram, S. M. Gopisetty, and R. F. Stengel. Nonlinear rule-based detection and identification of control system failures. In *American Control Conference*, pages 1600 – 1604, Philadelphia, PA, USA, 1998.
46. G. Schram, G. J. C. Copinga, P. M. Bruijn, and H. B. Verbruggen. Failure-tolerant control of aircraft; a fuzzy logic approach. In *American Control Conference*, pages 2281 – 2285, Philadelphia, PA, USA, 1998.

MEDICAL APPLICATIONS OF FUZZY LOGIC CONTROL

D.A.LINKENS AND M.F.ABBOD

Department of Automatic Control and Systems Engineering University of Sheffield Sheffield S1 3JD, United Kingdom
E-mail : D.Linkens@sheffield.ac.uk

The complexity and measurement problems posed in monitoring and control of the human body makes fuzzy logic an obvious paradigm for investigation. Only in recent years, however, has it emerged in the life sciences as a viable engineering technique, as evidenced by the Roadmap produced by ERUDIT (the European Network of Excellence for Fuzzy Logic and Uncertainty Modelling). One medical specialism where fuzzy control has been utilised extensively is that of anaesthesia. In this paper, a detailed description will be given of development of fuzzy control for unconsciousness management for patients undergoing surgery.

1 Introduction

A major positive driving force behind the introduction of advanced techniques into the area of medicine is that of improved patient care. This applies to all levels of healthcare, including primary care and high dependency hospital units. A negative force is that of financial escalating cost, enhanced by the essential safety critical nature of the tasks involved. Given these counteracting factors it is not surprising that the take-up of advanced engineering techniques is often slower in medicine than in industry. This does not fully explain, however, the very slow penetration of fuzzy logic control principles into medicine, especially since the complexity of human body dynamical functions should make it an ideal area for the use of fuzzy linguistic reasoning.

A recent Roadmap survey produced by ERUDIT [6] has considered the presence (or absence!) of fuzzy logic applications in 34 sub-specialisms in medicine. Although limited to a literature search area over a 10 year period, a number of tentative statements were made:

1. Internal medicine, anaesthesia, radiology, electrophysiology, pharmacokinetics, and neuromedicine already use fuzzy logic methods to a considerable amount. They use fuzzy expert systems, fuzzy control, fuzzy signal and image processing, fuzzy modelling, and fuzzy neural simulation.
2. No specific application has until now developed in the surgical disciplines, dental medicine, general practice and for nursing.
3. Most of the regionally defined medical disciplines participate from the methods developed or used in the neighbouring disciplines.

4. Papers from the medical information basic sciences are appearing at a high rate. Innovations in these sectors will keep the development in the other sectors growing.

As observed above, the specialism of anaesthesia has attracted considerable attention as an application suitable for fuzzy reasoning and control paradigms.

2 Anaesthesia

Anaesthesia represents one of the major areas in modern medicine which provides a significant challenge to the utilisation of systems engineering techniques. These techniques cover the fields of measurement, signal conditioning, decision support, on-line control and fault diagnosis. The three major sub-areas of unconsciousness (depth of anaesthesia), analgesia (pain relief) and muscle relaxation each offer particular challenges in the above areas. While this paper does not concentrate on measurement details, it is well recognised that reliable methods of instrumentation are vital in successful employment of computerised assistance in anaesthesia management. The highly inferential nature of measurements in this field provide additional challenges, which are more severe than conventional industrial environments.

Muscle relaxation provides a relatively simple method of measurement based on evoked potentials in the arm. Much work has been conducted in this area in the authors' laboratory, including recent work with both fixed and self-organising fuzzy logic controllers for automated on-line infusion of modern drugs. Thus, in one of the earliest published papers on fuzzy control in medicine, Linkens and Mahfouf [14] described a simulation-based study into the design of fuzzy control algorithms using hand-crafting of expert rules from hospital operating theatre data and a training simulator. Incorporating pharmacokinetic/dynamics drug models. Later work began to use self-organising (SOFLC) methods for regulation aspects [13,8], together with a hierarchical architecture for supervisory control [9]. Following detailed simulation studies, SOFLC was implemented successfully in clinical trials [19] using a fast-acting drug, atracurium. Using a system engineering methodology for introducing these advanced techniques into clinical practice [20], many trials have been undertaken using a range of modern relaxant drugs [21].

Multi-variable SO has been compared with GPC (Generalised Predictive Control) in extensive simulation studies involved simultaneous regulation of muscle relaxation and unconsciousness (inferred from cardiovascular measurements) [15].

3 Monitoring and Control of Depth of Anaesthesia

The field of Depth of Anaesthesia (DOA) is a challenging area for fuzzy control since direct measurements are unavailable. Thus, a hierarchical structure containing fuzzy reasoning at each level has being developed [16]. This provides self-organising fuzzy modelling (SOFM) at the lowest level for inferential monitoring of anaesthetic depth. At the next level, regulatory control is achieved for the anaesthesia maintenance phase via a self-organising fuzzy logic controller (SOFLC). This hierarchy is extended to higher several levels for adaptation, supervision, surgical evaluation and alarm monitoring. Thus, it demonstrates a totally fuzzy-logic-based architecture for manipulating complex process (i.e. the patient) [25].

An existing fuzzy logic-based programme named "SADA" provides good control algorithms with self-learning capabilities, but only limited measurement inference based mostly on cardiovascular indicators (i.e. changes in blood pressure (BP) and heart rate (HR)) [16,17]. However, these changes are subject to moderation by drugs used in anaesthesiology (e.g. opioids and anticholinergics), by surgical situation (e.g. body position) or patient's circumstance (e.g. fluid loss and blood loss), so they are unreliable. In the search for a reliable monitor of DOA the electroencephalogram (EEG) is an obvious signal to investigate. The EEG, generated from within the central nervous system (CNS), is not affected by neuromuscular blockers, and the raw signal has been known for some time to show graded changes with increasing concentration of anaesthetics. However, these changes vary for different agents and no unifying feature has been identified for measurement of DOA in spite of extensive signal processing.

Attention, therefore, has turned to evoked potentials (EP), in the search for a signal which reflects the subjective clinical signs that anaesthesiologists have used over the years to assess their patients during anaesthesiology. The signal should show similar graded changes with anaesthetic concentration for different agents; show appropriate changes with surgical events, and indicate awareness or very light anaesthesia. The EPs, which are changes in electrical potential evoked by auditory, somatosensory or visual stimuli, have two advantages over the EEG in the study of anaesthesia: (a) the EP is an indication of the responsiveness of the CNS whereas the EEG reflects the resting level; (b) the EPs have anatomical significance. The auditory evoked responses (AER) has been tested with the above criteria in mind [27]. Thus, the mid-latency auditory evoked response (MLAER) has been validated extensively as a measure of DOA. The main features of MLAER which reflect the anaesthetic depth are the changes in latencies and amplitudes of its waves due to drug changes and surgical stimuli. A digital signal processing (DSP) technique was carried out to smooth MLAER and improve the signal to noise ratio (S/N). Feature extraction has been implemented to extract the factors describing the changes in amplitudes and latencies of MLAER waves which are related to DOA. These factors have been merged together using fuzzy logic to create a reliable index for DOA.

Previous research on closed-loop control has indicated that patients can be modelled and controlled well with Artificial Intelligence techniques such as fuzzy logic and neural networks [7,12]. Recent research on the measurements of DOA has concentrated on use of the AER signal [22,24,10]. More recently, neural networks and fuzzy logic have been used to analyse the signal and measure the DOA [28]. In recent work by the authors, neural networks and fuzzy logic [11] have been used to control the amount of infused drug using a rule-based fuzzy logic controller.

This paper describes the structure of a real-time measuring system based on fuzzy logic. The system uses neuro-fuzzy and multiresolution wavelet analysis for monitoring the DOA based on AER signals. The measuring system uses a DSP signal processing card hosted in a PC, providing averaging and analysis using multiresolution wavelet analysis. The analysed signal is fed to a neuro-fuzzy system where the inference takes place to obtain a measure for DOA. Another measure for DOA is based on the cardiovascular system status using a rule-based fuzzy logic classifier. The two measures are fused together using rule-based fuzzy logic to decide the final DOA. Based on the classified DOA, a target concentration is decided by a rule-based fuzzy logic controller which feeds the target to a Target Controlled Infusion (TCI) algorithm. The patient is simulated using a mathematical model with numerical simulation. The system forms a closed-loop controller for monitoring the DOA for patients undergoing surgical operation.

4 Monitoring Depth of Anaesthesia

A multi-sensor fusion system structure using cardiovascular indicators and AER signals to monitor DOA has been constructed as shown in Fig. 1. The first measurement system uses cardiovascular data which can be logged on-line to the system, such as systolic arterial pressure (SAP), heart rate (HR), while the second uses the AER system for assessing DOA based on the evoked potentials. The two measures are fused together to obtain a final measure for DOA.

4.1 Rule-Base of CV_DOA

The first level estimates the primary depth of anaesthesia (*CV_DOA*) from on-line cardiovascular signals such as SAP and HR. At this level, the rules can come either from anaesthetists' experience, in which case we apply a simple fuzzy modelling system to model the *CV_DOA*, or they may be derived from self-learning via clinical trials.

The *CV_DOA* classifier is based on HR and SAP and the output is the DOA. The input variables are classified into three linguistic levels for each variable, *low, medium*, and *high*. *Medium* means the input variable is within the normal range, *low* and *high* mean below or above the normal range respectively.

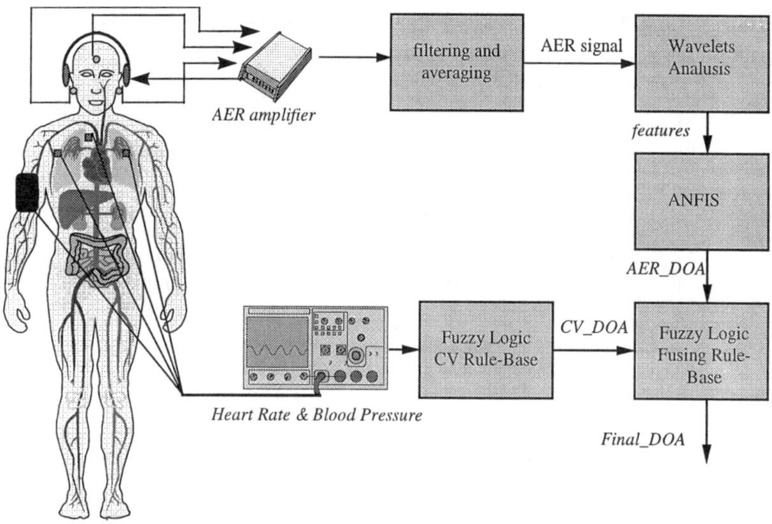

Figure 1: Block diagram of the depth of anaesthesia measuring system.

In contrast, the output variable (DOA) has six linguistic levels, *deep, ok_deep, ok, ok_light, light,* and *awake*. The rule-base was derived from anaesthetists' experience as shown in Table 1.

Verification of the rule base output can be done by comparing the linguistic rules interpreted from the anaesthetist with data logged from actual operations (SAP, HR, assessed DOA). In order to fine tune the anaesthetist's rules, genetic algorithm (GA) optimisation was used to train the fuzzy logic rule-base in terms of tuning the positions and width of the fuzzy sets. Logged data from the operations were used as training data.

Table 1: *CV_DOA* classifier rule-base.

		SAP		
	CV_DOA	*high*	*medium*	*low*
HR	*high*	awake	ok_light	ok
	medium	light	ok	ok_deep
	low	ok	ok_deep	deep

4.2 Rule-Base for AER_DOA

Depth of anaesthesia is also estimated via the evoked brain potentials which are measured using one or more scalp electrodes. The AER signals are the responses in the EEG to clicks applied to the ear. It can be recorded in operating theatre and extracted from the EEG by computer averaging. Fig. 2 shows a typical AER signal. Assessment of DOA was based on analysing the AER signal using wavelets for extracting its features, then feeding the features to a neurofuzzy system for final assessment of DOA

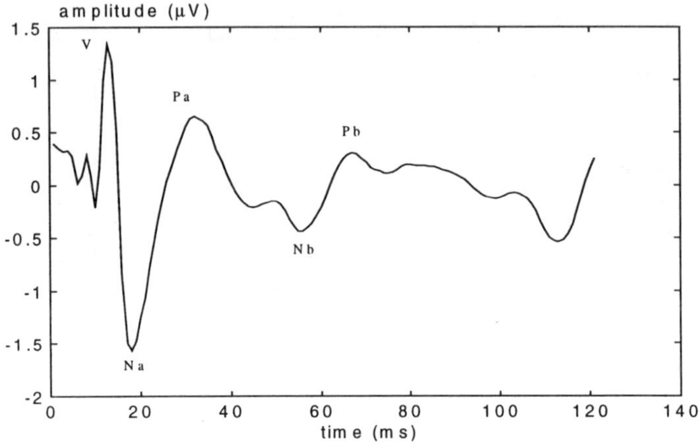

Figure 2: Auditory evoked potential response

One method for analysing the AER signal is that of Fourier Transforms (FT). It has been used for the analysis of MLAER to find the region corresponding to the maximum power in the frequency spectrum as a means to identify the major oscillatory components in the signal. It has been shown that an increase in the DOA yields a shift in the power spectrum of the oscillatory component along with an attenuation in the response. The shortcoming of FT is that it cannot represent the transient nature of the peaks in the signal. However, the Wavelet Transform [2] has the ability to deal with non-stationary signals, as well as the ability to decompose the signal into its frequency components, allowing the identification of major oscillatory components.

Multiresolution wavelet analysis was used to obtain the coefficients of the *detail* components of the AER. A 5-point weighted moving average was used to smooth the data; it was observed that this weighted moving average did not add any noticeable delay in the signal. These wavelet coefficients were then used to calculate the average energy in each of 6 components (Table 2). The notes taken down during the course of the surgical procedure were then used to label the features. These features were furthermore labelled with the anaesthetist's expert opinion (based on the clinical signs of anaesthesia and his experience) on the DOA at different stages of the surgical procedure. Thus, labelling of the data could be carried out. Six labels, corresponding to distinct anaesthetic depths were used: *awake, light, ok_light, ok, ok_deep*, and *deep*. The data outliers were removed and the remaining data were normalised to zero mean and unit variance for each of the six categories. The data from nine patients were used to create a training and a validation set.

The Adaptive Network Based Fuzzy Inference System (ANFIS) architecture is based on a fuzzy inference system implemented in the framework of an adaptive network [5]. Using a hybrid learning procedure, ANFIS can learn an input-output mapping based on human knowledge (in the form of if-then fuzzy rules).

The ANFIS architecture has been employed to model non-linear functions, identify non-linear components on-line in a control system, and predict chaotic time series. MRA was used to obtain the coefficients of the detail components D1-D4 of the AEP. In this particular case, ANFIS has been used to model the variation on the MRA to reflect the DOA. The system was trained and validated using logged data form patients.

Table 2: The table shows the wavelet coefficients used to calculate the energy for each detail component. The early and late coefficients, corresponding to the Brainstem and late cortical responses are not used in the computation. D3 has three separate components representing different time positions in the MLAER response. Each of the D3 components in the table can be likened to a frequency content in the signal at these time positions, a time-frequency decomposition.

Detail Component	Number of coefficients	coefficients used
D1	64	9-56
D2	32	5-28
D3_1	16	3-6
D3_2	16	7-10
D3_3	16	11-13
D4	8	2-7

4.3 Fusing CV_DOA and AER_DOA

Using the SAP and HR measurements to decide *CV_DOA* and the AER measurements to decide *AER_DOA*, the two are fused together to decide the final DOA. There are six states of DOA i.e., *awake, light, ok_light, ok, ok_deep*, and *deep*. From extensive discussion with anaesthetists, the rule-base to decide the final DOA has been elicited as shown in Table 3.

Table 3: *F_DOA* classifier rule-base.

	AER_DOA					
CV_DOA	*awake*	*light*	*ok_light*	*ok*	*ok_deep*	*deep*
awake	awake	awake	light	light	ok_light	ok
light	awake	light	ok_light	ok_light	ok	ok
ok_light	light	ok_light	ok_light	ok	ok	ok_deep
ok	ok_light	ok	ok	ok	ok	ok_deep
ok_deep	ok_light	ok	ok	ok	ok_deep	deep
deep	ok	ok	ok_deep	ok_deep	deep	deep

5 Target Controlled Infusion

The TCI system was first introduced by Alvis and co-workers [1] who controlled fentanyl infusion during anaesthesia. The idea behind the system is to keep the plasma concentration level constant. In order to do that, the infusion rate has to be determined. Once the desired level is reached, the system maintains the concentration in an open-loop manner.

The TCI system used in this study is based on a three-compartment patient model as shown in Fig. 3 and uses population average pharmacokinetic parameters. The three compartment pharmacokinetic model of the patient as well as the pharmacokinetic parameters (in min^{-1}) describes the flow rate of the drug between the various compartments. K_{10} is the rate constant for elimination from the central compartment, K_{12} is the rate constant for drug transfer from the central compartment to compartment 2, K_{13} is the rate constant for drug transfer from the central compartment to compartment 3, K_{21} is the rate constant for drug transfer from the compartment 2 to the central compartment and K_{31} is the rate constant for drug transfer from the compartment 3 to the central compartment. Table 4 lists typical values for the transfer rates.

Table 4: V_l is used to derive V_c, the volume of the central compartment 1 (also sometimes referred to as the volume of distribution) from the mass of the patient using the equation V_c = mass × 0.159. These values were used by Glass et al [4].

V_1 (lt Kg^{-1})	K_{10}(min^{-1})	K_{12}(min^{-1})	K_{13}(min^{+})	K_{21}(min^{-1})	K_{31}(min^{-1})
0.159	0.152	0.207	0.040	0.092	0.0048

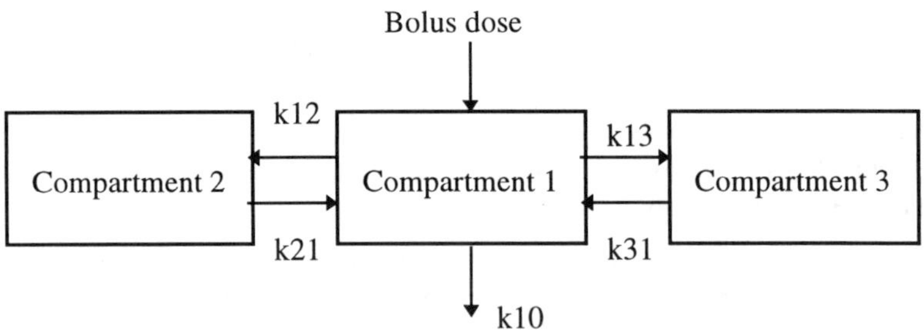

Figure 3: Diagram presenting three compartment model for the drug propofol.

6 Patient Model for the Drug Propofol

Propofol is an anaesthetic agent which is popular due to its properties such as its good solubility, short time onset, and quick recovery. It is rapidly metabolised in the liver by conjugation of glucuronide and sulphate. Numerous investigators have evaluated the pharmacokinetics of propofol following a wide range of doses as well as continuous infusion [18,3]. The kinetics have been described by two or three compartment models. The three compartment model has been estimated to have a distribution phase with a half-life of 30 to 60 minutes and an elimination half-life of 4 to 7 hours. This long elimination life is due to the effect of a deep compartment with limited perfusion that results in slow return of propofol to the central compartment. The volume of distribution of the central compartment has been calculated as 20 to 40 litres. These parameters may be altered by a variety of factors; e.g. gender, weight, pre-existing disease, age and concomitant medication.

The cardiovascular effects of propofol have been evaluated during its use in induction and maintenance of anaesthesia [23,26]. The effect of injection of propofol on arterial pressure, independently of the presence of cardiovascular disease, is to drop the arterial pressure. An induction of 2 to 2.5 mg/Kg produces a 25 to 40 % reduction of systolic pressure, similar changes being seen in the diastolic pressure. The cardiac output will decrease in association with the arterial pressure.

It has been widely reported [29] that a blood level of propofol of 2.5 to 6 µg/ml is required for the maintenance stage which enables the use of a continuous infusion system for regulating the level of anaesthesia.

6.1 *Pharmacokinetics and Pharmacodynamics Model*

The three compartment propofol model shown in Fig. 4 can be described by mathematical equations representing the central compartment, elimination and inter-compartmental distribution. The transfer function for plasma concentration can be formulated as follows:

$$G(s) = \frac{K_1(1+T_{c5}s)(1+T_{c6}s)e^{-\tau s}}{(1+T_{c1}s)(1+T_{c2}s)(1+T_{c3}s)(1+T_{c4}s)}$$

where

$K_1 = 6.58$
$T_{c1} = 2.21$ min
$T_{c2} = 25.14$ min
$T_{c3} = 268.1$ min
$T_{c4} = 10.0$ min
$T_{c5} = 10.87$ min
$T_{c6} = 208.3$ min
$\tau = 0.5$ min

The relation between the change in mean arterial pressure and the plasma concentration in the effect compartment can be modelled by the use of a *Hill Equation* as shown by:

$$E = E_{\max} \frac{CP_{ss}^{\alpha}}{CP_{ss}^{\alpha} + CP_{ss}(50)^{\alpha}}$$

Experimental tests provided the parameters associated with this equation [9] as shown in Table 5.

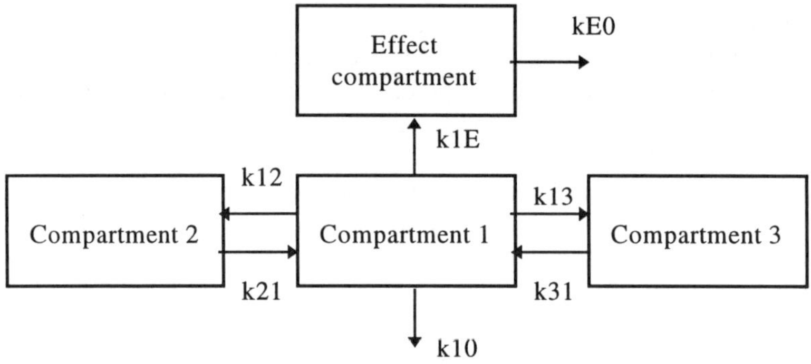

Figure 4: Diagram presenting three compartment model for drug propofol with an additional effect compartment.

Table 5: Table summarising the *Hill Equation* values.

	α	$CP_{ss}(50)$ (ug/ml)	E_{max} (mmHg)
Mean	4	1.298	-44
SD	0.45	0.219	-9.70

7 Feed-Back Control and Simulation

The patient model and the DOA measurements system were combined to make a simulation platform for testing the measurements system and the TCI algorithm as a DOA controller. A block diagram of the system is shown in Fig. 5.

The patient model was simulated using a 4th order Runge-Kutta algorithm in two parts, the pharmacokinetic model and the pharmacodynamic model. The output of the pharmacokinetic model is the plasma concentration which is fed to two paths, the first path being to the pharmacodynamic model block (effect compartment) and then to the non-linear effect block (Hill equation). This output is the mean arterial pressure which is fed in turn to two more blocks, to calculate the associated systolic pressure, and the heart rate. The second path is to the AER signal simulation block which consists of a look-up table for the AER signal classified based on the plasma concentration from nine patients. According to the plasma concentration calculated by the model, a certain group is selected which is related to the present plasma concentration, then a single AER trace is selected randomly from ten stored traces. Each trace consists of 120 data points sampled at 1 msec. The selected trace is used as an output AER signal. The AER signal is synchronised to the trigger signal from the AER system amplifier which is used in reality for the earphone trigger.

The input to the simulator is the infused drug rate calculated by the controller. The final output of the simulator is SAP and HR at 30 second update intervals. The AER signal is sampled at 120msec for each trace.

After collecting the AER signal, it is fed to the wavelet analysis block to analyse the required components of the signal which are then fed to the ANFIS block where the *AER_DOA* is calculated. In the meantime the recorded HR and SAP signals are fed to the fuzzy logic rule-base for analysis of *CV_DOA*. Prior to that both SAP and HR are filtered using a weighted moving average filter. The last stage in DOA analysis is to feed *CV_DOA* and *AER_DOA* to the final DOA analysis block. The calculated *F_DOA* is fed to the next block in the system which is the plasma concentration controller block. Based on the current target concentration and the measured DOA, a new target concentration is calculated using a fuzzy logic rule-based controller. The target concentration controller consists of three rule-bases. One rule-base is selected depending on the required DOA level. The three levels are *ok_light, ok*, and *ok_deep*.

The plasma target concentration is fed to the TCI block which is based on the initial information supplied. It calculates the drug infusion rate and feeds it to the syringe driver which in our case is represented in the simulated patient's model. The initial settings for the TCI algorithm are based on the patient weight, maximum infusion rate and the initial infusion rate.

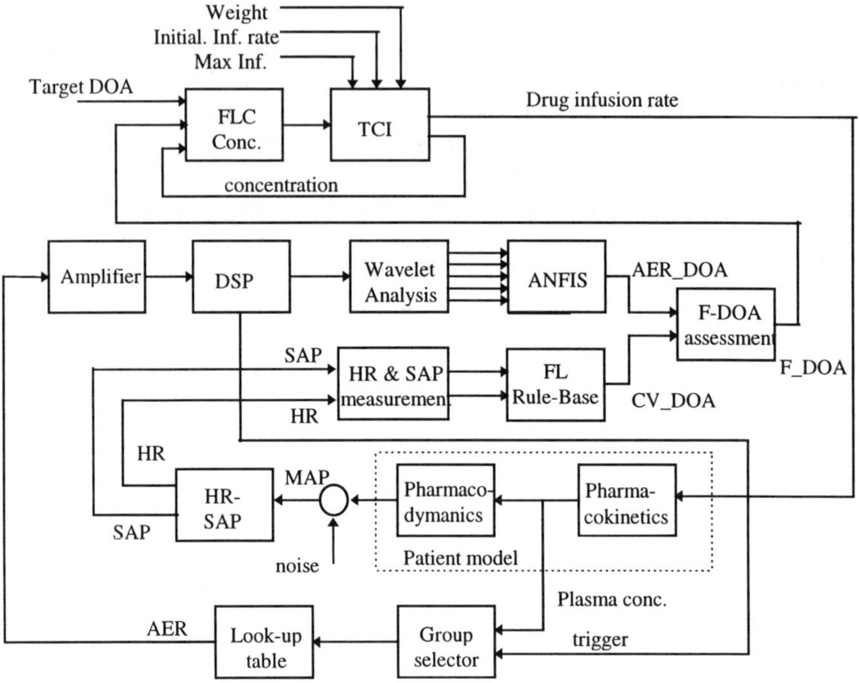

Figure 5: Block diagram of the closed-loop control system

The simulator was used to validate the system. Simulation runs were conducted for different patients: nominal patients with three different weights, a sensitive patient, and a resistive patient The type of AER signal used in the simulator and the random selection gives the phenomenon of a random component in the AER signal, as well as the inherited noise through the amplified AER. The simulator program is hosted in a Windows-based environment to provide user interface and graphical presentation of the control and patient parameters.

8 Simulation Results

In this section, results are reported for different types of simulation runs; nominal patients with three weight cases, sensitive patient, and resistive patient. The following parameters were used during the simulation runs reported here:

Patient weight	50, 70, and 90 Kg
Propofol Concentration	10 $\mu g\ ml^{-1}$
Maximum Infusion Rate	6000 $mg\ hr^{-1}$

HR base line 70 BPM.
SAP base line 140 mmHg.
Patient Sensitivity (gain) Normal 6.58
Sensitive 9.32
Resistive 4.68
5 point moving average filter used for *HR, SAP, AER_DOA*, *CV_DOA*, and *F_DOA*.

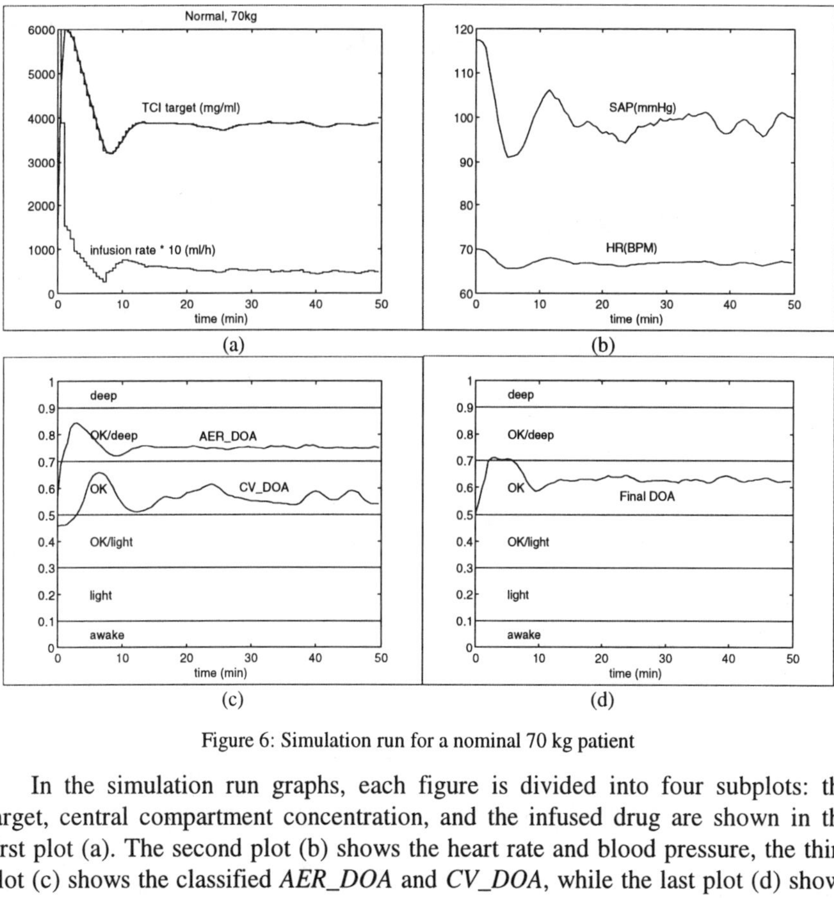

(a) (b) (c) (d)

Figure 6: Simulation run for a nominal 70 kg patient

In the simulation run graphs, each figure is divided into four subplots: the target, central compartment concentration, and the infused drug are shown in the first plot (a). The second plot (b) shows the heart rate and blood pressure, the third plot (c) shows the classified *AER_DOA* and *CV_DOA*, while the last plot (d) shows the final DOA as the two intermediate DOAs are fused together.

8.1 Normal Sensitivity Patient

Three simulation runs carried out for nominal different weight patients, namely 50, 70, and 90 Kg. Basically, the algorithm should perform the same for all the cases since one of its prime inputs is the patient's weight on which the infusion is based.

However, for the three cases, the controller performance is not the same in each case. This is due to the fact that the selection of AER signals is chosen randomly. Since the signal contained noise and different operation stimuli, the random selection factor appears to affect the behaviour of the patient in each case. This gives the model the facility to behave differently each time the simulator runs, which makes it look more like a real variety of patients. The average infusion rate during the maintenance stage was about 4000 mg/ml in each case. The 70 Kg simulation runs are shown in Fig. 6 where the final DOA was maintained steadily in the OK region, in turn the drug infusion was maintained steadily.

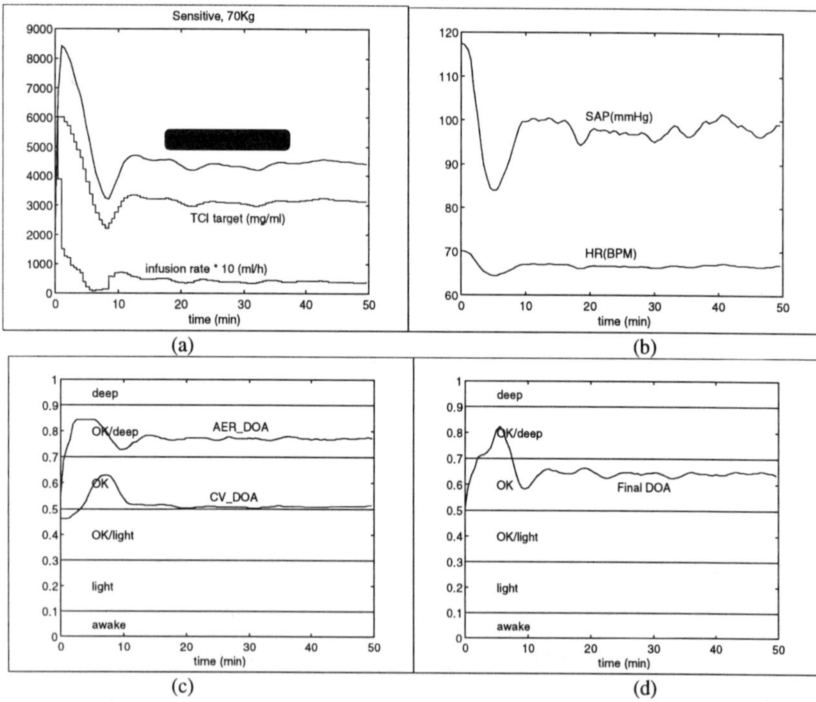

Figure 7: Simulation run for a sensitive 70 kg patient

8.2 *High Sensitivity Patient*

A sensitive patient is known to consume a lower amount of drug to reach the same level of DOA as for the nominal patient. This is due to the high gain to the drug. Such phenomena can be simulated by altering the patient's model gain to a higher level. For a nominal patient a gain of 6.58 was used while for the sensitive patient this was increased to 9.32. This is an increase of 41% which means 28.5%

reduction in the drug consumption. The simulation run is shown in Fig. 7. In the first part, where the target TCI and plasma concentration is shown, the target TCI appears lower than the plasma concentration, this being due to the high drug sensitivity of the patient. In contrast, in Fig. 6, the plasma concentration follows closely the target TCI which is part of the TCI algorithm. In the case of the sensitive patient, the model used in the TCI algorithm is the nominal parameters model, accordingly the open-loop TCI target would have been kept about the nominal target (4000 ml/mg). However, due to the correction by the feedback loop, the actual plasma concentration was controlled at about 4300 ml/mg.

8.3 Low Sensitivity Patient

The final case is the resistive patient which is described as a high drug consumer compared to the nominal case. The model gain selected in this case was 4.68 which means 41% lower gain. The simulation run is shown in Fig. 8 where the target TCI is higher than the plasma concentration. The same behaviour is obtained since the feedback control compensates for the low patient's gain by increasing the target TCI so the plasma concentration reaches the required nominal level of 4000 ml/mg. Again, the TCI algorithm did not follow the exact target TCI as it would in an open-loop mode with respect to a nominal model.(4000 ml/mg by covering the TCI target).

9 Clinical Trials

The system is being used on-line in the operating theatre for clinical trials in the Royal Hallamshire Hospital, Sheffield, UK. This procedure required connecting the system to a DATEX device for recording HR and BP, as well as a Graseby 3400 syringe infusion pump via the RS232 serial port.

A sample clinical trial shown here was carried out using the system based on continuous infusion of the drug propofol. The logging events are shown:

Patient Name : DB
Date : 6-July-1998
Sex : Female
Age : 27
Blood Pressure : 131/70
Heart Rate : 92
Weight : 81 Kg
Operation Type : Small incision cholecystectomy plus operative cholangiogram
Anaesthesia : start with 250mg fentanyl, propofol was started at 4000 ml/mg.

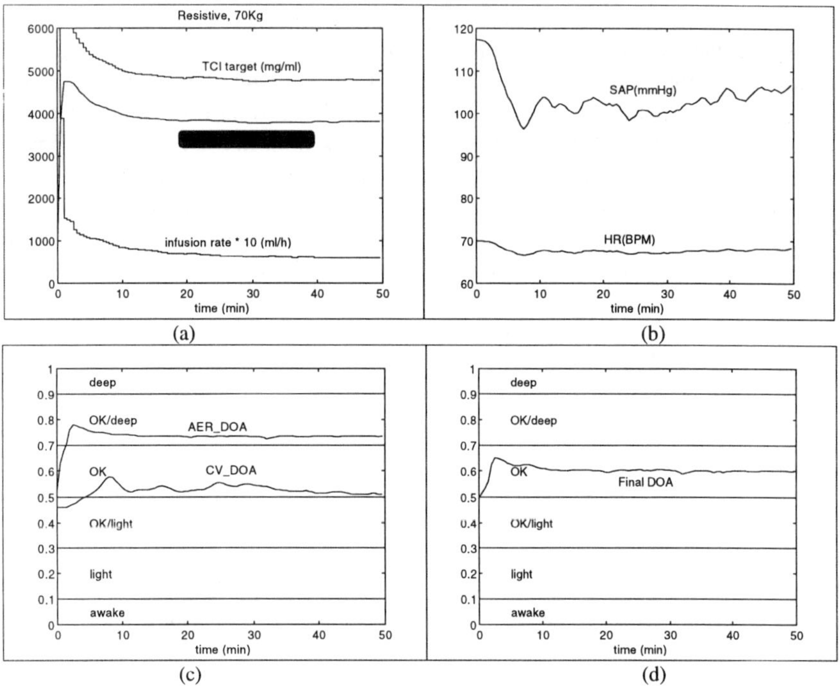

Figure 8: Simulation run for a high resistivity 70 kg patient

The time response is shown in Fig. 9 in four parts, (a) shows the TCI and drug infusion, (b) shows SAP and HR, (c) shows assessed *AER_DOA* and *CV_DOA*, (d) shows the final DOA by merging both DOA. During the operation the *AER_DOA* was kept at *ok-deep* level with some spikes that indicates a low level of DOA. This, in fact, was due to the effect of diathermy which gives large electrical artifacts. On the other hand, *CV_DOA* was kept at *ok_light* which is due to the effect of the muscle relaxant drugs given to the patient which interferes with the clinical signs. The fused final measure of DOA was kept successfully in the *ok* region.

10 Conclusions

A closed loop-control system for measuring and control of depth of anaesthesia has been developed based on two different measures, one based on the brain and the other on cardiovascular system. The system is aimed at providing more reliable measurements of DOA.

The system has been tested under three different patient drug sensitivities: nominal, sensitive, and resistive. A simulation study using a numerical model of the patient shows that the closed-loop system effectively maintains the patient at a clinically acceptable anaesthetic depth even when the noxious stimuli reflect those of a real surgical procedure.

Furthermore, the system has been tested successfully in clinical trials for measuring the depth of anaesthesia in operating theatre. It has been proposed to use the system in full closed-loop control after more validation tests have been completed.

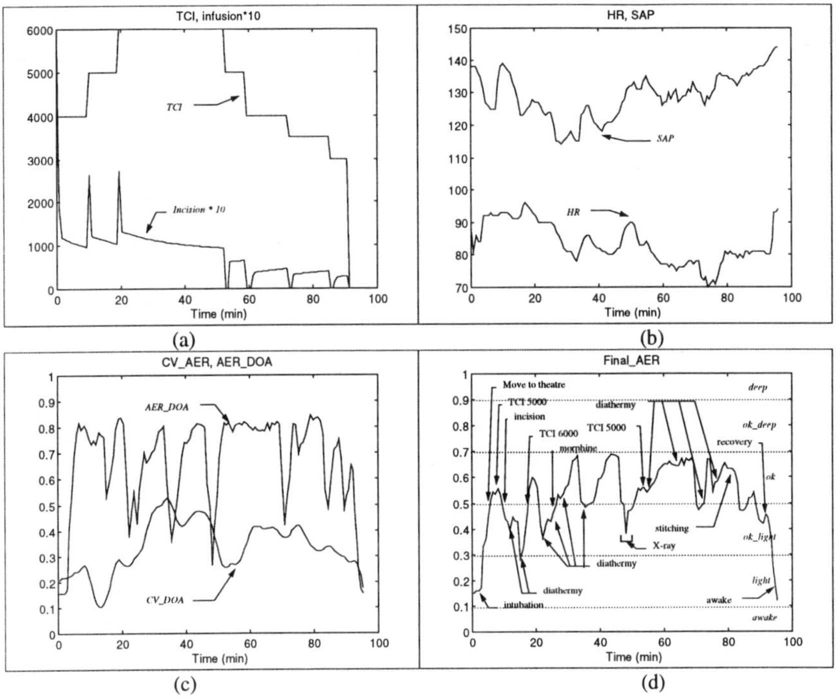

Figure 9: Time response of a clinical trial

11 Acknowledgement

The authors acknowledge financial support from the UK Engineering and Physical Science Research Council, and the University of Sheffield Centre for Research into Anaesthetic Mechanisms (CRAM). Clinical trials have been undertaken with the help of Dr. J.E. Peacock, Royal Hallamshire Hospital, Sheffield.

References

1. Alvis J. M., Reves J. G. , Spain J. A. and Sheppard L. C., Computer assisted continuous infusion of the intravenous analgesic fentanyl during general anaesthesia - an interactive system. IEEE Trans. Biomed. Eng. **BME-32,** 5 (1985) pp. 323-329.
2. Daubechies I., Orthonormal bases of compactly supported wavelets. Communications on Pure and Applied Mathematics **41** (1988) pp. 909-996.
3. Gepts E., Jonckheer K., Maes V., Sonck W. and Camu F., Disposition kinetics of propofol during alfentanil anaesthesia. Anaesthesia **43** (1988) pp. 8-13.
4. Glass P. S., Goodman D. K., Ginsberg B., Reeves J. G. and Jacobs J. R., Accuracy of pharmacokinetic model-driven infusion of propofol. Anesthesiology **71**, 3A (1989).
5. Jang J. R., ANFIS: adaptive-network, based fuzzy inference system. IEEE Trans Systems, Man, and Cybernetics **23**, 3 (1993) pp. 665-685.
6. Keyserlingk D. G., Roadmap for ERUDIT, February (1998).
7. Linkens D. A., (editor). Intelligent Control in Biomedicine (Taylor and Francis, 1994).
8. Linkens D. A. and Abbod M. F., Self-organising fuzzy logic control and the selection of its scaling factors. Trans. Inst. MC **14** (1992) pp. 114-125.
9. Linkens D. A. and Abbod M. F., Supervisory intelligent control using a fuzzy logic hierarchy. Trans. Inst. MC **15** (1993) pp. 112-132.
10. Linkens D. A., Abbod M. F. and Backory J.K., Closed-loop control of depth of anaesthesia: a simulation study using auditory evoked responses. Control Engineering Practice **5**, 12, (December 1997) pp. 1717-1725.
11. Linkens D. A., Abbod M. F. and Backory J. K., Closed-loop control of anaesthesia using a wavelet-based fuzzy logic system. EUFIT'97, Aachen, Germany (September 8-12 1997).
12. Linkens D. A. and Abbod M. F., (Editors). Journal of Biomedical Engineering - Applications, Basis & Communications, Special Issue on Control Methods in Anaesthesia, Taiwan (August 1998).
13. Linkens D. A. and Hasnain S. B., Self-organising fuzzy logic control and its application to muscle relaxation anaesthesia. Proc. IEE Pt.D. **138** (1990) pp. 274-284.
14. Linkens D. A. and Mahfouf M., Fuzzy logic knowledge-based control of muscle relaxant anaesthesia. IFAC symposium - Modelling and Control in Biomedicine, Venice, Italy (1988) pp. 185-190.
15. Linkens D. A., Mahfouf M. and Abbod M. F., Self-adaptive and self-organising control applied to non-linear multi-variable anaesthesia: a model based comparative study. IEE Proc. Pt.D. **139**, 4 (1992) pp. 381-394.

16. Linkens D. A., Shieh J. S., Elkfafi M. and Peacock J. E., Hierarchical fuzzy-based support system for anaesthesiology monitoring and control. IEEE International Conference on Fuzzy Systems, New Orleans (Sep. 8-11, 1996).
17. Linkens D. A., Shieh J. S. and Peacock J. E., Hierarchical fuzzy modelling for monitoring depth of anaesthesia. Fuzzy Sets and Systems **70** (1996) pp. 43-57.
18. Marsh B., White M., Morton N. and Kenny G. N. C., Pharmacokinetic model driven infusion of propofol in children, Brit. J. Anaesthesia **67** (1991) pp. 41-48.
19. Mason D. G., Edwards N. D., Linkens D. A. and Reilly C. S., Performance assessment of a fuzzy controller for atracurium-induced neuromuscular block. Brit. J. Anaesthesia **76** (1996) pp. 47-57.
20. Mason D. G., Linkens D. A., Edwards N. D. and Reilly C. S., Development of portable closed-loop atracurium infusion system: system methodology and safety issues. Int. J. Clin. Mon. and Comput. **13** (1996) pp. 243-252.
21. Mason D. G., Edwards N. D., Ross J. J., Linkens D. A. and Reilly C. S., Self-learning fuzzy control of atracurium-induced neuromuscular block during surgery. IFAC Symposium on Modelling and Control in Biomedical Systems, Warwick (23-26 March 1997).
22. Newton D. E. F., Thornton C., Konieczko K. M., Jordan C., Webster N. R., Luff N. P., Frith C. D. and Dore C. J., Auditory evoked response and awareness: a study in volunteers at sub-MAC concentration of isoflurane. Brit. J. Anaesthesia **69** (1992) pp. 122-129.
23. Peacock J. E., Lewis R. P., Reilly C. S. and Mimmo S., Effect of different rates of infusion of propofol for induction of anaesthesia in elderly patients. Brit. J. Anaesthesia **65** (1990) pp. 346-352.
24. Schwender D., Klasing S., Conzen P., Finsterer U., Poppel E. and Peter K., Midlatency auditory evoked potentials during anaesthesia with increasing endexpiratory concentrations of desflurane. Acta. Anaesthesiol. Scand. **40**, 2 (1996) pp. 171-176.
25. Shieh J. S., Linkens D. A. and Peacock J. E., Hierarchical rule-based and self-organising fuzzy logic control for Depth of Anaesthesia. Trans. IEEE Systems Man and Cybernetics **29**, 1 (1999).
26. Tackley R. M., Lewis G. T. R., Prys-Roberts, Boaden R. W., Dixon J. and Harvey T. J., Computer controlled infusion of propofol", Brit. J. Anaesthesia **62** (1989) pp. 46-53.
27. Thornton C., Creagh-Barry P., Jordan C. and Newton D. E. F., Anaesthetic depth and auditory evoked potentials. Acta. Anaesthesiol. Scand. **37,** S100 (1993) pp. 105-108.
28. Webb A., Allen R. and Smith D., Closed-loop control of depth of anaesthesia. Measurement + Control **29** (1996) pp. 211-215.
29. White M. and Kenny G. N. C., Intravenous propofol anaesthesia using a computerised infusion system. Anaesthesia **45** (1990) pp. 204-209.

LINEAR DESIGN APPROACH TO A FUZZY CONTROLLER

J. JANTZEN
Technical University of Denmark, Dept. of Automation, Building 326, DK-2800 Lyngby, DENMARK
E-mail: jj@iau.dtu.dk

A ball-balancer, basically an inverted pendulum problem, is stabilised by a linear controller. With certain design choices, a fuzzy controller is equivalent to a summation; thus it can replace the linear controller. It can be claimed, that the fuzzy controller performs at least as well as the linear controller, since the linear controller is contained in the fuzzy controller. The approach makes it somewhat easier to design a fuzzy controller.

1 Introduction

The ball-balancer consists of a moving cart, on top of which a steel ball rolls between two parallel, arched pipes. The cart rolls on tracks horizontally mounted on a heavy support (Fig. 1). The control objective is to balance the ball on the top, and, at the same time, position the cart in the middle of the track.

There is no design procedure in fuzzy control such as root-locus design, frequency response design, or pole placement design. Since a fuzzy controller is in general nonlinear, it is difficult to give it a simple mathematical description. The question is whether there are clues from linear control theory that can be used to organise the design procedure. The philosophy in this paper is to stay in the linear domain as long as possible, and let a linear controller serve as a reference for subsequent nonlinear designs. More specifically, the procedure is

1. to design a linear controller,

Figure 1: Laboratory rig.

2. to replace it with a linear fuzzy controller, and

3. to make the fuzzy controller nonlinear in a gradual manner.

The linear fuzzy controller in the second step should perform in exactly the same way as the linear controller in the first step.

2 Problem description

The cart-ball system was built for teaching electrical engineers about automatic control, originally with a focus on state-space control theory. It is educational, because the laboratory rig is sufficiently slow for visual inspection of different control strategies, and the mathematical model is sufficiently complex to be challenging. The rig was built during an M.Sc. project, and later the mathematical model was published in an educational journal.[1] A simulator of the same system was built in Matlab much later for a course on the Internet.[2]

The controller design will be based on a linearised model of the system,

$$\begin{aligned} \dot{\mathbf{x}} &= \mathbf{A}\mathbf{x} + \mathbf{B}u \\ \mathbf{y} &= \mathbf{C}\mathbf{x} \end{aligned} \tag{1}$$

The model is a linear state-space model, where $\mathbf{x}$ is a vector of state variables, $\dot{\mathbf{x}}$ is the vector time derivative $d\mathbf{x}/dt$, u is the input to the system, and $\mathbf{y}$ is a vector of output variables. The matrices $\mathbf{A}, \mathbf{B}$, and $\mathbf{C}$ are matrices of appropriate dimensions containing real numbers.

By pushing the cart left and right manually, it is possible to get the ball on top of the arc, but it is impossible to position the cart at a particular position at the same time. An automatic control system can do that, however. The cart position and ball angle from vertical are measured variables, and the manipulated variable is the horizontal force acting on the cart.

The ball rolls on curved pipes, one of which is made of aluminium while the other is a coil of resistance wire. The ball's angle from vertical is determined by its position on the pipes. The ball, being made of steel, connects the pipes electrically, and acts as a voltage divider producing a voltage proportional to the position (Fig. 2). The cart position is measured the same way using a brass roller, mounted on the cart, which rolls on a coil alongside the rails.

The rails are cylindrical bars mounted on the support, and the cart wheels are small, low-friction ball-bearings which roll on the bars. A wire pulls the cart, passing over a pulley in one end and a wire drum in the other end, both attached to the support. The wire drum is driven by a current-driven direct current (DC) print-motor.

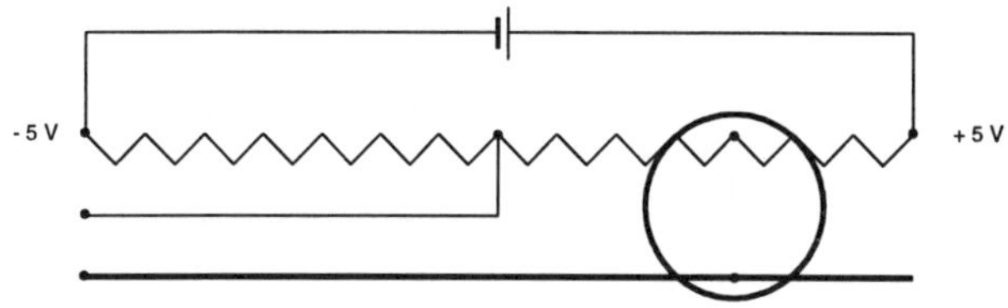

Figure 2: Ball position measurement.

Although the motor is current-driven, we assume the voltage is proportional to the current and in turn that the force is proportional to the current. This is an approximation, but it is a relatively fast DC motor with small electrical and mechanical time-constants. The numerical data of the rig are given by Jorgensen.[1]

The current to the motor is essentially a function of the controlled variables $\left(y, \dot{y}, \varphi, \dot{\varphi}\right)$ in which y is the position of the cart and φ is the angular deviation from vertical of the ball position. The velocity signals are obtained by differentiation in operational amplifiers.

Newton's laws and various relationships bring out a model consisting of two coupled, nonlinear differential equations.[1] After linearisation they can be put on the state space form (1), with

$$\mathbf{A} = \begin{bmatrix} 0 & 1 & 0 & 0 \\ 0 & 0 & a & 0 \\ 0 & 0 & 0 & 1 \\ 0 & 0 & c & 0 \end{bmatrix} \qquad \mathbf{B} = \begin{bmatrix} 0 \\ b \\ 0 \\ d \end{bmatrix}$$
$$\mathbf{C} = \begin{bmatrix} 1 & 0 & 0 & 0 \\ 0 & 0 & 1 & 0 \end{bmatrix} \tag{2}$$

The state vector $\mathbf{x}$ is the vector $\left(y, \dot{y}, \varphi, \dot{\varphi}\right)^T$ and the output vector is

$$\mathbf{y} = \begin{bmatrix} y \\ \varphi \end{bmatrix} \tag{3}$$

The input vector consists of just one input, the horizontal force acting on the cart, which can be substituted by the voltage u to the motor, as their relationship is 1 volt to 1 Newton; that is how the motor and the gear were designed. The physical constants are

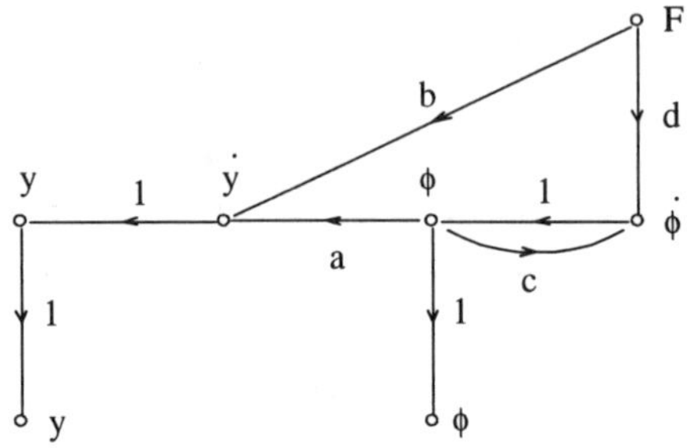

Figure 3: Signal flow in the state-space model.

$$(a, b, c, d) = (-1.34, 0.301, 14.3, -0.386)\,. \tag{4}$$

An overview of the model is provided by a *signal flow graph*. Given a state space model with matrices containing zeros and non-zero elements, the flow of the signals can be mapped into a directed graph, or *digraph*; the digraph is a picture of the couplings in the model (Fig. 3). The node set is given by an input node, four state nodes, and two output nodes; the arc set is given by the non-zero entries in the matrices. That is, if $a_{ij} \neq 0$, then there exists an arc from the jth state node to the ith state node; if $b_{ij} \neq 0$, then there exists an arc from the jth input node to ith state node; and if $c_{ij} \neq 0$, then there exists an arc from the jth state node to the ith output node. The numbers a_{ij}, b_{ij}, c_{ij} are assigned to the arcs, so-called *weights*.

If a designer decides to add a feedback connection from the output node φ to the input node F, he will create a loop $\varphi - F - \dot{\varphi} - \varphi$. If, alternatively, he adds a feedback connection from the output node y to the input node F, he will create a larger loop, $y - F - \dot{\varphi} - \varphi - \dot{y} - y$, as well as another loop, $y - F - \dot{y} - y$. In the first case, there is no feedback from the cart to the ball, while in the second case, there is feedback through the ball into the cart. Therefore it is easier to design a ball controller than a cart controller; the ball controller can be designed and tuned independently of how the cart behaves, while a cart controller will be influenced by the ball behaviour. The digraph is a useful qualitative tool in the design of any state space system.[3]

Since state space models are not unique – a given physical plant may be modelled by several state-space models – the digraph reflects the flow of signals in the *model*, rather than the physical system itself. Another pendulum problem, say, a rod balancing on a cart, can be modelled such that the digraph is the same, only the constants (a, b, c, d) differ.

3 Linear design approach

With a linear model and a linear controller it is relatively easy to mathematically analyse the system, and the stability of the closed loop system can be checked conveniently in Matlab. Two particular configurations for a controller come into mind: a state- or output-feedback controller, or two proportional-derivative (PD) controllers in a cascade. The former lends itself to mathematical analysis, while the latter is more intuitive with regard to manual tuning. In the linear domain, the two are equivalent, but here we have chosen the former possibility and a pole placement design procedure.

3.1 Pole placement design

To stabilise the system, we could first try to use output feedback. Referring to Fig. 3, two feedback gains are to be added, one from the cart position y to the input u and another from the angle φ to the input u, i.e.,

$$u = \mathbf{F}\mathbf{y} \tag{5}$$

where

$$\mathbf{F} = \begin{bmatrix} f_1 & f_3 \end{bmatrix}, \mathbf{y} = \begin{bmatrix} y \\ \varphi \end{bmatrix} \tag{6}$$

It turns out, however, from inspection of the closed loop equations, that there will always be poles in the right half of the complex plane, or multiple poles on the imaginary axis. In other words, output feedback is insufficient.

We therefore try (full) state feedback, from the states to the input, i.e.,

$$u = \mathbf{F}\mathbf{x}, \tag{7}$$

where

$$\mathbf{F} = [f_1 \; f_2 \; f_3 \; f_4], \mathbf{x} = \begin{bmatrix} y \\ \dot{y} \\ \varphi \\ \dot{\varphi} \end{bmatrix} \tag{8}$$

The stability of the system is governed by the poles of the closed loop system, or the eigenvalues of the closed-loop system matrix, i.e., $eig(\mathbf{A}+\mathbf{BF})$. The idea in a pole placement design is to choose the matrix $\mathbf{F}$ such that the eigenvalues are placed in the left half of the complex plane, perhaps even in certain desirable locations. A necessary condition for arbitrary pole assignability is that the system is controllable, i.e., that the controllability matrix

$$\mathbf{Q} = \left[\mathbf{B}\ \mathbf{AB}\ \mathbf{A}^2\mathbf{B} \ldots \mathbf{A}^{n-1}\mathbf{B}\right] \tag{9}$$

has full rank, i.e., *rank* $(\mathbf{Q}) = n$, the order of the system. Roughly speaking, a system is controllable if the state variables can be moved in any desired direction by a suitable sequence of control signals. It turns out that the rank of $\mathbf{Q}$ is 4, the system is controllable.

After insertion into the state space equations (1), the closed loop characteristic polynomial can be found,

$$|s\mathbf{I} - \mathbf{A} - \mathbf{BF}| = \det \begin{bmatrix} s & -1 & 0 & 0 \\ -f_1 b & s - f_2 b & -a - f_3 b & -f_4 b \\ 0 & 0 & s & -1 \\ -f_1 d & -f_2 d & -c - f_3 d & s - f_4 d \end{bmatrix} \tag{10}$$

$$= s^4 - (f_2 b + f_4 d)\, s^3 - (f_3 d + c + f_1 b)\, s^2 + (-f_2 da + f_2 bc)\, s + f_1 bc - f_1 da \tag{11}$$

In this polynomial it is possible to arbitrarily assign its coefficients by adjusting the feedback gains f_1, f_2, f_3, and f_4 (four linear equations with four unknowns); since the coefficients of the polynomial can be assigned arbitrarily, so can the roots. This means, we can place all poles in any location we desire.

Given the desired poles $\mathbf{p} = (p_1, p_2, p_3, p_4)^T$, the characteristic polynomial is also

$$(s - p_1)(s - p_2)(s - p_3)(s - p_4) = s^4 + r_1 s^3 + r_2 s^2 + r_3 s^1 + r_4 \tag{12}$$

By equating coefficients in (11) and (12), the feedback vector $\mathbf{F}$ can be found. Since we have a system with one control input, it is also possible to use Ackermann's formula (the function 'acker' in the Matlab Control Toolbox); it calculates the feedback gains given a set of desired pole locations.[4]

The next step is to decide on the desired closed-loop pole locations. A useful rule of thumb is: the farther away from the origin, the larger is the magnitude of the required control signal. An easy choice is the so-called *prototype response* poles, which can be looked up in a table (Franklin et al., p 388, Table 6.1b).[4] The recommended poles are, for a fourth order system,

$$\mathbf{p} = \begin{bmatrix} -0.6573 - i0.8302 \\ -0.6573 + i0.8302 \\ -0.9047 - i0.2711 \\ -0.9047 + i0.2711 \end{bmatrix} \tag{13}$$

These refer to a prototype response with little overshoot for a normalised system. The four poles can be multiplied by a constant w in order to obtain some variation in the speed of the response.

3.2 Fuzzy linear controller

The state feedback controller (7) can be written out,

$$u = f_1 y + f_2 \dot{y} + f_3 \varphi + f_4 \dot{\varphi} \tag{14}$$

It can be split into two controllers, one for the cart and one for the ball,

$$u = \left(f_1 y + f_2 \dot{y}\right) + \left(f_3 \varphi + f_4 \dot{\varphi}\right) \tag{15}$$

Next step is to replace the two controllers by linear fuzzy controllers F_1 and F_2, acting like summations, such that

$$u_1 = \left(f_1 y + f_2 \dot{y}\right) = F_1(y, \dot{y}) \tag{16}$$

$$u_2 = \left(f_3 \varphi + f_4 \dot{\varphi}\right) = F_2\left(\varphi, \dot{\varphi}\right) \tag{17}$$

The fuzzy controlled system should show exactly the same step response; this is a check that the implementation is correct.

There are several sources of nonlinearity in a fuzzy controller. The position, shape and number of fuzzy sets as well as nonlinear input scaling cause nonlinear transformations. The rules often express a nonlinear control strategy, and if the connectives **and** and **or** are implemented as for example **min** and **max** respectively, they are nonlinear. Several defuzzification methods are nonlinear. It is possible, however, to construct a rule base with a linear input-output mapping that acts like a summation.[5,6,7,8]

> *Input universes.* The input universes must be large enough for the inputs to stay within the limits (no *saturation*). Each input family should contain a number of terms, designed such that the sum of membership values for each input is 1. This can be achieved when the sets are triangular and cross their neighbour sets at the membership value $\mu = 0.5$; their peaks will thus be equidistant. Any input value can thus be a member of at most two sets, and its membership of each is a linear function of the input value.
>
> *Number of rules.* The number of terms in each family determines the number of rules, as they must be the **and** combination (*outer*

product) of all terms to ensure completeness. The output sets should preferably be singletons s_i equal to the sum of the peak positions of the input sets – the output sets may also be triangles, symmetric about their peaks, but singletons make defuzzification simpler.

Aggregation. We must choose multiplication for the connective **and**. Using the weighted average of rule contributions for the control signal (corresponding to *centre of gravity for singletons* defuzzification, *COGS*), the denominator vanishes, because all firing strengths add up to 1.

The following checklist summarises the design choices for achieving a fuzzy rule base equivalent to a summation:

1. Use triangular input sets that cross at $\mu = 0.5$;
2. use multiplication (*) for the **and** connective;
3. the rule base must be the **and** combination of all input families;
4. use output singletons, positions determined by the sum of the peak positions of the input sets;
5. use the defuzzification method *COGS*.

With these design choices the control surface degenerates to a diagonal plane. A flexible fuzzy controller, that allows these choices, is two controllers in one so to speak. When linear, it has a transfer function and the usual methods regarding tuning and stability of the closed loop system apply.

3.3 Transferring gains to the fuzzy controller

Since a fuzzy controller is normally equipped with input gains as well as an output gain, the feedback gains from the previous section have to be transferred to the fuzzy controller gains. Including gains, the fuzzy controllers can be written as

$$u_1 = F_1\left(g_1 y, g_2 \dot{y}\right) * gu_1 \tag{18}$$

$$u_2 = F_2\left(g_3 \varphi, g_4 \dot{\varphi}\right) * gu_2 \tag{19}$$

where (g_1, g_2, g_3, g_4) are input gains, gu_1 is the output gain on the cart controller, and gu_2 the output gain on the ball controller. Since each fuzzy controller acts like a summation, the following relationship holds for the cart controller

$$\left(f_1 y + f_2 \dot{y}\right) = \left(g_1 y + g_2 \dot{y}\right) gu_1 \tag{20}$$

implying,

$$f_1 = g_1 gu_1 \tag{21}$$

$$f_2 = g_2 gu_1 \tag{22}$$

The left hand sides are known from the previous section, the task is to choose g_1, g_2, and gu_1 under the constraints (21)-(22). With three gains and two equations, there is one degree of freedom. In order to exploit at least one of the input universes as much as possible, we can choose either g_1 or g_2 freely. If, for instance, the input universes are chosen to be standard universes $[-100, 100]$, then we have the following further constraints, in order to avoid saturation in the universes,

$$|g_1 y|_{\max} \leq 100 \tag{23}$$

$$\left|g_2 \dot{y}\right|_{\max} \leq 100 \tag{24}$$

We can choose to maximise, say, g_1 under constraint (23), then gu_1 is given by (21), and subsequently g_2 is given by (22). It must be checked then, that (24) is still valid; if it turns out to be violated, a recalculation is necessary with g_2 chosen according to (24), and the rest according to (21)-(22). For the ball controller, the following holds

$$f_3 = g_3 gu_2 \tag{25}$$

$$f_4 = g_4 gu_2 \tag{26}$$

The choice of g_3, g_4, and gu_2 follow an analogous procedure.

3.4 Fuzzy nonlinear controller

The next step in the design procedure is to gradually make the linear fuzzy controller nonlinear. It is common practice to build a rule base from *terms* such as Pos, Zero, and Neg, representing *labels* of fuzzy sets. An *input family* may consist of those three terms. Consequently, with two inputs it is possible to build $3 \times 3 = 9$ rules. Nine rules is a manageable amount often used in practice.

The shape of the sets and the choice of rules affect the control strategy and the dynamics of the closed loop system. There are essentially four characteristic shapes of the control surface.

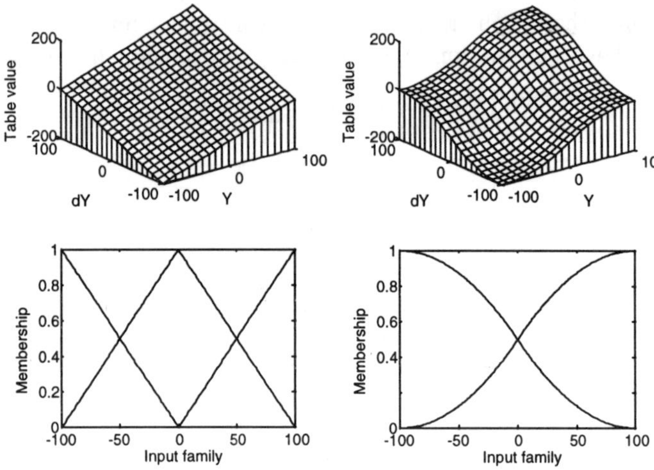

Figure 4: Linear surface (top left) and steep surface (top right) with the input families that generated them (bottom left and right respectively).

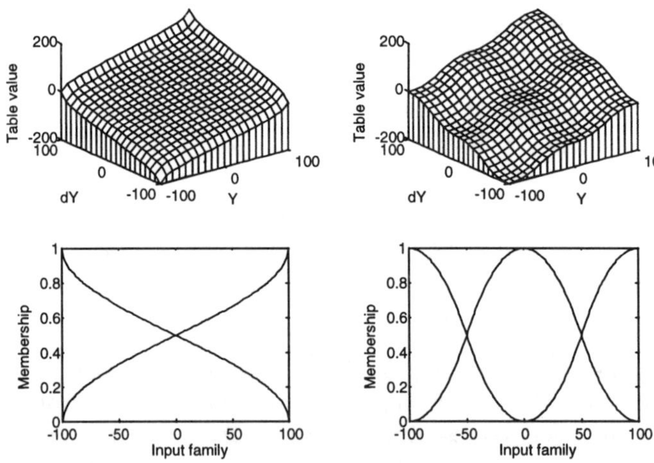

Figure 5: Gently sloping surface (top left) and bumpy surface (top right) with the input families that generated them (bottom left and right respectively).

A *linear* surface (Fig. 4, left) results from the rules below with triangular input sets,

$$\begin{array}{l} 1.\ \text{If } y \text{ is Neg and } \dot{y} \text{ is Neg then output is } -200 \\ 2.\ \text{If } y \text{ is Neg and } \dot{y} \text{ is Zero then output is } -100 \\ 3.\ \text{If } y \text{ is Neg and } \dot{y} \text{ is Pos then output is } 0 \\ 4.\ \text{If } y \text{ is Zero and } \dot{y} \text{ is Neg then output is } -100 \\ 5.\ \text{If } y \text{ is Zero and } \dot{y} \text{ is Zero then output is } 0 \\ 6.\ \text{If } y \text{ is Zero and } \dot{y} \text{ is Pos then output is } 100 \\ 7.\ \text{If } y \text{ is Pos and } \dot{y} \text{ is Neg then output is } 0 \\ 8.\ \text{If } y \text{ is Pos and } \dot{y} \text{ is Zero then output is } 100 \\ 9.\ \text{If } y \text{ is Pos and } \dot{y} \text{ is Pos then output is } 200 \end{array} \tag{27}$$

The surface in the figure is equivalent to a summation of the two inputs (cp. the values on the axes). Since the surface is equivalent to a summation, the controller is equivalent to (16).

The *steep* surface (Fig. 4, right) is built using only rules 1, 3, 7, and 9 together with the input sets shown in the figure; they are segments of cosine functions. Notice the absence of the centre rule (no. 5). That surface has a steeper slope, or higher gain, near the centre compared to the linear surface, but they have the same values pairwise in the four corners.

The *gentle slope* surface (Fig. 5, left) is built using the same four rules 1, 3, 7, and 9 but the input sets have been reflected. They are based on inverse trigonometric functions, i.e., $Pos(x) = 0.5 + \sin^{-1}(x/100)/\pi$, and $Neg(x) = \cos^{-1}(x/100)/\pi$. That surface has a more gentle slope, or lower gain, near the centre of the table than the linear surface has; but it too has the same values pairwise in the four corners.

The *bumpy* surface (Fig. 5, right) is a blend of the previous two surfaces. It is built using the set of nine rules (27) with nonlinear input sets as shown in the figure. This is often the default in commercial packages. It has a flat plateau near the centre and bumps in several other places. Even this surface has the same values as the other surfaces in the four corners.

For a study of how a response reacts to the various nonlinear surfaces, see Jantzen.[9]

3.5 Fine-tuning the nonlinear fuzzy controller

The final step in the design procedure is to fine-tune the gains, now that the fuzzy controller is nonlinear. The choice of gains in the fine-tuning phase is traditionally the result of intuition and experience. The pendulum responses usually contain two components, a slow one and a fast one. The slow frequency is due to the movement of the cart and the faster frequency is due to the movement of the ball. A few rules of thumb can be derived:

- *Gain* g_1. Influences mainly the amplitude and frequency of the slow component.
- *Gain* g_2. Influences mainly the damping of the slow component.
- *Gain* gu_1. Combines both of the previous effects up to a certain limit.
- *Gain* g_3. Influences mainly the amplitude and frequency of the fast component.
- *Gain* g_4. Influences mainly the damping of the fast component.
- *Gain* gu_2. Combines both of the previous effects up to a certain limit.

As with all nonlinear systems, the responses are amplitude dependent and thereby depend on the step size.

4 Simulation results and discussion

A test has been conducted on the Matlab simulator of the cart-ball system, available from the World Wide Web.[10] The implementation uses a fairly primitive Euler type simulation procedure, but it is good enough for the purpose. Throughout all simulations, a simulation step of 10 msecs is used, a compromise between accuracy and elapse time. The cart is initially positioned on the middle of the track, with the ball resting on the left end-stop. More precisely, the initial conditions are

$$\mathbf{x} = (0, 0, -0.2094, 0)^T$$

When started, the cart moves left to get the ball up and over the top, then it moves back toward the middle, perhaps with some overshoot.

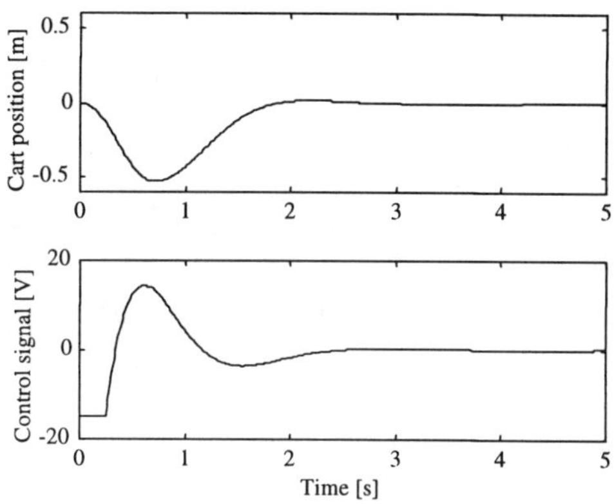

Figure 6: Cart position response with a linear pole placement controller.

4.1 Pole placement design

The parameter w can be seen as a convenient optimisation parameter, the larger the better. The control signal is physically limited to 15 volts, however, and if w is too high, the system will become unstable. In order to define a single parameter as a performance measure, we use the Integral Absolute Error (IAE) on the cart position,

$$IAE = \int_{t=0}^{5} |e(t)|\, dt$$

The error $e(t) = ref - y(t)$, where the reference $ref = 0$. Now w can be increased from $w = 1$ until IAE is minimal; this happens around $w = 3.2$. The corresponding step response is in Fig. 6 and the settings are in Table 1. Notice that all feedback gains are positive; in fact the cart controller has positive feedback in the sense that it tries to push the cart *away* from the centre (all directions are assumed positive towards the right). This may seem counter-intuitive, but it is necessary considering the situation where the cart is to the left of the middle ($y < 0$) with all other states equal to zero. The cart controller will pull left, the ball rolls right, and the ball controller will take over and guide the cart back towards the middle. The two controllers are constantly battling each other, but in most cases the ball controller dominates.

w	f_1	f_2	f_3	f_4	IAE
3.2	27.7	27.7	175	47.5	0.492

Table 1: Optimal feedback gains for pole placement controller.

Controller	g_1	g_2	gu_1	g_3	g_4	gu_2	IAE
Linear fuzzy	65	65	0.426	313	85	0.559	0.492
Nonlinear fuzzy	65	65	0.330	313	85	0.559	0.465

Table 2: Fuzzy controller settings.

4.2 Fuzzy linear controller

When replacing the pole placement controller by a fuzzy controller with a linear surface, the response is exactly the same, as long as the gain factors are chosen correctly. Implementation-wise, it is easier to use a look-up table, rather than a rule base, with a finite resolution of, say, 21-by-21 cells. It saves computing the inference at each sampling. To avoid discretisation problems, the table values are interpolated using bilinear interpolation.

4.3 Transferring gains to the fuzzy controller

Plots from the previous run indicate that g_2 and g_4 should be chosen to avoid saturation in the velocity universes; the rest of the gains follow from equations (25)-(26), see Table 2. The performance is the same, as expected (although there can be differences on the fourth decimal of IAE).

4.4 Fuzzy nonlinear controller

Closer analysis of the state feedback controller shows that all four feedback gains must be positive, not zero, in order to stabilise the system. The bumpy surface has a horizontal tangent in the centre, this is obviously a zero gain. The linear table has a nonzero gradient in the centre, and evidently it works better. The steep surface (Fig. 4, right) has been chosen in the hopes that it will work better. The response is shown in Fig. 7; it is more oscillatory and the performance measure is worse.

4.5 Fine-tuning the nonlinear fuzzy controller

The steep surface has a larger gain than the linear controller everywhere except for the corners; it is thus to be expected that the gains will have to be de-tuned. The first place to start is to decrease gu_1, because it is more efficient. Figure 8

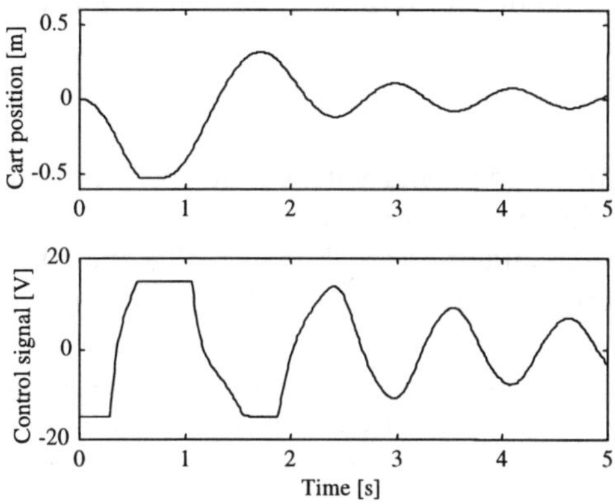

Figure 7: Cart response with the steep control surface.

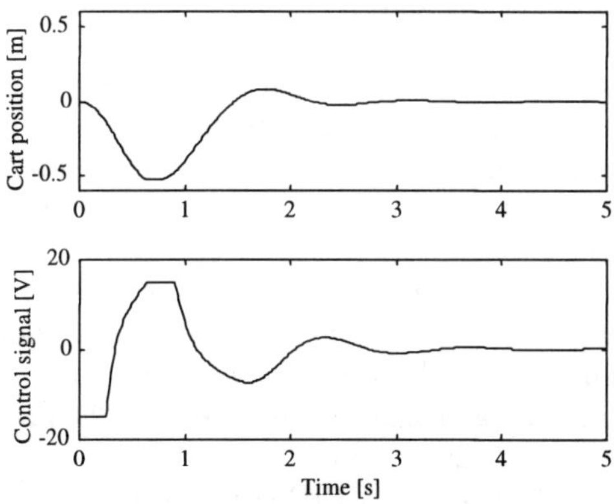

Figure 8: Response with steep surface after fine-tuning.

shows the response with the settings from Table 2. The performance measure improved about five percent relative to the best linear controller. The input gains g_1 and g_2 can be further adjusted, but this has been skipped here.

5 Conclusions

It seems that the fuzzy controller with the steep surface in the cart controller works better than the linear controller, at least in the sense of the performance criterion IAE. One should be aware, however, that there is no formal guarantee that the linear controller performs optimally; the comparison may turn out to be unfair if a better linear controller can be found. It can nevertheless be said, without a doubt, that the fuzzy controller performs at least as well, because it contains a linear controller within. The nonlinear options in the fuzzy controller give it the potential for performing better, if the designer is able to exploit them to his advantage.

There is a large potential for the linear design approach in fuzzy PID controllers, due to the widespread application of PID control and the well-known tuning rules.

References

1. V. Jorgensen. A ball-balancing system for demonstration of basic concepts in the state-space control theory. *Int.J.Elect.Enging Educ.*, 11:367–376, 1974.
2. Jan Jantzen. Fuzzy control course on the internet. http://www.iau.dtu.dk/~ jj/learn, 1996.
3. K. J. Reinschke. *Multivariable Control: A Graph-Theoretic Approach*, volume 108 of *Lecture Notes in Control and Information Sciences*. Springer-Verlag, 1988.
4. Gene F. Franklin, J. David Powell, and Abbas Emami-Naeini. *Feedback Control of Dynamic Systems.* Electrical and Computer Engineering: Control Engineering. Addison-Wesley, second edition, 1991.
5. W. Siler and H. Ying. Fuzzy control theory: The linear case. *Fuzzy Sets and Systems*, 33:275–290, 1989.
6. M. Mizumoto. Realization of PID controls by fuzzy control methods. In IEEE, editor, *First Int. Conf. On Fuzzy Systems*, number 92CH3073-4, pages 709–715, San Diego, 1992. The Institute of Electrical and Electronics Engineers, Inc.
7. W.Z. Qiao and M. Mizumoto. PID type fuzzy controller and parameters adaptive method. *Fuzzy Sets and Systems*, 78:23–35, 1996.

8. Jan Jantzen. Design of fuzzy controllers. Technical Report 98-E-864 (design), Technical University of Denmark: Dept. of Automation, http://www.iau.dtu.dk/~ jj/pubs, 1998. Lecture notes, 27 p.
9. Jan Jantzen. A robustness study of fuzzy control rules. In EUFIT, editor, *Proceedings Fifth European Congress on Fuzzy and Intelligent Technologies*, pages 1222–1227, Promenade 9, D-52076 Aachen, 1997. ELITE Foundation.
10. Petur Snaeland, Hördur Kvaran, and Jan Jantzen. The cart-ball simulator. http://www.iau.dtu.dk/~ jj/pubs, 1998. v 3.